Decision Engineering

Series Editor

Professor Rajkumar Roy
Department of Enterprise Integration School of Industrial and Manufacturing Science
Cranfield University
Cranfield
Bedford
MK43 0AL
UK

Other titles published in this series

Cost Engineering in Practice
John McIlwraith

IPA – Concepts and Applications in Engineering
Jerzy Pokojski

Strategic Decision Making
Navneet Bhushan and Kanwal Rai

Product Lifecycle Management
John Stark

From Product Description to Cost: A Practical Approach
Volume 1: The Parametric Approach
Pierre Foussier

From Product Description to Cost: A Practical Approach
Volume 2: Building a Specific Model
Pierre Foussier

Decision-Making in Engineering Design
Yotaro Hatamura

Composite Systems Decisions
Mark Sh. Levin

Intelligent Decision-making Support Systems
Jatinder N.D. Gupta, Guisseppi A. Forgionne and Manuel Mora T.

Knowledge Acquisition in Practice
N.R. Milton

Global Product: Strategy, Product Lifecycle Management and the Billion Customer Question
John Stark

Enabling a Simulation Capability in the Organisation
Andrew Greasley

Network Models and Optimization
Mitsuo Gen, Runewei Cheng and Lin Lin

Management of Uncertainty
Gudela Grote

Introduction to Evolutionary Algorithms
Xinjie Yu and Mitsuo Gen

Yong Yin · Ikou Kaku · Jiafu Tang · JianMing Zhu

Data Mining

Concepts, Methods and Applications in Management and Engineering Design

Springer

Yong Yin, PhD
Yamagata University
Department of Economics
and Business Management
1-4-12, Kojirakawa-cho
Yamagata-shi, 990-8560
Japan
yin@human.kj.yamagata-u.ac.jp

Ikou Kaku, PhD
Akita Prefectural University
Department of Management Science
and Engineering
Yulihonjo, 015-0055
Japan
ikou_kaku@akita-pu.ac.jp

Jiafu Tang, PhD
Northeastern University
Department of Systems Engineering
110006 Shenyang
China
jftang@mail.neu.edu.cn

JianMing Zhu, PhD
Central University
of Finance and Economics
School of Information
Beijing
China
tyzjm65@163.com

ISBN 978-1-4471-6118-9 ISBN 978-1-84996-338-1 (eBook)
DOI 10.1007/978-1-84996-338-1
Springer London Dordrecht Heidelberg New York

British Library Cataloguing in Publication Data
A catalogue record for this book is available from the British Library

Cover design: eStudioCalamar, Girona/Berlin

Printed on acid-free paper

Springer is part of Springer Science+Business Media (www.springer.com)

Preface

Today's business can be described by a single word: turbulence. Turbulent markets have the following characteristics: shorter product life cycles, uncertain product types, and fluctuating production volumes (sometimes mass, sometimes batch, and sometimes very small volumes).

In order to survive and thrive in such a volatile business environment, a number of approaches have been developed to aid companies in their management decisions and engineering designs. Among various methods, data mining is a relatively new approach that has attracted a lot of attention from business managers, engineers and academic researchers. Data mining has been chosen as one of ten emerging technologies that will change the world by *MIT Technology Review*.

Data mining is a process of discovering valuable information from observational data sets, which is an interdisciplinary field bringing together techniques from databases, machine learning, optimization theory, statistics, pattern recognition, and visualization.

Data mining has been widely used in various areas such as business, medicine, science, and engineering. Many books have been published to introduce data-mining concepts, implementation procedures and application cases. Unfortunately, very few publications interpret data-mining applications from both management and engineering perspectives.

This book introduces data-mining applications in the areas of management and industrial engineering. This book consists of the following: Chapters 1–6 provide a focused introduction of data-mining methods that are used in the latter half of the book. These chapters are not intended to be an exhaustive, scholarly treatise on data mining. It is designed only to discuss the methods commonly used in management and engineering design. The real gem of this book lies in Chapters 7–14, where we introduce how to use data-mining methods to solve management and industrial engineering design problems. The details of this book are as follows.

In Chapter 1, we introduce two simple but widely used methods: decision analysis and cluster analysis. Decision analysis is used to make decisions under an uncertain business environment. Cluster analysis helps us find homogenous objects, called clusters, which are similar and/or well separated.

Chapter 2 interprets the association rules mining method, which is an important topic in data mining. Association rules mining is used to discover association relationships or correlations among a set of objects.

Chapter 3 describes fuzzy modeling and optimization methods. Real-world situations are often not deterministic. There exist various types of uncertainties in social, industrial and economic systems. After introducing basic terminology and various theories on fuzzy sets, this chapter aims to present a brief summary of the theory and methods on fuzzy optimization and tries to give readers a clear and comprehensive understanding of fuzzy modeling and fuzzy optimization.

In Chapter 4, we give an introduction of quadratic programming problems with a type of fuzzy objective and resource constraints. We first introduce a genetic algorithms based interactive approach. Then, an approach is interpreted, which focuses on a symmetric model for a kind of fuzzy nonlinear programming problem by way of a special genetic algorithm with mutation along the weighted gradient direction. Finally, a non-symmetric model for a type of fuzzy nonlinear programming problems with penalty coefficients is described by using a numerical example.

Chapter 5 gives an introduction of basic concepts and algorithms of neural networks and self-organizing maps. The self-organizing maps based method has many practical applications, such as semantic map, diagnosis of speech voicing, solving combinatorial optimization problems, and so on. Several numerical examples are used to show various properties of self-organizing maps.

Chapter 6 introduces an important topic in data mining, privacy-preserving data mining (PPDM), which is one of the newest trends in privacy and security research. It is driven by one of the major policy issues of the information era: the right to privacy. Data are distributed among various parties. Legal and commercial concerns may prevent the parties from directly sharing some sensitive data. How parties collaboratively conduct data mining without breaching data privacy presents a grand challenge. In this chapter, some techniques for privacy-preserving data mining are introduced.

In Chapter 7, decision analysis models are developed to study the benefits from cooperation and leadership in a supply chain. A total of eight cooperation/leadership policies of the leader company are analyzed by using four models. Optimal decisions for the leader company under different cost combinations are analyzed.

Using a decision tree, Chapter 8 characterizes the impact of product global performance on the choice of product architecture during the product development process. We divide product architectures into three categories: modular, hybrid, and integral. This chapter develops analytic models whose objectives are obtaining global performance of a product through a modular/hybrid/integral architecture. Trade-offs between costs and expected benefits from different product architectures are analyzed and compared.

Chapter 9 reviews various cluster analysis methods that have been applied in cellular manufacturing design. We give a comprehensive overview and discussion for similarity coefficients developed to date for use in solving the cell formation problem. To summarize various similarity coefficients, we develop a classification system to clarify the definition and usage of various similarity coefficients in designing cellular manufacturing systems. Existing similarity (dissimilarity) coefficients developed so far are mapped onto the taxonomy. Additionally, production information-based similarity coefficients are discussed and a historical evolution of these similarity coefficients is outlined. We compare the performance of twenty well-known similarity coefficients. More than two hundred numerical cell formation problems, which are selected from the literature or generated deliberately, are used for the comparative study. Nine performance measures are used for evaluating the goodness of cell formation solutions.

Chapter 10 develops a cluster analysis method to solve a cell formation problem. A similarity coefficient is proposed, which incorporates alternative process routing, operation sequence, operation time, and production volume factors. This similarity coefficient is used to solve a cell formation problem that incorporates various real-life production factors, such as the alternative process routing, operation sequence, operation time, production volume of parts, machine capacity, machine investment cost, machine overload, multiple machines available for machine types and part process routing redesigning cost.

In Chapter 11, we show how to use a fuzzy modeling approach and a genetic-based interactive approach to control a product's quality. We consider a quality function deployment (QFD) design problem that incorporates financial factor and plan uncertainties. A QFD-based integrated product development process model is presented firstly. By introducing some new concepts of planned degree, actual achieved degree, actual primary costs required and actual planned costs, two types of fuzzy nonlinear optimization models are introduced in this chapter. These models not only consider the overall customer satisfaction, but also the enterprise satisfaction with the costs committed to the product.

Chapter 12 introduces a key decision making problem in a supply chain system: inventory control. We establish a new algorithm of inventory classification based on the association rules, in which by using the support-confidence framework the consideration of the cross-selling effect is introduced to generate a new criterion that is then used to rank inventory items. Then, a numerical example is used to explain the new algorithm and empirical experiments are implemented to evaluate its effectiveness and utility, comparing with traditional ABC classification.

In Chapter 13, we describe a technology, surface mountain technology (SMT), which is used in the modern electronics and electronic device industry. A key part for SMT is to construct master data. We propose a method of making master data by using a self-organizing maps learning algorithm and prove such a method is effective not only in judgment accuracy but also in computational feasibility. Empirical experiments are invested for proving the performance of the indicator. Consequently, the continuous weight is effective for the learning evaluation in the process of making the master data.

Chapter 14 describes applications of data mining with privacy-preserving capability, which has been an area gaining researcher attention recently. We introduce applications from various perspectives. Firstly, we present privacy-preserving association rule mining. Then, methods for privacy-preserving classification in data mining are introduced. We also discuss privacy-preserving clustering and a scheme to privacy-preserving collaborative data mining.

Yamagata University, Japan
December 2010

Yong Yin
Ikou Kaku
Jiafu Tang
JianMing Zhu

Contents

Chapter 1
Decision Analysis and Cluster Analysis

In this chapter, we introduce two simple but widely used methods: decision analysis and cluster analysis. Decision analysis is used to make decisions under an uncertain business environment. The simplest decision analysis method, known as a decision tree, is interpreted. Decision tree is simple but very powerful. In the latter half of this book, we use decision tree to analyze complicated product design and supply chain design problems.

Given a set of objects, cluster analysis is applied to find subsets, called clusters, which are similar and/or well separated. Cluster analysis requires similarity coefficients and clustering algorithms. In this chapter, we introduce a number of similarity coefficients and three simple clustering algorithms. In the second half of this book, we introduce how to apply cluster analysis to design complicated manufacturing problems.

1.1 Decision Tree

Today's volatile business environment is characterized by short product life cycles, uncertain product types, and fluctuating production volumes (sometimes mass, sometimes batch, and sometimes very small volumes.) One important and challenging task for managers and engineers is to make decisions under such a turbulent business environment. For example, a product designer must decide a new product type's architecture when future demand for products is uncertain. An executive must decide a company's organization structure to accommodate an unpredictable market.

An analytical approach that is widely used in decision analysis is a decision tree. A decision tree is a systemic method that uses a tree-like diagram. We introduce the decision tree method by using a prototypical decision example.

Wata Company's Investment Decision

Lee is the investment manager of Wata, a small electronics components company. Wata has a product assembly line that serves one product type. In May, the board of executive directors of Wata decides to extend production capacity. Lee has to consider capacity extension strategy. There are two possible strategies.

1. Construct a new assembly line for producing a new product type.
2. Increase the capacity of existing assembly line.

Because the company's capital is limited, these two strategies cannot be implemented simultaneously. At the end of May, Lee collects related information and summarizes them as follows.

1. Tana, a customer of Wata, asks Wata to supply a new electronic component, named Tana-EC. This component can bring Wata \$150,000 profit per period. A new assembly line is needed to produce Tana-EC. However, this order will only be good until June 5. Therefore, Wata must decide whether or not to accept Tana's order before June 5.
2. Naka, another electronics company, looks for a supplier to provide a new electronic component, named Naka-EC. Wata is a potential supplier for Naka. Naka will decide its supplier on June 15. The probability that Wata is selected by Naka as a supplier is 70%. If Wata is chosen by Naka, Wata must construct a new assembly line and obtain a \$220,000 profit per period.
3. The start day of the next production period is June 20. Therefore, Wata can extend the capacity of its existing assembly line from this day. Table 1.1 is an approximation of the likelihood that Wata would receive profits. That is, Lee estimates that there is roughly a 10% likelihood that extended capacity would be able to bring a profit with \$210,000, and that there is roughly a 30% likelihood that extended capacity would be able to bring a profit with \$230,000, *etc.*

Table 1.1 Distribution of profits

Profit from extended capacity	Probability
\$210,000	10%
\$230,000	30%
\$220,000	40%
\$250,000	20%

Using information summarized by Lee, we can draw a decision tree that is represented chronologically as Figure 1.1. Lee's first decision is whether to accept Tana's order. If Lee refused Tana's order, then Lee would face the uncertainty of whether or not Wata could get an order from Naka. If Wata receives Naka's order, then Wata would subsequently have to decide to accept or to reject Naka's order. If Wata were to accept Naka's order, then Wata would construct a new assembly line for Naka-EC. If Wata were to instead reject the order, then Wata would extend the capacity of the existing assembly line.

A decision tree consists of nodes and branches. Nodes are connected by branches. In a decision tree, time flows from left to right. Each branch represents a decision or a possible event. For example, the branch that connects nodes A and B is a decision, and the branch that connects nodes B and D is a possible event with a probability 0.7. Each rightmost branch is associated with a numerical value that is the outcome of an event or decision. A node that radiates decision branches is called a decision node. That is, the node has decision branches on the right side. Similarly, a node that radiates event branches is called an event node. In Figure 1.1, nodes A and D are decision nodes, nodes B, C, and E are event nodes.

Expected monetary value (EMV) is used to evaluate each node. EMV is the weighted average value of all possible outcomes of events. The procedure for solving a decision tree is as follows.

Step 1 Start from the rightmost branches, compute each node's EMV. For an event node, its EMV is equal to the weighted average value of all possible outcomes of events. For a decision node, the EMV is equal to the maximum EMV of all branches that radiate from it.

Step 2 The EMV of the leftmost node is the EMV of a decision tree.

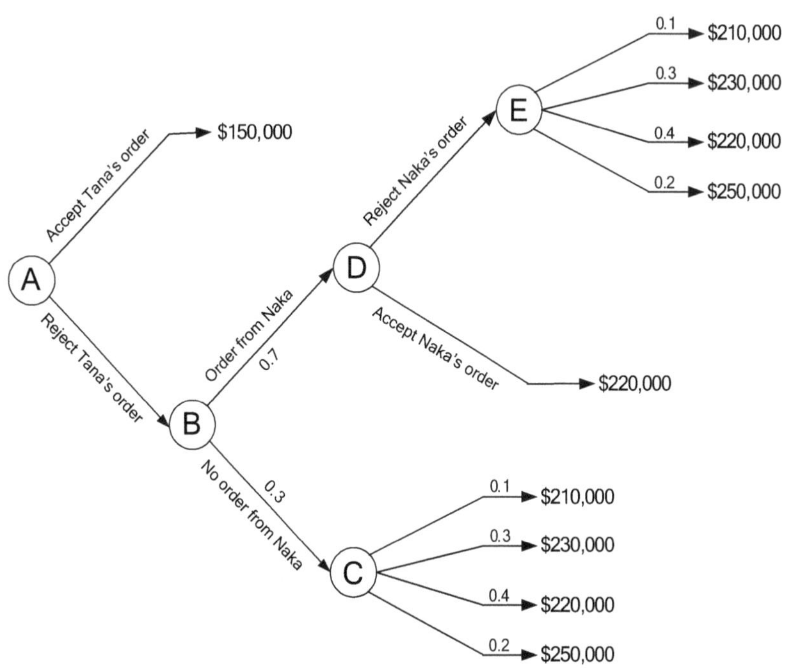

Figure 1.1 The decision tree

Following the above procedure, we can solve the decision tree in Figure 1.1 as follows.

Step 1 For event nodes C and E, their EMVs, EMV_C and EMV_E, are computed as follows:

$$210{,}000 \times 0.1 + 230{,}000 \times 0.3 + 220{,}000 \times 0.4 + 250{,}000 \times 0.2 = 228{,}000 .$$

The EMV of decision node D is computed as

$$\mathrm{Max}\,\{\mathrm{EMV}_E, 220{,}000\} = \mathrm{EMV}_E = 228{,}000 .$$

The EMV of event node B is computed as

$$0.3 \times \mathrm{EMV}_C + 0.7 \times \mathrm{EMV}_D = 0.3 \times 228{,}000 + 0.7 \times 228{,}000 = 228{,}000 .$$

Finally, the EMV of decision node A is computed as

$$\mathrm{Max}\,\{\mathrm{EMV}_B, 150{,}000\} = \mathrm{EMV}_B = 228{,}000 .$$

Step 2 Therefore, the EMV of the decision tree is 228,000.

Based on the result, Lee should make the following decisions. Firstly, he would reject Tana's order. Then, even if he receives Naka's order, he would reject it. Wata would expand the capacity of the existing assembly line.

1.2 Cluster Analysis

In this section, we introduce one of the most used data-mining methods: cluster analysis. Cluster analysis is widely used in science (Hansen and Jaumard 1997; Hair *et al.* 2006), engineering (Xu and Wunsch 2005; Hua and Zhou 2008), and the business world (Parmar *et al.* 2009).

Cluster analysis groups individuals or objects into clusters so that objects in the same cluster are more similar to one another than they are to objects in other clusters. The attempt is to maximize the homogeneity of objects within the clusters while also maximizing the heterogeneity between the clusters (Hair *et al.* 2006).

A similarity (dissimilarity) coefficient is usually used to measure the degree of similarity (dissimilarity) between two objects. For example, the following coefficient is one of the most used similarity coefficient: the Jaccard similarity coefficient.

$$S_{ij} = \frac{a}{a+b+c}, \qquad 0 \le S_{ij} \le 1 ,$$

where S_{ij} is the similarity between machine i and machine j.

S_{ij} is the Jaccard similarity coefficient between objects i and j. Here, we suppose an object is represented by its attributes. Then, a is the total number of attributes, which objects i and j both have, b is the total number of attributes belonging only to object i, and c is the total number of attributes belonging only to object j.

Cluster analysis method relies on similarity measures in conjunction with clustering algorithms. It usually follows a prescribed set of steps, the main ones being:

Step 1 Collect information of all objects. For example, the objects' attribute data.

Step 2 Choose an appropriate similarity coefficient. Compute similarity values between object pairs. Construct a similarity matrix. An element in the matrix is a similarity value between two objects.

Step 3 Choose an appropriate clustering algorithm to process the values in the similarity matrix, which results in a diagram called a tree, or dendrogram, that shows the hierarchy of similarities among all pairs of objects.

Step 4 Find clusters from the tree or dendrogram, check all predefined constraints such as the number of clusters, cluster size, *etc.*

For a lot of small cluster analysis problems, step 3 could be omitted. In step 2 of the cluster analysis procedure, we need a similarity coefficient. A large number of similarity coefficients have been developed. Table 1.2 is a summary of widely used similarity coefficients. In Table 1.2, d is the total number of attributes belonging to neither object i nor object j.

In step 3 of the cluster analysis procedure, a clustering algorithm is required to find clusters. A large number of clustering algorithms have been proposed in the literature. Hansen and Jaumard (1997) gave an excellent review of various clustering algorithms. In this section, we introduce three simple clustering algorithms: single linkage clustering (SLC), complete linkage clustering (CLC), and average linkage clustering (ALC).

Table 1.2 Definitions and ranges of selected similarity coefficients

Similarity coefficient	Definition S_{ij}	Range
1. Jaccard	$a/(a+b+c)$	0–1
2. Hamann	$[(a+d)-(b+c)]/[(a+d)+(b+c)]$	−1–1
3. Yule	$(ad-bc)/(ad+bc)$	−1–1
4. Simple matching	$(a+d)/(a+b+c+d)$	0–1
5. Sorenson	$2a/(2a+b+c)$	0–1
6. Rogers and Tanimoto	$(a+d)/[a+2(b+c)+d]$	0–1
7. Sokal and Sneath	$2(a+d)/[2(a+d)+b+c]$	0–1
8. Rusell and Rao	$a/(a+b+c+d)$	0–1
9. Baroni-Urbani and Buser	$[a+(ad)^{1/2}]/[a+b+c+(ad)^{1/2}]$	0–1
10. Phi	$(ad-bc)/[(a+b)(a+c)(b+d)(c+d)]^{1/2}$	−1–1
11. Ochiai	$a/[(a+b)(a+c)]^{1/2}$	0–1
12. PSC	$a^2/[(b+a)*(c+a)]$	0–1
13. Dot-product	$a/(b+c+2a)$	0–1
14. Kulczynski	$1/2[a/(a+b)+a/(a+c)]$	0–1
15. Sokal and Sneath 2	$a/[a+2(b+c)]$	0–1
16. Sokal and Sneath 4	$1/4[a/(a+b)+a/(a+c)$ $+d/(b+d)+d/(c+d)]$	0–1
17. Relative matching	$[a+(ad)^{1/2}]/[a+b+c+d+(ad)^{1/2}]$	0–1

SLC algorithm is the simplest algorithm based on the similarity coefficient method. Once similarity coefficients have been calculated for object pairs, SLC groups two objects (or an object and an object cluster, or two object clusters) which have the highest similarity. This process continues until the predefined number of object clusters has been obtained or all objects have been combined into one cluster. SLC greatly simplifies the grouping process. Because, once the similarity coefficient matrix has been formed, it can be used to group all objects into object groups without any further revision calculation. SLC algorithm usually works as follows:

Step 1 Compute similarity coefficient values for all object pairs and store the values in a similarity matrix.

Step 2 Join the two most similar objects, or an object and an object cluster, or two object clusters, to form a new object cluster.

Step 3 Evaluate the similarity coefficient value between the new object cluster formed in step 2 and other remainder object clusters (or objects) as follows:

$$S_{tv} = \underset{\substack{i \in t \\ j \in v}}{\text{Max}}\{S_{ij}\} \tag{1.1}$$

where object i is in the object cluster t, and object j is in the object cluster v.

Step 4 When the predefined number of object clusters is obtained, or all objects are grouped into a single object cluster, stop; otherwise go to step 2.

CLC algorithm does the reverse of SLC. CLC combines two object clusters at minimum similarity level, rather than at maximum similarity level as in SLC. The algorithm remains the same except that Equation 1.1 is replaced by

$$S_{tv} = \underset{\substack{i \in t \\ j \in v}}{\text{Min}}\{S_{ij}\}$$

SLC and CLC use the "extreme" value of the similarity coefficient to form object clusters. ALC algorithm, developed to overcome this deficiency in SLC and CLC, is a clustering algorithm based on the average of pair-wise similarity coefficients between all members of the two object clusters. The ALC between object clusters t and v is defined as follows:

$$S_{tv} = \frac{\sum_{i \in t} \sum_{j \in v} S_{ij}}{N_t N_v}$$

where N_t is the total number of objects in cluster t, and N_v is the total number of objects in cluster v.

In the remainder of this section, we use a simple example to show how to perform cluster analysis.

Step 1 Attribute data of objects.

The input data of the example is in Table 1.3. Each row represents an object and each column represents an attribute. There are 5 objects and 11 attributes in this

Table 1.3 Object-attribute matrix

		Attribute										
		1	2	3	4	5	6	7	8	9	10	11
Object	1	1	1			1	1	1	1		1	
	2	1			1	1				1		1
	3			1			1		1		1	
	4		1		1			1		1		1
	5			1			1		1		1	

example. An element 1 in row i and column j means that the ith object has the jth attribute.

Step 2 Construct similarity coefficient matrix.

The Jaccard similarity coefficient is used to calculate the similarity degree between object pairs. For example, the Jaccard similarity value between objects 1 and 2 is computed as follows:

$$S_{12} = \frac{a}{a+b+c} = \frac{2}{2+5+3} = 0.2\,.$$

The Jaccard similarity matrix is shown in Table 1.4.

Table 1.4 Similarity matrix

		Objekt				
		1	2	3	4	5
Object	1		0.2	0.375	0.2	0.375
	2			0	0.429	0
	3				0	1
	4					0

Step 3 For this example, SLC gives a dendrogram shown in Figure 1.2.

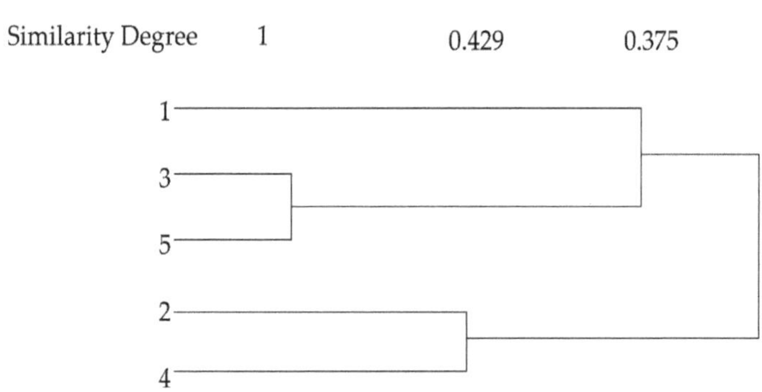

Figure 1.2 The dendrogram from SLC

Step 4 Based on similarity degree, we can find different clusters from the dendrogram. For example, if we need to find clusters that consist of the same objects, *i.e.*, Jaccard similarity values between object pairs equal to 1, then we have 4 clusters as follows:

Cluster 1: object 1
Cluster 2: objects 3 and 5
Cluster 3: object 2
Cluster 4: object 4

If similarity values within a cluster must be larger than 0.374, we obtain 2 clusters as follows:

Cluster 1: objects 1, 3 and 5
Cluster 2: objects 2 and 4

SLC is very simple, but it does always produce a satisfied cluster result. Two objects or object clusters are merged together merely because a pair of objects (one in each cluster) has the highest value of similarity coefficient. Thus, SLC may identify two clusters as candidates for the formation of a new cluster at a certain threshold value, although several object pairs possess significantly lower similarity coefficients. The CLC algorithm does just the reverse and is not good as the SLC. Due to this drawback, these two algorithms sometimes produce improper cluster analysis results.

In later chapters, we introduce how to use the cluster analysis method for solving manufacturing problems.

References

Hair JF, Black WC, Babin BJ, Anderson RE, Tatham RL (2006) Multivariate Data Analysis, 6th edn. Prentice Hall, Upper Saddle River, NJ

Hansen P, Jaumard B (1997) Cluster analysis and mathematical programming. Math Program 79:191–215

Hua W, Zhou C (2008) Clusters and filling-curve-based storage assignment in a circuit board assembly kitting area. IIE Trans 40:569–585

Parmar D, Wu T, Callarman T, Fowler J, Wolfe P (2010) A clustering algorithm for supplier base management. Int J Prod Res 48(13):3803–3821

Xu R, Wunsch II D (2005) Survey of clustering algorithms. IEEE Trans Neural Networks 16(3): 645–678

Chapter 2
Association Rules Mining in Inventory Database

Association rules mining is an important topic in data mining which is the discovery of association relationships or correlations among a set of items. It can help in many business decision-making processes, such as catalog design, cross-marketing, cross-selling and inventory control. This chapter reviews some of the essential concepts related to association rules mining, which will be then applied to the real inventory control system in Chapter 12. Some related research into development of mining association rules are also introduced.

This chapter is organized as follows. In Section 2.1 we begin with explaining briefly the background of association rules mining. In Section 2.2, we outline certain necessary basic concepts of association rules. In Section 2.3, we introduce the Apriori algorithm, which can search frequent itemsets in large databases. Section 2.4 introduces some research into development of mining association rules in an inventory database. Finally, we summarize this chapter in Section 2.5.

2.1 Introduction

Data mining is a process of discovering valuable information from large amounts of data stored in databases. This valuable information can be in the form of patterns, associations, changes, anomalies and significant structures (Zhang and Zhang 2002). That is, data mining attempts to extract potentially useful knowledge from data. Therefore data mining has been treated popularly as a synonym for knowledge discovery in databases (KDD). The emergence of data mining and knowledge discovery in databases as a new technology has occurred because of the fast development and wide application of information and database technologies.

One of the important areas in data mining is association rules mining. Since its introduction in 1993 (Agrawal *et al.* 1993), the area of association rules mining has received a great deal of attention. Association rules mining finds interesting association or correlation relationships among a large set of data items. With massive amounts of data continuously being collected and stored, many industries are becoming interested in mining association rules from their database. The discovery of interesting association relationships among huge amounts of business transaction records can help in many business decision-making processes, such as catalog design, cross-marketing, cross-selling and inventory control. How can we find association rules from large amounts of data, either transactional or relational? Which association rules are the most interesting? How can we help or guide the mining procedure to discover interesting associations? In this chapter we will explore each of these questions.

A typical example of association rules mining is market basket analysis. For instance, if customers are buying milk, how likely are they to also buy bread on the same trip to the supermarket? Such information can lead to increased sales by helping retailers to selectively market and plan their shelf space, for example, placing milk and bread within single visits to the store. This process analyzes customer buying habits by associations between the different items that customers place in their shopping baskets. The discovery of such associations can help retailers develop marketing strategies by gaining insight into which items are frequently purchased together by customers. The results of market basket analysis may be used to plan marketing or advertising strategies, as well as store layout or inventory control. In one strategy, items that are frequently purchased together can be placed in close proximity in order to further encourage the sale of such items together. If customers who purchase milk also tend to buy bread at the same time, then placing bread close to milk may help to increase the sale of both of these items. In an alternative strategy, placing bread and milk at opposite ends of the store may entice customers who purchase such items to pick up other items along the way (Han and Micheline 2001). Market basket analysis can also help retailers to plan which items to put on sale at reduced prices. If customers tend to purchase coffee and bread together, then having a sale on coffee may encourage the sale of coffee as well as bread.

If we think of the universe as the set of items available at the store, then each item has a Boolean variable representing the presence or absence of the item. Each basket can then be represented be a Boolean vector of values assigned to these variables. The Boolean vectors can be analyzed for buying patterns that reflect items that are frequently associated or purchased together. These patterns can be represented in the form of association rules. For example, the information that customers who purchase milk also tend to buy bread at the same time is represented in association rule as following form:

$$buys(X, "milk") \Rightarrow buys(X, "bread")$$

where X is a variable representing customers who purchased such items in a transaction database. There are a lot of items can be represented by the form above however most of them are not interested. Typically, association rules are considered interesting if they satisfy several measures that will be described below.

2.2 Basic Concepts of Association Rule

Agrawal *et al.* (1993) first developed a framework to measure the association relationship among a set of items. The association rule mining can be defined formally as follows.

$I = \{i_1, i_2, \cdots i_m\}$ is a set of items. For example, goods such as milk, bread and coffee for purchase in a store are items. $D = \{t_1, t_2, \cdots t_n\}$ is a set of transactions, called a *transaction database*, where each transaction t has an identifier *tid* and a set of items *t-itemset*, *i.e.*, t = (*tid*, *t-itemset*). For example, a customer's shopping cart going through a checkout is a transaction. X is an *itemset* if it is a subset of I. For example, a set of items for sale at a store is an itemset.

Two measurements have been defined as *support* and *confidence* as below. An itemset X in a transaction database D has a *support*, denoted as $sp(X)$. This is the ratio of transactions in D containing X,

$$sp(X) = \frac{|X(t)|}{|D|}$$

where $X(t) = \{t \text{ in } D | t \text{ contains } X\}$.

An itemset X in a transaction database D is called frequent if its support is equal to, or greater than, the threshold minimal support (*min_sp*) given by users. Therefore support can be recognized as frequencies of the occurring patterns.

Two itemsets X and Y in a transaction database D have a confidence, denoted as $cf(X \Rightarrow Y)$. This is the ratio of transactions in D containing X that also contain Y.

$$cf(X \Rightarrow Y) = \frac{|(X \cup Y)(t)|}{|X(t)|} = \frac{sp(X \cup Y)}{sp(X)}.$$

Because the confidence is represented as a conditional probability of both X an Y, having been purchased under the condition if X had been purchased, then the confidence can be recognized as the strength of the implication of the form $X \Rightarrow Y$.

An *association rule* is the implication of the form $X \Rightarrow Y$, where $X \subset I$, $Y \subset I$, and $X \cap Y = \phi$. Each association rule has two quality measurements, *support* and *confidence*, defined as: (1) the support of a rule $X \Rightarrow Y$ is $sp(X \cup Y)$ and (2) the confidence of a rule $X \Rightarrow Y$ is $cf(X \Rightarrow Y)$.

Rules that satisfy both a minimum support threshold (*min_sp*) and a minimum confidence threshold (*min_cf*), which are defined by users, are called strong or valid. Mining association rules can be broken down into the following two subproblems:

1. Generating all itemsets that have support greater than, or equal to, user specified minimum support. That is, generating all frequent itemsets.
2. Generating all rules that have minimum confidence in the following simple way: for every frequent itemset X, and any $B \subset X$, let $A = X - B$. If the confidence of a rule $A \Rightarrow B$ is greater than, or equal to, the minimum confidence (*min_cf*), then it can be extracted as a valid rule.

Table 2.1 A transaction database

tid	Items
$t001$	I_2, I_3, I_5
$t002$	I_1, I_2, I_3, I_5
	I_2, I_3, I_5
$t003$	I_1, I_2, I_3, I_5
$t004$	I_2, I_5

To demonstrate the support-confidence framework, we use a database example from the supermarket shown in Table 2.1.

In Table 2.1, let the item universe be $I = \{I_1, I_2, I_3, I_4, I_5\}$ and transaction universe be $tid = \{t001, t002, t003, t004\}$. Therfore, *tid* identifies uniquely a transaction in which several *items* have been purchased. Association rule $X \Rightarrow Y$ has a support $sp(X \cup Y)$ equal to the frequencies of X and Y items purchased together in transaction database D, and a confidence $cf(X \Rightarrow Y)$ equal to the ratio of how many times X items were purchased with Y items. There are many association rules that can be constructed by combining various items. For example, using only 10 items can make about 57,000 association rules (numbers of association rules can be calculated by $\sum_{k=2}^{m} C_m^k \cdot (2^k - 2)$, where selecting k items from m items produces association rules). However, not all of the association rules are interesting in practice; only parts of the association rules are valid. Under the consideration of the support-confidence framework, the association rules which have larger frequencies (their supports are larger than *min_sp*) and stronger relationships (their confidences are larger than *min_cf*) can be recognized as being valid. For example, let *min_sp* = 50% (to be frequent, an itemset must occur in at least two transactions in the above transaction database), and *min_cf* = 60% (to be a high-confidence or valid rule, at least 60% of the time you find the antecedent of the rule in the transactions; you must also find the consequence of the rule there). We can generate all frequent itemsets in Table 2.1 as follows. By scanning the database D, item $\{I_1\}$ occurs in the two transactions, $t001$ and $t003$. Its frequency is 2, and its support $sp\{I_1\} = 50\%$ is equal to $min\_sp = 50\%$. Therefore $\{I_1\}$ is a frequent item. Similarly, we can find that $\{I_2\}, \{I_3\}$ and $\{I_5\}$ are frequent items but $\{I_4\}$ is not a frequent item (because $sp\{I_1\} = 25\%$ is smaller than *min_sp*).

Now consider two-item sets in Table 2.1, where 8 two-item sets exist (because the combination of 2 items from all items (5 items) is 10, there should be 10 two-item sets, however 2 two-item sets do not appear in the transaction database). For example, $\{I_1, I_2\}$ occurs in the one transaction ($t003$). Its frequency is 1, and its support, $sp\{I_1 \cup I_2\} = 25\%$, which is less than $min\_sp = 50\%$. Therefore $\{I_1, I_2\}$ is not a frequent itemset. On the other hand, itemset $\{I_1, I_3\}$ occurs in the two transactions ($t001$ and $t003$), its frequency is 2, and its support, $sp\{I_1 \cup I_3\} = 50\%$, which is equal to $minsupp = 50\%$. Therefore $\{I_1, I_3\}$ is a frequent itemset. Similarly, we can find that $\{I_2, I_3\}, \{I_2, I_5\}$ and $\{I_3, I_5\}$ are frequent itemsets but $\{I_1, I_4\}, \{I_1, I_5\}$, $\{I_2, I_5\}$ and $\{I_3, I_4\}$ are not frequent itemsets.

We also need to consider three-item sets in Table 2.1, where 5 three-item sets exist (because the combination of 3 items from all items (5 items) is 10, there should be 10 three-item sets, however five two-item sets do not appear in the transaction database). For example, $\{I_1, I_2, I_3\}$ occurs in the one transaction ($t003$). Its frequency is 1, and its support, $sp\{I_1 \cup I_2 \cup I_3\} = 25\%$, which is less than $min\_sp = 50\%$. Therefore $\{I_1, I_2, I_3\}$ is not a frequent itemset. On the other hand, itemset $\{I_2, I_3, I_5\}$ occurs in the two transactions ($t002$ and $t003$), its frequency is 2, and its support, $sp\{I_2 \cup I_3 \cup I_5\} = 50\%$, which is equal to $minsupp = 50\%$. Therefore $\{I_2, I_3, I_5\}$ is a frequent itemset. In the same way, we can find that all three-item sets are not frequent except itemset $\{I_2, I_3, I_5\}$.

Similarly, four-items and five-items set also should be considered in Table 2.1. Only one four-item set $\{I_1, I_2, I_3, I_5\}$ exists but it is not frequent, and there are no five-item sets.

According to the above definition, $\{I_1\}, \{I_2\}, \{I_3\}, \{I_5\}, \{I_1, I_3\}, \{I_2, I_3\}, \{I_2, I_5\}, \{I_3, I_5\}$ and $\{I_2, I_3, I_5\}$ in Table 2.1 are frequent itemsets.

When the frequent itemsets are determined, generating association rules from the frequent itemsets is easy. For example, consider frequent three-item set $\{I_2, I_3, I_5\}$, because

$$cf\{I_2 \cup I_3 \Rightarrow I_5\} = \frac{sp(I_2 \cup I_3 \cup I_5)}{sp(I_2 \cup I_3)} = \frac{2}{2} = 100\% \,.$$

Since this is greater than $min\_cf = 60\%$, $I_2 \cup I_3 \Rightarrow I_5$ can be extracted as a valid rule. In the same way, $I_2 \cup I_5 \Rightarrow I_3$ and $I_3 \cup I_5 \Rightarrow I_2$ are also valid association rules but the relationships among items are different (because $cf\{I_2 \cup I_5 \Rightarrow I_3\} = 66.7\%$ and $cf\{I_3 \cup I_5 \Rightarrow I_2\} = 100\%$). Also because

$$cf\{I_2 \Rightarrow I_3 \cup I_5\} = \frac{sp(I_2 \cup I_3 \cup I_5)}{sp(I_2)} = \frac{2}{3} = 66.7\% \,.$$

This is greater than $min\_cf = 60\%$, and so $I_2 \Rightarrow I_3 \cup I_5$ can be extracted as a valid rule. In the same way, $I_2 \Rightarrow I_5 \cup I_3$ and $I_3 \Rightarrow I_5 \cup I_2$ are also valid association rules.

Similarly, we can find that $I_1 \Rightarrow I_3$ and $I_3 \Rightarrow I_1$ are valid from the frequent itemset $\{I_1, I_3\}$. Considering the association rules exist over two-item itemsets, generating all over two-item frequent itemsets, we can obtain all of association rules.

It is not clear which itemsets are frequent, and then which and how many association rules are valid if we do not investigate all of frequent itemsets and association rules. As mentioned above, there are too many candidate itemsets to search all of the itemsets. Therefore mining all valid frequent itemsets automatically from a transaction database is very important. In fact, it is a main research topic in data mining studies. Agrawal *et al.* (1993) have built an Apriori algorithm for mining association rules from databases. This algorithm has since become a common model for mining association rules. In this chapter, we just introduce the Apriori algorithm to show how to reduce the search space of frequent itemsets. In fact, there is a lot of research on mining association rules. Detailed overviews of mining association rule algorithms can be found in Zhang and Zhang (2002), Zhang *et al.* (2004) and Han and Micheline (2001).

2.3 Mining Association Rules

Identifying frequent itemsets is one of the most important issues faced by the knowledge discovery and data mining community. There have been a number of excellent algorithms developed for extracting frequent itemsets in very large databases. Apriori is a very famous and widely used algorithm for mining frequent itemsets (Agrawal *et al.* 1993). For efficiency, many variations of this approach have been constructed. To match the algorithms already developed, we first present the Apriori algorithm, and then present selectively several related algorithms of mining special association rules.

2.3.1 The Apriori Algorithm: Searching Frequent Itemsets

Apriori is an influential algorithm for mining frequent itemsets. The algorithm employs an iterative approach known as a level-wise search, where k-itemsets are used to explore $(k+1)$-itemsets. To improve the efficiency of the level-wise generation of frequent itemsets, an important property called Apriori property is used to reduce the search: all non-empty subsets of a frequent itemset must also be frequent. This property belongs to a special category of properties called antimonotone, in the sense that if a set cannot pass a test, all of its supersets will fail the same test as well. It is called antimonotone because the property is monotonic in the context of failing a test. That is, there are itemsets I and J where $I \subset J$ then $sp(I) \geq sp(J)$. That means if itemset I is not frequent then (for example in Table 2.1, $\{I_4\}$ is not a frequent item) itemset J, which included I, is also not frequent (for example in Table 2.1 $\{I_1, I_4\}$ and $\{I_3, I_4\}$ are not frequent itemsets).

By using the Apriori property, first the set of frequent one-item sets are found. This set is denoted L_1. L_1 can be used to find L_2, the set of frequent two-item sets in the following way. A set of candidate one-item sets is generated by joining L_1 with itself, which is denoted C_1. That is, C_1 is a superset of L_1, its members may or may not be frequent but all of the frequent one-item sets are included in C_1. A scan of the database to determine the count of each candidate in C_1 would result in the determination of L_1. Any one-item set that is not frequent cannot be a subset of a frequent two-item set. It is true for any k-itemsets. Hence, if any $(k-1)$-subset of a candidate k-itemset is not in L_{k-1}, then the candidate cannot be frequent either and so can be removed from C_k. Then L_2 can be used to find L_3, and so on, until no more frequent k-itemsets can be found. Therefore, we need to scan fully the database k times when searching for frequent k-itemsets.

The following algorithm in Figure 2.1 is used to generate all frequent itemsets in a given database D. This is the Apriori algorithm.

In the Apriori algorithm shown in Figure 2.1, step 1 finds the frequent one-item sets, L_1. Then L_{k-1} is used to generate candidates C_k in order to find L_k in steps 2 through 5. The *apriori_gen* procedure generates the candidates and then uses the apriori property to eliminate those having a subset that is not frequent, which is shown in detail in Figure 2.2.

Algorithm Apriori (D, *min_sp*)
Input: D, *min_sp*
Output: Answer
Begin
1) $L_1 = \{\text{large one-item sets}\}$;
2) For ($k = 2; L_{k-1} \neq \theta; k++$) do {
3) C_k = *apriori_gen* (L_{k-1}); // New candidates
4) $L_k = \{c \in C_k \mid c.\text{support} \geq min\_sp\}$
5) }
6) Answer $= \cup L_k$;
End

Figure 2.1 Main program of the Apriori algorithm

Function *apriori_gen*(L_{k-1})
Input: L_{k-1}
Output: C_k
Begin
For (each pair $p, q \in L_{k-1}$) do {
If
$p.item_1 = q.item_1 \wedge \cdots \wedge p.item_{k-1} = q.item_{k-1} \wedge p.item_{k-1} < q.item_{k-1}$
Then
$c = p \cup q.item_{k-1}$
$C_k = C_k \cup \{c\}$
}
For (each $p \in C_k$) do {
For (each k-subsets s of c) do {
If $c \notin L_{k-1}$ Then
Delete c
}
}
return C_k
End

Figure 2.2 Algorithm of function *apriori_gen*(L_{k-1})

In Figure 2.2, the *apriori_gen* function performs two kinds of actions. The first action is the join component in which L_{k-1} is joined with L_{k-1} to generate potential candidates. The second action employs the Apriori property to remove candidates that have a subset that is not frequent. An example to show the action of *apriori_gen* can be illustrated as follows with Table 2.1. Let $L_1 = \{I_1\}, \{I_2\}, \{I_3\}, \{I_5\}$, which consists of all frequent one-item sets. In order to join with L_1 to generate a candidate set of two-item sets, each pair of $\{I_1\}, \{I_2\}, \{I_3\}, \{I_5\}$ are combined as $\{I_1, I_2\}, \{I_1, I_3\}, \{I_1, I_5\}, \{I_2, I_3\}$, $\{I_2, I_5\}$ and $\{I_3, I_5\}$ as C_2. Consider the Apriori property that all subsets of a frequent itemset must also be frequent, we can determine that those six candidates may remain in C_2. Then the set of frequent two-item sets, L_2, is determined, consisting of $\{I_1, I_3\}, \{I_2, I_3\}$, $\{I_2, I_5\}, \{I_3, I_5\}$, which have support greater than or equal to *min_sp*. Similarly, the set of candidate three-item sets, C_2, can also be generated with L_2. Each pair of $\{I_1, I_3\}, \{I_2, I_3\}$,

$\{I_2, I_5\}, \{I_3, I_5\}$ are combined as $\{I_1, I_2, I_3\}, \{I_1, I_2, I_5\}, \{I_1, I_3, I_5\}$ and $\{I_2, I_3, I_5\}$ as C_3. However, we can find that $\{I_1, I_2\}$ and $\{I_1, I_5\}$ are not frequent two-item sets. Therefore we remove them from C_3, to save the effort of unnecessarily obtaining their counts during the subsequent scan of the database to determine L_3 because C_3 only includes $\{I_2, I_3, I_5\}$. Note that when given a candidate k-itemset, we only need to check if its $(k-1)$-subsets are frequent since the Apriori algorithm uses a level-wise search strategy. This is because there is only one itemset in L_3 so that the *apriori_gen* function is completed.

2.3.2 Generating Association Rules from Frequent Itemsets

Once the frequent itemsets have been searched, it is straightforward to generate association rules. Notice that $sp(I) \geq sp(J)$ where itemsets I are the non-empty subsets of J (*i.e.*, $I \subset J$); if J is a frequent itemset, then for every subset I the association rule $I \Rightarrow J - I$ can be established if $cf\{I \Rightarrow J - I\} \geq \min\_cf$. According to this property, the following algorithm in Figure 2.3 can be used to generate all of association rules.

```
Generating association rule
Input: Frequent itemsets
Output: Association rules
Begin
1. For each frequent itemsets l_k   (k ≥ 2) do {
2.      H_1:={h ∈ l_k}|cf(l_k − {h} ⇒ {h}) ≥ min_cf}
3.      Call Ap-Genrule(l_k, H_1);
4. }

5. Procedure Ap-Genrule(l_k, H_m){
6.    If k > m + 1){
7.          H_{m+1} = apriori_gen(H_m);
8.          For each h_{m+1} ∈ H_{m+1} {
9.               cf = sp((l_k)/ sp(l_k − h_{m+1})
10.              If cf ≥ min_cf then
11.                    Output (l_k − h_{m+1}) ⇒ h_{m+1};
12.              Else
13.                    H_{m+1} := H_{m+1} − {h_{m+1}};
14.              }
15.              Ap-Genrule(l_k, H_{m+1});
16.   }
17. }
End
```

Figure 2.3 Algorithm to generate association rules

In Figure 2.3, H_m is the consequent set of m which has larger confidence than *min_cf*. First in step 2, the consequent sets of one item should be made. For example, by using the data from Table 2.1, $\{I_2, I_3\}$ is a two-item frequent itemset. Suppose

$min\_cf = 60\%$, then

$$cf(I_2 \Rightarrow I_3\} = 66.7\% > \min\_cf$$
$$cf(I_3 \Rightarrow I_2\} = 66.7\% > \min\_cf \ .$$

For generating larger H_{m+1}, we can use the function of the *apriori_gen* (H_m) in step 7, and calculate the confidence of each. If their confidence is larger than *min_cf* then output them as association rules; else withdraw them from H_{m+1} (in steps 8–14). We can also use an example to illustrate the algorithm. Because $\{I_2, I_3\}$ is a frequent itemset then *apriori_gen* $(\{I_2, I_3\})$ makes several three-item sets, but only $\{I_2, I_3, I_5\}$ is frequent. Calculate the confidence of the subset of itemset $\{I_2, I_3, I_5\}$ to give

$$cf(I_2 \cup I_3 \Rightarrow I_5\} = 100\%$$
$$cf(I_2 \cup I_5 \Rightarrow I_3\} = 66.7\%$$
$$cf(I_3 \cup I_5 \Rightarrow I_2\} = 100\%$$
$$cf(I_2 \Rightarrow I_3 \cup I_5\} = 66.7\%$$
$$cf(I_3 \Rightarrow I_2 \cup I_5\} = 66.7\%$$
$$cf(I_5 \Rightarrow I_2 \cup I_3\} = 66.7\% \ .$$

Because the confidence of the six subsets of $\{I_2, I_3, I_5\}$ is larger than *min_cf*, they then become association rules. Moreover, there are no frequent itemsets of four items in Table 2.1 so that the procedure of generating association rules is completed.

2.4 Related Studies on Mining Association Rules in Inventory Database

Since the Apriori is a very basic algorithm for mining frequent itemsets in large databases, many variations of the Apriori have been proposed that focus on improving the efficiency of the original algorithm and on the developing the algorithm for other research fields. There are many excellent publications that summarize this topic, for example, Zhang and Zhang (2002), Zhang *et al.* (2004) and Han and Micheline (2001). In this section, we discuss two typical approaches of mining association rules established in current literature: mining multidimensional association rules from relational databases (Han and Micheline 2001, Fukuda *et al.* 1996a, 1996b, 1999, 2001), and mining association rules with a time-window (Xiao *et al.* 2009), which are considered to be very important in inventory management.

2.4.1 Mining Multidimensional Association Rules from Relational Databases

We have studied association rules that imply a single predicate, for example, the predicate buys. Hence we can refer to the association rules with a form of $A \Rightarrow B$

as a one-dimensional association rule because it contains a single distinct predicate (for example buys) with multiple occurrences. Such association rules can commonly be mined from transactional data.

However, rather than using a transactional database, sales and related information are stored in a relational database. Such a store of data is multidimensional, by definition. For instance, in addition to keeping track of the items purchased in sales transactions, a relational database may record other attributes associated with the items, such as the quantity purchased or the price, or the branch location of the sales. Additional relational information regarding the customers who purchased the items, such as customer age, occupation, credit rating, income, and address, may also be stored. Considering each database attribute as a predicate, it can therefore be interesting to mine association rules containing multiple predicates. Association rules that involve two or more dimensions or predicates can be referred to as multidimensional association rules, such as

$$age(X, "20...29") \wedge occupation(X, "student") \Rightarrow buys(X, "laptop")$$

where X is a variable representing customers who purchased items in relational databases. There are three predicates (age, occupation, and buys) in the above association rules. Note that these predicates can be categorical (nominal attributes) or quantitative (numeric attributes). Here, we just consider a simple case in which quantitative attributes are separated using predefined concept hierarchies, namely optimized association rules, which is proposed by Fukuda *et al.* (1996b, 1999, 2001). This rule has a simple form

$$A \in [v_1, v_2] \Rightarrow B$$

which states that customers who buy item A in the range between v_1 and v_2 are likely to buy item B. If the resulting task-relevant data are stored in a relational table, then the Apriori algorithm requires just a slight modification so as to find all frequent ranges rather than frequent itemsets. Also if an instance of the range is given, the confidence of this rule can be calculated easily. In practice, however, we want to find a range that yields a confident rule. Such a range is called a *confident* range. Unfortunately, a confident range is not always unique, and we may find a confident range that contains only a very small number of items. If the confident range has a maximum support then the association rule is called an *optimized support* rule. This range captures the largest cluster of customers that are likely to buy item B with a probability no less than the given *min_cf*. Instead of the optimized support rule it is also interesting to find the frequent range that has a maximum confidence. Such association rules are called *optimized confidence* rules. This range clarifies a cluster of more than, for instance, 10% of customers that buy item B with the highest confidence.

Tables 2.2 and 2.3 show the examples of the optimized support rule and optimized confidence rule. Suppose $min\_sp = 10\%$ and $min\_cf = 50\%$. We may have many instances of ranges that yield confident and ample rules, shown in Tables 2.2 and 2.3.

Table 2.2 Examples of confidence rules

Range	[1000, 10 000]	[5000, 5500]	[500, 7000]
Support	20%	2%	15%
Confidence	50%	55%	52%

Among those ranges in Table 2.2, range [1000, 10 000] is a candidate range for an optimized support rule.

Table 2.3 Examples of ample rules

Range	[1000, 5000]	[2000, 4000]	[3000, 8000]
Support	13%	10%	11%
Confidence	65%	50%	52%

Among those ranges in Table 2.3, range [1000, 5000] is a candidate range for an optimized confidence rule. It can be considered that although [1000, 5000] is a superset of [2000, 4000], the confidence of the rule of the former range is greater than that of the latter range, but observation will confirm that corresponding situations could really occur.

Fukuda *et al.* (1996b, 1999, 2001) proposed two asymptotically optimal algorithms to find optimized support rules and optimized confidence rules, on the assumption that data are sorted with respect to the numeric attributes. Sorting the database, however, could create a serious problem if the database is much larger than the main memory, because sorting data for each numeric attribute would take an enormous amount of time. To handle giant databases that cannot fit in the main memory, they presented other algorithms for approximating optimized rules, by using randomized algorithms. The essence of those algorithms is that they generated thousands of almost equidepth buckets (*i.e.*, buckets are made by dividing the value of the attribute into a sequence of disjointed ranges, and the size of any bucket is the same), and then combines some of those buckets to create approximately optimized ranges. In order to obtain such almost equidepth buckets, a sample of data is first created that fits into the main memory, thus ensuring the efficiency with which the sample is sorted.

2.4.2 Mining Association Rules with Time-window

Most of the early studies on mining association rules focus on the quantitative property of transaction. The time property of transaction, however, is another important feature that has also attracted many studies in recent years. Transaction time is believed to be valuable for discovering a customer's purchasing patterns over time, *e.g.*, finding periodic association rules. Periodic association rules were first studied by Ramaswamy and Siberschatz (1998) to discover the association rules that repeat in every cycle of a fixed time span. Li *et al.* (2003) presented a level-wise Apriori-based algorithm, named Temporal-Apriori, to discover the calendar-based periodic association rules, where the periodicity is in the form of a calendar, *e.g.*, day, week or month. Lee and Jiang (2008) relaxed the restrictive crisp periodicity of the periodic association rule to a fuzzy one, and developed an algorithm for mining fuzzy periodic association rules.

The periodic association rule is based on the assumption that people always do the same purchasing activities after regular time intervals so that some association rules might not hold on the whole database but on a regularly partitioned database. However, the key drawback of the mining periodic association rule is that the periodicities have to be user-specified, *e.g.*, days, weeks or months, which limit the ability of algorithms on discovering a customer's arbitrary time-cycled activities with unknown interval. Moreover, a customer's purchasing patterns over time are more complex than just periodically repeated behaviors. Many association rules might appear occasionally or repeat asynchronously (Huang and Chang 2005) or with changing periodicities. Finding all of these association rules which appear in different time segments, and then using them to discover all potential patterns of a customer is challenging work.

In practice, there are many candidates of association rules that do not satisfy the thresholds of *min_sp* and *min_cf* over the whole database, but satisfy them in segments of the database partitioned by specified time-windows, which are also called *part-time* association rules. These *part-time* association rules reflect the customer's purchasing pattern at different time phases, which is useful information for timely market management because market managers need to know the behaviors of customers on different time phases. However, these rules cannot be discovered by all of the existing algorithms, including those algorithms for mining periodic association rule. To discover *part-time* association rules, we first need to introduce the notion of the association rule with time-windows (ARTW), which is formally described as follows.

Let $I = \{i_1, i_2, \ldots, i_n\}$ be a set of items. $D = \{t_1, t_2, \cdots t_n\}$ is a set of transactions, where each transaction t has an identifier *tid* and a set of items *t-itemset*, *i.e.*, t = (*tid*, *t-itemset*). A time stamp t_t indicating the time when the transaction occurred, and a set of items t_c such that $t_c \in I$. Let T be the total time span of the real-time database such that $t_t \in T$ for all transactions. Let (t_1, t_2) be a time-window such that $t_1, t_2 \in T$. Let $|D^{(t_1,t_2)}|$ be the number of transactions that occur in (t_1, t_2), where $|D(X)^{(t_1,t_2)}|$ is the number of transactions containing itemset X occurring in (t_1, t_2), and $|(t_1, t_2)|$ is the width of the time-window. An ARTW is an implication of the form $X \Rightarrow^{(t_1,t_2)} Y$, where $X \subseteq I, Y \subseteq I, X \cap Y = \phi$, and $(t_1, t_2) \in T$. This rule has support $s\%$ in time-window (t_1, t_2) if and only if

$$\frac{|D(X)^{(t_1,t_2)}|}{|D^{(t_1,t_2)}|} \geq s\% \ ,$$

and has conference $c\%$ in time-window (t_1, t_2) if and only if

$$\frac{|D(X \cup Y)^{(t_1,t_2)}|}{|D(X)^{(t_1,t_2)}|} \geq c\% \ .$$

As usual, two thresholds, *i.e.*, the *min_sp* and the *min_cf*, are required to determine whether an ARTW holds or not. Besides, another threshold, namely, the minimum time-window, noted as *min_win*, is also required in determining an ARTW. Generally, an association rule with too narrow of a time-window is often meaningless to market management because it does not reflect a stable purchase pattern of customer. Therefore, we use the following criteria to determine whether an ARTW: $X \Rightarrow^{(t_1,t_2)} Y$ holds or not:

1. Its support $s\%$ is greater than or equal to the predefined *min_sp*.
2. Its conference $c\%$ is greater than or equal to the predefined *min_cf*.
3. The width of time-window $|(t_1, t_2)|$ is greater than or equal to the predefined *min_win*.

In addition, to avoid the algorithm tramping in the situation of *one transaction is frequent*, the *min_win* is required to be much greater than $1/min\_sp$.

An ARTW might have many pairs of disjointed time-windows with different widths and different intervals. For example, the association rule may hold in the time-window of (t_1, t_2), (t_3, t_4), ... and so on. Therefore, the periodic association rules are just a subset of the ARTW that are with fixed time-window width and fixed length of interval. Especially, when the *min_win* is equal to the total time span of the whole database, the found ARTW are exactly the traditional notion of association rules that hold on the whole database.

An important index of ARTW, *i.e.*, the *tc%* named time-coverage rate, is employed to represent the strength of ARTW on time length, which is defined as follows:

$$tc\% = \frac{|\,(t_1, t_2)_{A \Rightarrow B}\,|}{|T|} \times 100\% \ ,$$

where $|(t_1, t_2)|_{A \Rightarrow B}$ is the length of the time-window of ARTW $A \Rightarrow^{(t_1,t_2)} B$, and $|T|$ is the total time span of the database. Therefore, an ARTW holds if and only if its time-coverage rate is equal to or greater than the $min\_win/|T|$. Since an association rule may be associated with many different and disjointed time-windows, the same association rule with different time-windows may be recognized as different ARTW by the algorithm. So we merge those ARTW with identical itemset A and itemset B into one. Therefore, an ARTW always indicates an association rule $A \Rightarrow^{\{(t_1,t_2)\}} B$ that holds on a set of disjointed time-windows. The time-coverage of ARTW is consequently changed to:

$$tc\% = \frac{\sum |\,(t_1, t_2)_{A \Rightarrow B}\,|}{|T|} \times 100\% \ .$$

Especially, when $tc\% = 100\%$, the ARTW is degraded to a traditional *full-time* association rule. If we relax the time-coverage rate from 100% to lower values, like 80% or 50%, more association rules are expected to be discovered.

In order to discover all ARTW from a real-time database, as usual, two subproblems are involved. The first is to find all of the frequent itemsets with time-window (FITW), which accordingly indicates the itemsets that are frequent only in a specified time-window. The second is to find out, from the set of FITW discovered in step one, all of ARTW in the database. Since the solution to the second subproblem is straightforward, our research efforts mainly focus on the first subproblem to search all FITW from a real-time database with respect to the specified thresholds of *min_sp*, *min_cf* and *min_win*, efficiently and completely, which can be described as follows.

For a pair of time-windows $(t_1, t_2) \in T$, let $|D^{(t_1,t_2)}|$ be the number of transactions occurring in (t_1, t_2), let $|D(X)^{(t_1,t_2)}|$ be the number of transactions occurring in (t_1, t_2) while containing itemset X, and let $|(t_1, t_2)|$ be the width of time-window. The support of itemsets X in time-window (t_1, t_2) can be defined as

$$\text{support}\left(X^{(t_1,t_2)}\right) = \frac{|D(X)^{(t_1,t_2)}|}{|D^{(t_1,t_2)}|} \times 100\% \; .$$

Therefore, an itemset X is said to be frequent if and only if its support is greater than or equal to the predefined *min_sp* and the width of its time-window is greater than or equal to the predefined *min_win*, *i.e.*, $|(t_1, t_2)| > min\_win$. It is also noticeable that an itemset X may be frequent in many pairs of disjointed time-windows. For this case, all of them will be noted as independent FITW, such as $X^{(t_1,t_2)}$, $X^{(t_3,t_4)}, \ldots$, and so on.

The traditional way to discover frequent itemsets is to use the knowledge that all subsets of a frequent itemsets are also frequent. Analogously, if an itemset X in time-window (t_1, t_2) has a support $s\%$, then all of its subsets will have supports equal to or greater than $s\%$ in the same time-window (t_1, t_2), which is straightforward. Therefore, we develop the similar knowledge that all subsets of an FITW are also FITW. This insight will help us in developing an Apriori-based algorithm to quickly discover all FITW from a large database.

2.5 Summary

Mining association rules from huge amounts of data is useful in business. Market basket analysis studies the buying habits of customers by searching for sets of items that are frequently purchased together. Association rule mining consists of first finding frequent itemsets. Agrawal *et al.* (1993) have proposed a support-confidence framework for discovering association rules. This framework is now widely accepted as a measure of mining association rules. Han and Micheline (2001), Zhang and Zhang (2002), Zhang *et al.* (2004), and Fukuda *et al.* (2001) summarize the topics on various technologies of mining association rules. Because mining association rules in inventory databases focus on the discovery of association relationships among large items data which have some special characteristics of inventory management, we have summarized and discussed briefly the topics related to mining association rules in inventory databases in this chapter. The key points of this chapter are:

1. basic concepts for dealing with association rules;
2. support-confidence framework concerning association rule mining; and
3. discussion of two topics related to inventory data mining: multidimensional association rules and association rules with time-windows.

Acknowledgements The work of mining association rules with a time-window is research conducted by Dr. Yiyong Xiao, Dr. Renqian Zhang and Professor Ikou Kaku. Their contribution is appreciated.

References

Agrawal R, Imilienski T, Swami A (1993) Mining association rules between sets of items in large datasets. In: Proceedings of SIGMOD, pp 207–216

Fukuda T, Morimoto Y, Morishita S, Tokuyama T (1996a) Data mining using two-dimensional optimized association rules: scheme, algorithms, and visualization. In: Proceedings of the ACM SIGMOD Conference on Management of Data, pp 13–23

Fukuda T, Morimoto Y, Morishita S, Tokuyama T (1996b) Mining optimized association rules for numeric attributes. In: Proceedings of the ACM SIGACT-SIGMOD-SIGART Symposium on Principles of Database Systems, pp 182–191

Fukuda T, Morimoto Y, Morishita S, Tokuyama T (1999) Mining optimized association rules for numeric attributes. J Comput Syst Sci 58(1):1–15

Fukuda T, Morimoto Y, Tokuyama T (2001) Data mining in data science series. Kyouritu, Tokyo

Han J, Micheline K (2001) Data mining: concepts and techniques. Morgan Kaufmann, San Fransisco, CA, pp 226–269

Huang KY, Chang CH (2005) SMCA: a general model for mining asynchronous periodic patterns in temporal databases. IEEE Trans Know Data Eng 17(6):774–785

Lee SJ, Jiang YJ (2008) Mining fuzzy periodic association rules. Data Know Eng 65:442–462

Li Y, Ning P, Wang XS, Jajodia S (2003) Discovering calendar-based temporal association rules. Data Know Eng 44(2):193–218

Ramaswamy BS, Siberschatz A (1998) Cyclic association rules. In: Proceedings of the 14th International Conference on Data Engineering, pp 412–421

Xiao YY, Zhang RQ, Kaku I (2009) A framework of mining association rules with time-window on real-time transaction database. In: Proceedings of the 2nd International Conference on Supply Chain Management, pp 412–421

Zhang C, Zhang S (2002) Association rule mining: models and algorithms, chaps 1–2. Springer, Berlin Heidelberg New York, pp 1–46

Zhang C, Zhang S, Wu X (2004) Knowledge discovery in multiple databases, chap 2. Springer, Berlin Heidelberg New York, pp 27–62

Chapter 3
Fuzzy Modeling and Optimization: Theory and Methods

After introducing definitions, properties and the foundation of fuzzy sets theory, this chapter aims to present a brief summary of the theory and methods of fuzzy optimization. We also attempt to give readers a clear and comprehensive understanding of knowledge, from the viewpoint of fuzzy modeling and fuzzy optimization, classification and formulation for the fuzzy optimization problems, models, and some well-known methods. The importance of interpretation of the problem and formulation of an optimal solution in a fuzzy sense are emphasized.

3.1 Introduction

Traditional optimization techniques and methods have been successfully applied for years to solve problems with a well-defined structure/configuration, sometimes known as hard systems. Such optimization problems are usually well formulated by crisply specific objective functions and a specific system of constraints, and solved by precise mathematics. Unfortunately, real world situations are often not deterministic. There exist various types of uncertainties in social, industrial and economic systems, such as randomness of occurrence of events, imprecision and ambiguity of system data and linguistic vagueness, *etc.*, which come from many sources (Simon 1995) including errors of measurement, deficiency in history and statistical data, insufficient theory, incomplete knowledge expression, and the subjectivity and preference of human judgments, *etc.* As pointed out by Zimmermann (1991), various kinds of uncertainties can be categorized as stochastic uncertainty and fuzziness.

Stochastic uncertainty relates to the uncertainty of occurrences of phenomena or events. Its characteristics lie in the fact that descriptions of information are crisp and well defined, however they vary in their frequency of occurrence. Systems with this type of uncertainty are called stochastic systems, which can be solved by stochastic optimization techniques using probability theory. In some other situations, the decision maker (DM) does not think the commonly used probability distribution is always appropriate, especially when the information is vague, relating to human language and behavior, imprecise/ambiguous system data, or when the information could not be described and defined well due to limited knowledge and deficiency in its understanding. Such types of uncertainty are categorized as fuzzy, which can be further classified into ambiguity or vagueness. Vagueness here is associated with the difficulty of making sharp or precise distinctions, *i.e.*, it deals with the situation where the information cannot be valued sharply or cannot be described clearly in linguistic term, such as preference-related information. This type of fuzziness is usually represented by a membership function, which reflects the decision maker's subjectivity and preference on the objects. Ambiguity is associated with the situation in which the choice between two or more alternatives is left unspecified, and the occurrence of each alternative is unknown owing to deficiency in knowledge and tools. It can be further classified into preference-based ambiguity and possibility-based ambiguity from the viewpoint of where the ambiguity comes from. The latter is sometimes called imprecision. If the ambiguity arises from the subjective knowledge or objective tools, *e.g.*, "the processing time is around 2 min," it is a preference-based ambiguity, and is usually characterized by a membership function. If the ambiguity is due to incompleteness, *e.g.*, "the profit of an investment is about $2 or $1.9 to $2.1," it is a possibility-based ambiguity and is usually represented by ordinary intervals, and hence it is characterized by a possibility distribution, which reflects the possibility of occurrence of an event or an object. A system with vague and ambiguous information is called a soft one in which the structure is poorly defined and it reflects human subjectivity and ambiguity/imprecision. It cannot be formulated and solved effectively by traditional mathematics-based optimization techniques nor probability-based stochastic optimization approaches. However, fuzzy set theory (Zadeh 1965, 1978), developed by Zadeh in the 1960s and fuzzy optimization techniques (Zimmermann 1991; Lai and Hwang 1992a, 1992b) provide a useful and efficient tool for modeling and optimizing such systems. Modeling and optimization under a fuzzy environment is called fuzzy modeling and fuzzy optimization.

The study of the theory and methodology of the fuzzy optimization has been active since the concepts of fuzzy decision and the decision model under fuzzy environments were proposed by Bellman and Zadeh in the 1970s (Bellman *et al.* 1970). Various models and approaches to fuzzy linear programming (Fang *et al.* 1999; Fang and Li 1999; Hamacher *et al.* 1978; Han *et al.* 1994; Ishibuchi *et al.* 1994; Rommelfanger 1996; Ramik and Rommelfanger 1996; Tanaka and Asai 1984a,b; Wang 1997; Wang and Fang 1997), fuzzy multi-objective programming (Sakawa and Yano 1989, 1994) fuzzy integer programming (Chanas and Kuchta 1998; Stoica *et al.* 1984), fuzzy dynamic programming (Kacprzyk and Esogbue 1996), possibility linear programming (Dubois 1987; Lai and Hwang 1992a,b; Ramik and Rimanek 1985; Rommelfanger *et al.* 1989; Tanaka and Asai 1984a,b), and fuzzy nonlinear programming (Liu and Fang 2001; Tang and Wang 1997a,b; Tang *et al.* 1998; Trappey *et al.* 1988) have been developed over the years by many researchers. In the meantime, fuzzy ranking (Bortolan *et al.* 1985), fuzzy set operation, sensitivity analysis (Ostermark 1987) and fuzzy dual theory (Verdegay 1984a,b), as well as the application of fuzzy optimization to practical problems also represent important topics. Recent surveys on the advancement of the fuzzy optimization has been found in Delgado *et al.* (1994), Fedrizzi *et al.* (1991a,b), Inuiguchi (1997), Inuiguchi and Ramik (2000), Kacprzyk and Orlovski (1987), and Luhandjula (1989), and especially the systematic survey on the fuzzy linear programming made by Rommelfanger and Slowinski (1998). The survey on other topics of fuzzy optimization like discrete fuzzy optimization and fuzzy ranking have been investigated by Chanas and Kuchta (1998) and Bortolan (1985), respectively. The classification of uncertainties and of uncertain programming has been made by Liu (1999, 2002). The latest survey on fuzzy linear programming is provided by Inuiguchi and Ramik (2000) from a practical point of view. The possibility linear programming is focused and its advantages and disadvantages are discussed in comparison with the stochastic programming approach using examples. There are fruitful literatures and broad topics in this area, it is not easy to embrace them all in one chapter, hence the above surveys serve as an introduction and summarize some advancement and achievements of fuzzy optimization under special cases.

3.2 Basic Terminology and Definition

3.2.1 Definition of Fuzzy Sets

Let X be a classical set of objects, called the universe, whose generic elements are denoted by x. If A is a crisp subset of X, then the membership of x in A can be viewed as the characteristic function $\mu_A(x)$ from X to $\{0, 1\}$ such that

$$\mu_A(x) = \begin{cases} 1 & \text{if and only if} \quad x \in A \\ 0 & \text{otherwise} \end{cases} \tag{3.1}$$

If $\{0, 1\}$ is allowed to be the real interval [0,1], A is called a fuzzy set proposed by Zadeh, and $\mu_A(x)$ denotes the degree of membership of X in A. The closer the value of $\mu_A(x)$ is to 1, the more x belongs to A.

Generally speaking, a fuzzy set A denoted by $\tilde{A}$ is characterized by the set of ordered pairs (Zimmermann 1991):

$$\tilde{A} = \{(x, \mu_{\tilde{A}}(x))|x \in X\}\,.$$

Of course, the characteristic function can be either a membership function or a possibility distribution. If the membership function is preferred, then the characteristic function is usually denoted by $\mu(x)$. On the other hand, if the possibility distribution is preferred, the characteristic function will be specified as $\pi(x)$.

Along with the expression of $\tilde{A} = \{(x, \mu_{\tilde{A}}(x))|x \in X\}$, the following notation may be used in some cases. If $X = \{x_1, x_2, \cdots, x_n\}$ is a finite numerable set, then a fuzzy set $\tilde{A}$ is then expressed as

$$\tilde{A} = \mu_{\tilde{A}}(x_1)/x_1 + \mu_{\tilde{A}}(x_2)/x_2 + \cdots + \mu_{\tilde{A}}(x_n)/x_n\,.$$

For example let $X = \{\text{Li, Wang, Huang, Tang, Zhang}\}$ be a set of five girls; the heights for the five girls are given as follows:

Li: 165 cm Wang: 172 cm Huang: 159 cm Tang: 175 cm Zhang: 168 cm.

Then we select $\tilde{A}$ to be a set of "tall" girls, where "tall" is a linguistic proposition. How do we formulate this set $\tilde{A}$? Which girl belongs to $\tilde{A}$? To deal with those problems, we develop the following procedure.

Step 1 Determine the preferred level, for example we consider that a girl is absolutely tall if her height is 175 cm. On the other hand if her height is less than 160 cm, she is not tall at all. Of course, the preferred value should be accepted due to common sense. Obviously, Tang is absolutely tall, and Huang is absolutely not tall, but how about Zhang, Li and Wang?

Step 2 Calculate the reference value, namely the degree belongs to "tall." From common sense, the higher the height, the greater degree to which the girl belongs to "tall." Generally, we draw a degree-scaled line linearly proportional to the previous height line in order to represent the degree of membership indicating that a girl belongs to A. As a result, the following degrees may be available:
Degree (Li is tall) = 0.3;
Degree (Wang is tall) = 0.8;
Degree (Zhang is tall) = 0.5333.
Therefore $\tilde{A}$ may be characterized by

$$\tilde{A} = \{0.3/\text{Li}, 0.8/\text{Wang}, 0/\text{Huang}, 1/\text{Tang}, 0.5333/\text{Zhang}\}.$$

Certainly, if X is a real set, $\tilde{A}$ may be characterized by a continuous membership function.

3.2.2 Support and Cut Set

The support of a fuzzy set $\tilde{A}$, denoted by *supp* $\tilde{A}$ is a crisp subset of X and is presented as:

$$supp\,\tilde{A} = \{x | \mu_{\tilde{A}}(x) > 0 \quad \text{and} \quad x \in X\}.$$

The α-level (α-cut) set of a fuzzy $\tilde{A}$ is a crisp subset of X and is denoted by

$$A_\alpha = \{x | \mu_{\tilde{A}}(x) > \alpha \quad \text{and} \quad x \in X\}.$$

From the definition of support and α-level cut set, A_α is a subset of *supp* $\tilde{A}$.

3.2.3 Convexity and Concavity

Definition 3.1. A fuzzy set $\tilde{A}$ in X is a convex fuzzy set if and only if for every pair of points x_1 and x_2 in X, the membership of $\tilde{A}$ satisfies the inequality:

$$\mu_{\tilde{A}}(\delta x_1 + (1-\delta)x_2) \geqslant \min\{\mu_{\tilde{A}}(x_1), \mu_{\tilde{A}}(x_2)\}$$

where, $\delta \in [0, 1]$.

Alternatively, a fuzzy set is convex if all α-level sets are convex. Similarly, we can define a concave fuzzy set.

3.3 Operations and Properties for Generally Used Fuzzy Numbers

3.3.1 Fuzzy Inequality with Tolerance

Fuzzy inequality with tolerance has the following general form:

$$y \tilde{\leqslant} 0 \tag{3.2}$$

where $\tilde{\leqslant}$ and $\tilde{\geqslant}$ means tolerance in the inequality.

Equation 3.2 may be viewed as a fuzzy subset denoted by $\tilde{I}_1 = \{y | y \tilde{\leqslant} 0\}$, characterized by $\mu_{\tilde{I}}(y)$ which satisfies:

$$\begin{cases} \mu_{\tilde{I}_1}(y) = 1 & y \in (-\infty, 0] \\ 0 < \mu_{\tilde{I}_1}(y) < 1 \quad \text{and is monotonic non-increasing} & y \in (0, \infty) \\ \mu_{\tilde{I}_1}(-\infty) = 0 & \end{cases}$$

Definition 3.2. A fuzzy set $y\, tilde \leqslant 0$ is a fuzzy inequality with positive tolerance if the membership functions $\mu_{\tilde{I}_1}(y)$ satisfy the above properties.

Similarly, we may define a fuzzy inequality with negative tolerance $y \tilde{\geqslant} 0$ as a fuzzy set $\tilde{I}_2$, which the membership function $\mu_{\tilde{I}_2}(y)$ satisfies:

$$\begin{cases} \mu_{\tilde{I}_2}(y) = 1 & y \in [0, +\infty] \\ 0 < \mu_{\tilde{I}_2}(y) < 1 \quad \text{and is monotonic non-increasing} & y \in (-\infty, 0) \\ \mu_{\tilde{I}_2}(-\infty) = 0 & \end{cases}$$

$\mu_{\tilde{I}_1}(y)$ and $\mu_{\tilde{I}_2}(y)$ stand for the degree of satisfaction with the tolerances that exist in the constraints $y \tilde{\leqslant} 0$ and $y \tilde{\geqslant} 0$, respectively.

Generally, formulate $\mu_{\tilde{I}_1}(y)$ and $\mu_{\tilde{I}_2}(y)$ as

$$\mu_{\tilde{I}_1}(y) = \begin{cases} 1 & y \leqslant 0 \\ \dfrac{1}{1 + y^{\delta}} & y > 0 \end{cases} \tag{3.3}$$

$$\mu_{\tilde{I}_2}(y) = \begin{cases} 1 & y \geqslant 0 \\ \dfrac{1}{1 + (-y)^{\delta}} & y < 0 \end{cases} \tag{3.4}$$

where δ is a positive constant. The term $y \tilde{\lessapprox} 0$ ($y \tilde{\gtrapprox} 0$) with membership functions as in Equations 3.3 and 3.4 is called an infinite tolerance type fuzzy inequality.

Contrarily, given the largest acceptable tolerance (LAT > 0), $y \tilde{\lessapprox} 0$ ($y \tilde{\gtrapprox} 0$) may then be characterized by

$$\mu_{\tilde{I}_1}(y) = \begin{cases} 1 & y \leqslant 0 \\ 1 - \left(\frac{y}{\text{LAT}}\right)^{\delta} & 0 < y \leqslant \text{LAT} \\ 0 & y > LAT \end{cases} \tag{3.5}$$

$$\mu_{\tilde{I}_2}(y) = \begin{cases} 1 & y \geqslant 0 \\ 1 - \left(\frac{-y}{\text{LAT}}\right)^{\delta} & -\text{LAT} < y < 0 \\ 0 & y \leqslant -\text{LAT} \end{cases} \tag{3.6}$$

The above fuzzy inequality with largest acceptable tolerance is called a definite tolerance type fuzzy inequality.

3.3.2 Interval Numbers

Let R be a real number domain.

Definition 3.3. A closed interval $[m, n] \in R(m \leqslant n)$ is called a closed interval number if the algebraic operations satisfy the extension principle.

Similarly, we may define an open interval number as (m, n), a semi-open/semiclosed interval number as $(m, n]$ or $[m, n)$.

Generally, we denote the following interval number by $|m, n|$. As a fuzzy number in real domain R, the interval number $|m, n|$ characterized by $\mu_{\tilde{R}}(x)$ is defined by

$$\mu_{\tilde{R}}(x) = \begin{cases} 1 & x \in |m, n| \\ L(x) & x < m \\ R(x) & x > n \end{cases} \tag{3.7}$$

where $L(x)$ and $R(x)$ are non-decreasing and non-increasing, respectively, and satisfy $\lim_{x \to -\infty} L(x) = 0$ and $\lim_{x \to \infty} R(x) = 0$.

Algebraic operations for an interval number satisfy the following:

Property 3.1. $|a, b| + |c, d| = |a + c, b + d|$

Property 3.2. $|a, b| - |c, d| = |a - c, b - d|$

Property 3.3. $|a, b| * |c, d| = |ac \wedge ad \wedge bc \wedge bd, ac \vee ad \vee bc \vee bd|$

Property 3.4.

$$|a, b| \div |c, d| = |a/c \wedge a/d \wedge b/c \wedge b/d, a/c \vee a/d \vee b/c \vee b/d|$$

where $\wedge$ and $\vee$ indicate the operations min and max, respectively.

3.3.3 L–R Type Fuzzy Number

Definition 3.4. A fuzzy number $\tilde{m}$ is of L-R type if there exist reference functions L and R, and scalars $\alpha > 0, \beta > 0$, such that the membership functions have the following type:

$$\mu_{\tilde{m}}(x) = \begin{cases} L\left(\dfrac{m-x}{\alpha}\right) & x \leqslant m \\ R\left(\dfrac{x-m}{\beta}\right) & x \geqslant m\,. \end{cases} \tag{3.8}$$

L-R type fuzzy number is denoted by $(m, \alpha, \beta)_{LR}$, where $L(x)$,$R(x)$ are left and right reference functions, and α and β are called left and right spreads, respectively.

Algebraic operations for L-R type fuzzy number satisfy the following: set $\tilde{m} = (m, \alpha, \beta)_{LR}, \tilde{n} = (n, \gamma, \delta)_{LR}$.

Property 3.5. $\tilde{m} + \tilde{n} = (m + n, \alpha + \gamma, \beta + \delta)_{LR}$

Property 3.6. $-\tilde{m} = (-m, \beta, \alpha)_{RL}$

Property 3.7. $\tilde{m} - \tilde{n} = (m - n, \alpha + \delta, \beta + \gamma)_{LR}$

Property 3.8. $\tilde{m} \leqslant \tilde{n} \Leftarrow m \leqslant n, \alpha \geqslant \gamma, \beta \leqslant \delta$

Let $\alpha = \beta = 0, (m, \alpha, \beta)_{LR} = m$, namely general real number m is a special $L - R$ type fuzzy number with left and right spreads of 0.

A function $L(x)$ is called a reference function (Zimmermann 1985) if

1. $L(-x) = L(x)$;
2. $L(0) = 1$;
3. $L(x)$ is decreasing on $[0, \infty)$.

In the following we introduce some commonly used reference function $L(x)$.

1. $L(x) = \max\{0, 1 - |x|^p\}$;
2. $L(x) = e^{-|x|^p}$;
3. $L(x) = \frac{1}{1+|x|^p}$;
4. $L(x) = \begin{cases} 1 & x \in [-p, +p] \\ 0 & \text{else} \end{cases}$.

3.3.4 Triangular Type Fuzzy Number

Definition 3.5. A fuzzy number $\tilde{m}$ is a triangular type fuzzy number if it has the membership function $\mu_{\tilde{m}}(x)$ defined by

$$\mu_{\tilde{m}}(x) = \begin{cases} 0 & x \leqslant m - \alpha \\ \dfrac{x - (m - \alpha)}{\alpha} & m - \alpha < x \leqslant m \\ \dfrac{m + \beta - x}{\beta} & m < x \leqslant m + \beta \\ 0 & x > m + \beta \end{cases} \tag{3.9}$$

Triangular type fuzzy number $\tilde{m}$ is a special $L-R$ type fuzzy number, denoted by (m,α,β).

Algebraic operations for triangular type fuzzy number satisfy the following.

Property 3.9. $(m,\alpha,\beta)+(n,\gamma,\delta)=(m+n,\alpha+\gamma,\beta+\delta)$.

Property 3.10. $-(m,\alpha,\beta)=(-m,\beta,\alpha)$.

Property 3.11. $(m,\alpha,\beta)-(n,\gamma,\delta)=(m-n,\alpha+\gamma,\beta+\delta)$.

Property 3.12. $(m,\alpha,\beta)\leqslant(n,\gamma,\delta)\Leftarrow m\leqslant n,\alpha\geqslant\gamma,\beta\leqslant\delta$.

Property 3.13. $(m,\alpha,\beta)\leqslant 0\Leftarrow m\leqslant 0,\alpha\geqslant 0,\beta=0$.

Property 3.14. If $m,n>0,(m,\alpha,\beta)*(n,\gamma,\delta)\simeq(mm,m\gamma+n\alpha,m\delta+n\beta)$.

Property 3.15. If $a,m>0,a*(m,\alpha,\beta)=(am,a\alpha,a\beta)$.

Property 3.16. If $a<0,m>0,a*(m,\alpha,\beta)=(-am,-a\beta,-a\alpha)$.

The *problem* is how to define and formulate $\tilde{M}\geqslant\tilde{N}(\tilde{M}\leqslant\tilde{N})$.

3.3.5 Trapezoidal Fuzzy Numbers

Definition 3.6. A fuzzy number $\tilde{M}$ is trapezoidal fuzzy number denoted by (m,n,α,β) if $\exists m,n\in R,m\neq n$ such that the membership functions $\mu_{\tilde{M}}(x)$ satisfy the following:

1. $\mu_{\tilde{M}}(x)=1,\quad$ *for* $x\in[m,n]$.
2. $\mu_{\tilde{M}}(x)=1-\frac{m-x}{\alpha},\quad$ *for* $m-\alpha\leqslant x\leqslant m$.
3. $\mu_{\tilde{M}}(x)=1-\frac{x-n}{\beta},\quad$ *for* $n<x\leqslant n+\beta$.
4. $\mu_{\tilde{M}}(x)=0,\quad$ *for else.*

Trapezoidal fuzzy number is a special case of interval number. Algebraic operations for trapezoidal numbers satisfy the following:

Set $\tilde{M}_1=(m_1,n_1,\alpha_1,\beta_1),\tilde{M}_2=(m_2,n_2,\alpha_2,\beta_2)$.

Property 3.17. $\tilde{M}_1+\tilde{M}_2=(m_1+m_2,n_1+n_2,\alpha_1+\alpha_2,\beta_1+\beta_2)$.

Property 3.18. $-\tilde{M}_1=(-n_1,-m_1,\beta_1,\alpha_1)$.

Property 3.19. $\tilde{M}_1-\tilde{M}_2=(m_1-m_2,n_1-n_2,\alpha_1-\alpha_2,\beta_1-\beta_2)$.

Property 3.20. $\tilde{M}_1\leqslant\tilde{M}_2\Leftarrow m_1\leqslant m_2,n_1\leqslant n_2,m_1-\alpha_1\leqslant m_2-\alpha_2,n_1+\beta_1\leqslant n_2+\beta_2$.

Generally, $\forall[a_1,a_2]\in\tilde{A},[b_1,b_2]\in\tilde{B}$ define $\mathrm{Max}\{\tilde{A},\tilde{B}\},\mathrm{Min}\{\tilde{A},\tilde{B}\}$, as:

$$\mathrm{Max}\{\tilde{A},\tilde{B}\}=[\max\{a_1,b_1\},\max\{a_2,b_2\}]\,.$$

$$\mathrm{Min}\{\tilde{A},\tilde{B}\}=[\min\{a_1,b_1\},\min\{a_2,b_2\}]\,.$$

The *problem* is how to define and formulate $\tilde{M}\geqslant\tilde{N}(\tilde{M}\leqslant\tilde{N})$.

3.4 Fuzzy Modeling and Fuzzy Optimization

To understand and solve a complex problem under a fuzzy environment effectively, two tasks should be accomplished, *i.e.*, fuzzy modeling and fuzzy optimization. Fuzzy modeling aims to build an appropriate model based upon the understanding of the problem and analysis of the fuzzy information. However, fuzzy optimization aims at solving the fuzzy model "optimally" by means of optimization techniques and tools based on the formulation of the fuzzy information in terms of their membership functions and/or possibility distribution functions, *etc.*. Generally speaking, these tasks represent two different processes, however, there are no precise boundaries between them. The whole process for applying fuzzy optimization to solve a complex problem can be decomposed into seven stages as follows:

Stage 1 Understand the problem. In this stage, the state, constraints and goals of the system, as well as the relationships among them are understood clearly and expressed by sets.

Stage 2 Fuzziness analysis. Based on understanding of the background of the problem, such questions, like which kind of fuzzy information (elements) is involved and what is the position (*e.g.*, fuzzy goal, fuzzy system of constraints, fuzzy coefficients) it takes, as well as the way (*e.g.*, ambiguity/imprecision in quantity, vagueness in linguistic) in which it is expressed, should be analyzed and summarized. In this stage the fuzzy information is usually expressed in a semantic way.

Stage 3 Development of fuzzy model. Based upon stages 1 and 2, an appropriate fuzzy optimization model will be built by adopting some mathematical tools, catering to the characteristics of the problem. There are two methods for developing fuzzy models, *i.e.*, using the principles of cause-and-effect and those of transition, and using ordinary equations to express the cause-and-effect relationships. During model development, sets and logic relationships are first established and justified. The optimization model may take the form of fuzzy linear programming, fuzzy nonlinear programming, fuzzy dynamic programming, fuzzy multi-objective programming, or positivistic linear programming.

Stage 4 Description and formulation of the fuzzy information. On the basis of the stage 2, the fuzzy information including ambiguity and vagueness has been distinguished. What remains is the task to quantify the information in terms of appropriate tools and theory using fuzzy mathematics. In light of the nature and the way the fuzzy information is expressed, a membership function or a possibility distribution function can be selected to formulate it. The membership function is subjectively determined, and preference-based, which reflects the decision maker's (DM) preference on the objects. It usually applies to the situations involving the human factor with all its vagueness of perception, subjectivity, goals and conception, *e.g.*, fuzzy goals with aspiration, fuzzy constraints with tolerance. Such goals and constraints are expressed vaguely without sharp thresholds to give the necessary flexibility and elasticity. Nevertheless, the possibility distribution function expresses the possibility measure of occurrence of an event or an object, and it can be constructed in an objective or subjective way. It usually applies to the cases where ambiguity in natural language and/or values is involved, *e.g.*, ambiguous coefficients/parameters in the objective function and/or the system of constraints. These coefficients are considered as positivistic variables restricted by a possibility distribution. The membership function or the possibility distribution function may take a linear or nonlinear form, reflecting the DM's preference and understanding of the problem. This stage takes care of the transition from fuzzy modeling to fuzzy optimization.

Stage 5 Transform the fuzzy optimization model into an equivalent or an approximate crisp optimization model. It consists of three procedures, *i.e.*, determination of types of the optimal solution, interpretation and transformation. First of all, the type of the optimal solution is determined, depending on the understanding of the problem and the preference of the DM. That is to say, selection of the type of the optimal solution to a fuzzy model depends absolutely on understanding and definition of the optimal solution in a fuzzy sense. The subsequent task is to propose an appropriate interpretation method and some new concepts to support the understanding and definition of the optimal solution, based on theories and principles on fuzzy mathematics, such as fuzzy ranking, the extension principle, fuzzy arithmetic, *etc.*. The interpretation procedure is important for the following procedures. Some well-known interpretations are reviewed in Inuiguchi and Ramik (2000). Finally, the fuzzy model is transformed into an equivalent or approximate crisp optimization model on the basis of the interpretation. For a fuzzy model, different forms of crisp optimization models may be built depending on different types of the optimal solution and interpretations applied.

Stage 6 Solve the crisp optimization model. In light of the characteristics of the crisp optimization model, such as linear or nonlinear, single objective or multiple objectives, decision variable with continuous, discrete or mixed mode, appropriate optimization techniques and algorithms, *e.g.*, traditional heuristic algorithm or intelligent optimization techniques like genetic algorithm (Wang and Fang 1997; Tang and Wang 1997a,b; Jivenez and Verdegay 1999), rule-based system approaches (Dubois and Prade 1994) or hybrid algorithms, can all be adopted or developed for solving the model.

Stage 7 Validity examination. As indicated in Inuiguchi and Ramik (2000), the obtained optimal/efficient solution in stage 6 is not always acceptable, so there is a need to check its validity. If the solution is unreasonable, the fuzzy modeling process and/or the subsequent optimization process should be improved iteratively.

Among the above stages, 4 through 6 indicate that the basic procedure of fuzzy optimization is to transform a fuzzy model into a deterministic/crisp optimization one, and the most important task is how to make this transformation. During the transformation, the first thing to do is to understand the problem and then to determine the type of optimal solution, *e.g.*, a deterministic solution or a fuzzy solution, according to the understanding. Then, an appropriate interpretation and some concepts for supporting the understanding and definition of the optimal solution are proposed, and finally a transformation approach can be developed based on the interpretation. The selection of a particular approach to a fuzzy optimization problem depends on several factors including the nature of the problems, the DM's preference and the ranking of the objective as well as its evaluation.

3.5 Classification of a Fuzzy Optimization Problem

An optimization problem consists of two fundamental elements, *i.e.*, a goal or a utility function and a set of feasible domains. As indicated by Dubois and Prade (1994), fuzzy optimization refers to the search for extremism of a real-valued function when the function is fuzzily valued, and/or when the domain is fuzzily bounded. With this understanding, a fuzzy optimization problem (FOP) can be described as follows.

Let universe $X = \{x\}$ be a set of alternatives, X_1 is a subset or a fuzzy subset of X, the objective/utility function is a mapping $f\colon X_1 \to L(R)$, where $L(R)$ is a subset or a class of fuzzy subsets of real-value set R. The feasible domain is described by a subset or a fuzzy set $C \subset X$, with a membership function $\mu_C(X) \in [0, 1]$, which denotes the degree of feasibility of x. In this case, a fuzzy optimization problem may be generally expressed as FOP (Fedrizzi *et al.* 1991a,b; Kacprzyk and Orlovski 1987):

$$f(x, r) \to \max_{x \in C} \tag{3.10}$$

where r is either a crisp constant or a fuzzy coefficient. Equation 3.10 says the following: find an x "belonging" to domain C such that $f(x, r)$ can reach a possible "maximum", in a fuzzy sense. This can be interpreted in various ways, *e.g.*, the way as explained by Zimmermann (1976).

How to interpret the terms "belonging" and "maximum" in a fuzzy sense in Equation 3.10 constitutes the diversity of the FOP, which will be clarified and focused on in the coming section. Hence, Equation 3.10 is just a sketch of the FOP.

Similar to deterministic optimization problems, in general, the FOP may be classified into two different types, namely, fuzzy extreme problems and fuzzy mathematical programming problems. This chapter mainly discusses the fuzzy mathematical programming problems.

3.5.1 *Classification of the Fuzzy Extreme Problems*

The fuzzy extreme problems, *i.e.*, extreme of a fuzzy function, are also known as unconstrained fuzzy optimization problems, in which the domain C equals X. The fuzzy extreme problems generally can be described in the following two forms, depending on the definition of the fuzzy function (Dubois and Prade 1980).

1. Fuzzy extreme based on the fuzzy function defined from a fuzzy domain to a fuzzy domain. It has the following form:

$$\tilde{Y} = f(\tilde{X}, r) \to \max / \min \,, \tag{3.11}$$

where $\tilde{X} \subset X$ is a fuzzy set in X. The term $f\colon X \to R$ is a classical real-valued function from the fuzzy domain $\tilde{X}$ to the fuzzy domain $\tilde{Y} \subset R$; and $f(\tilde{X}, r)$ is a fuzzy function, hence a subset of R. The membership function of the fuzzy function $f(\tilde{X}, r)$ satisfies:

$$\mu_{\tilde{Y}}(y) = \sup_{f(x,r)=y} \mu_{\tilde{X}}(x) \,. \tag{3.12}$$

Equation 3.12 says that there exists an x in the fuzzy domain $\tilde{X}$ of X, at which the crisp function obtains an "extreme"

$$\tilde{f}(x, r) \to \max / \min \,. \tag{3.13}$$

2. Fuzzy extreme based on the fuzzy function defined from a crisp domain to a fuzzy domain.

Let X, Y be the universes, $\tilde{P}(Y)$ is the set of all fuzzy sets in Y, and $\tilde{f}$: $X \to \tilde{P}(Y)$ is a fuzzy function, defined by the membership function $\mu_{\tilde{f}(x,r)}(y) = \mu_{\tilde{R}}(x, y)$, and $\mu_{\tilde{R}}(x, y), \forall (x, y) \in X * Y$ is the membership function of a fuzzy relation. Equation 3.13 aims to find an x in X such that the function $\tilde{f}(x, r)$ defined by a fuzzy relation reaches "maximum" or "minimum." The coefficient r in the fuzzy function is usually a fuzzy number, and the fuzziness of the function comes from the coefficient. Hence, this type of fuzzy function is denoted by $f(x, \tilde{r})$ in what follows for the sake of convenience.

In any form of the fuzzy extreme problems, the extreme of the function is not unique, and there are no unique relationships between the extreme of the objective function and the notion of the optimal decision. The solution to the fuzzy extreme problem depends on the ways in which the extreme of the function is interpreted. Possible interpretations of the fuzzy extreme can be found in Dubois and Prade (1980). The concepts of maximizing set (Zimmermann 1991), maximum and minimum of fuzzy numbers and some integral methods for fuzzy ranking can be applied to solve the fuzzy extreme problems.

3.5.2 Classification of the Fuzzy Mathematical Programming Problems

Fuzzy mathematical programming (FMP) problems are also known as constrained fuzzy optimization problems. It can be generally expressed in the following form:

$$\begin{aligned} & f(x, r) \to \max \\ & s.t.\ x \in C = \left\{x \in X \,|\, g_i(x, s) \,\tilde{\leqq}\, 0, i = 1, 2, \cdots m\right\} . \end{aligned} \tag{3.14}$$

In this case, the domain C may be formulated as a crisp system of constraints or fuzzy system of constraints in terms of fuzzy equations, fuzzy inequalities, inequalities/equations with fuzzy coefficients, whereas the $f(x,r)$ may be either a crisp objective function or an objective function with fuzzy coefficients. The goal of the problem, C_0, is expressed by $f(x,r) \to \max$, which may be a fuzzy goal denoting $m\tilde{a}x$ or a crisp one.

Recently many methods have been proposed for classifying fuzzy mathematical programming. Zimmermann (1985) classified the fuzzy mathematical programming into symmetric and asymmetric models. Luhandjula (1989) categorized the fuzzy mathematical programming into flexible programming, fuzzy stochastic programming and mathematical programming with the fuzzy coefficients. Inuiguchi and Ramik (2000) further classified the fuzzy mathematical programming into the following three categories in view of the kinds of uncertainties involved in the problems:

- fuzzy mathematical programming with vagueness, *i.e.*, flexible programming;
- fuzzy mathematical programming with ambiguity, *i.e.*, positivistic programming; and
- fuzzy mathematical programming with vagueness and ambiguity, *i.e.*, robust programming.

In the authors' opinion, the formulation and classification of the fuzzy mathematical programming problems depend in what and where the fuzziness are involved. The fuzziness may emerge in the following possible ways:

1. fuzzy goal, *i.e.*, the goal which is expressed vaguely, and usually with an aspiration level, and the target value of the objective function has some leeway, *e.g.*, the target value of the objective function $f(x,r)$ is achieved as maximum as possible;
2. fuzzy constraints, which represent the system of constraints with tolerances or elasticity in terms of $\tilde{\leqslant}, \tilde{\geqslant}$ or $\tilde{=}$; and
3. fuzzy coefficients in the objective function and/or the system of constraints.

From the viewpoint of the way the fuzziness emerged and the coefficients involved in the objective function and/or the system of constraints in the problems, fuzzy mathematical programming problems are classified into FMP with crisp coefficients and the FMP with fuzzy coefficients, including:

FMP1-FMP with fuzzy goals C_0 and fuzzy constraints C, *i.e.*,

$$\begin{cases} m\tilde{a}x\ f(x,r) \\ s.t.\ x \in C \end{cases} \tag{3.15}$$

FMP2-FMP with fuzzy constraints C, *i.e.*,

$$\begin{cases} \max f(x,r) \\ s.t.\ x \in C \end{cases} \tag{3.16}$$

FMP3-FMP with fuzzy constraints C and fuzzy coefficients in the objective function $f(x, \tilde{r})$, *i.e.*,

$$\begin{cases} \max f(x, \tilde{r}) \\ s.t.\ x \in C \end{cases} \tag{3.17}$$

FMP4-FMP with fuzzy goal C_0 and fuzzy coefficients in the system of constraints $C(x, \tilde{s})$, *i.e.*,

$$\begin{cases} \tilde{\max} f(x, r) \\ s.t.\ x \in C(x, \tilde{s}) \end{cases} \tag{3.18}$$

FMP5-FMP with fuzzy coefficients in the objective function $f(x, \tilde{r})$, *i.e.*,

$$\begin{cases} \max f(x, \tilde{r}) \\ s.t.\ x \in C(x, s) \end{cases} \tag{3.19}$$

FMP6-FMP with fuzzy coefficients in the system of constraints $C(x, \tilde{s})$, *i.e.*,

$$\begin{cases} \max f(x, r) \\ s.t.\ x \in C(x, \tilde{s}) \end{cases} \tag{3.20}$$

FMP7-FMP with fuzzy coefficients in the objective function $f(x, \tilde{r})$ and the system of constraints $C(x, \tilde{s})$, *i.e.*,

$$\begin{cases} \max f(x, \tilde{r}) \\ s.t.\ x \in C(x, \tilde{s}) \end{cases} \tag{3.21}$$

Here the problems FMP1, FMP3, FMP4 and FMP7 are referred to as symmetric, while the FMP2, FMP5 and FMP6 are asymmetric problems, with regards to fuzziness. That means a problem is classified as symmetric or asymmetric from the viewpoint of fuzziness involved in the goal (or objective function) and/or the system of constraints. The symbol $f(x, \tilde{r})$ representing the objective function with the fuzzy coefficients is used to distinguish them from the fuzzy goal C_0. The notation $C(x, \tilde{s})$ is used to represent the system of constraints with the fuzzy coefficients in order to distinguish them from the fuzzy constraints C. This classification is adopted in the rest of the chapter.

The fuzzy goal and fuzzy constraints are characterized by a preference-based membership function. In comparison with the category by Inuiguchi and Ramik (2000), FMP1 and FMP2 are in the category of flexible programming problems. The fuzzy coefficients in the objective function and in the system of constraints may be characterized by a preference-based membership function and a possibility distribution function. When a fuzzy coefficient is formulated by a possibility distribution function, it is viewed upon as a positivistic variable restricted by the possibility distribution. In this case, FMP5, FMP6 and FMP7 are so-called possibility programming problems, denoted by PMP5, PMP6 and PMP7, respectively, hereafter, and FMP3 and FMP4 are robust programming problems, denoted by PMP3 and PMP4, respectively.

3.5.3 Classification of the Fuzzy Linear Programming Problems

Owing to the simplicity of the linear programming formulation and the existence of some developed software for optimization, linear programming has been an important and most frequently applied operations research technique for real-world problems. Since the introduction of fuzzy sets theory into traditional linear programming problems by Zimmermann (1976) and the fuzzy decision concept proposed by Bellman and Zadeh (1970), the fuzzy linear programming (FLP) has been developed in a number of directions with successful applications. It has been an important area of fuzzy optimization. Hence, classification of the fuzzy linear programming problems is emphasized as follows. Traditional linear programming problems can be presented in the following general form:

$$\begin{aligned} &\max c^T x \\ &s.t.\ Ax \leqslant b.x \geqslant 0 \end{aligned} \tag{3.22}$$

where $c^T = (c_1, c_2, \cdots, c_n), A = (A_{ij})_{mn}, b = (b_1, b_2, \cdots, b_n)^T$, $x = (x_1, x_2, \cdots, x_n)^T$ are the benefit coefficient vector, technical coefficient matrix, resources vector and decision variable vector, respectively.

The formulation of a linear programming problem under fuzzy environment depends on in what and where the fuzziness is introduced. In general, fuzziness may be initiated in fuzzy linear programming problems in the following ways:

1. The fuzzy goal, *i.e.*, the maximum of the linear objective function is expressed vaguely and usually with an aspiration level, and it has flexibility, *e.g.*, the target value of the objective function $c^T x$ is deemed maximum as possible and pursues an aspiration level.
2. The fuzzy constraints, *i.e.*, linear system of constraints expressed by fuzzy relations in terms of fuzzy equations or/and fuzzy inequalities.
3. The objective function with the fuzzy benefit coefficients $\tilde{c}_i$.
4. The linear system of constraints with the fuzzy technical coefficients $\tilde{A}_{ij}$ and/or fuzzy resources/thresholds $\tilde{b}_i$.

Based on the above cases, the fuzzy linear programming problems can be categorized as follows:

Category I FLP with crisp coefficients. In this type of the FLP, the goal and/or the system of constraints is/are formulated by DMs in a vague and subjective way. The goal and the system of constraints are called the fuzzy goal and the fuzzy constraints, respectively. This type of FLP includes:

- FLP1-FLP with the fuzzy goals and the fuzzy constraints, *i.e.*, the goal of the objective function is formulated vaguely, *e.g.*, in terms of $\widetilde{\max}$, and the linear system of constraints are defined by fuzzy relations ($\tilde{\leqslant}$) with tolerances, *e.g.*, FLP as defined by Zimmermann (1991).

- FLP2-FLP with the fuzzy constraints, *i.e.*, the linear system of constraints are defined by fuzzy relation ($\tilde{\lesssim}$) with tolerances.

Category II FLP with fuzzy coefficients. In this type of the FLP, some or all of the coefficients are ambiguous, and can usually be expressed by fuzzy numbers. This type of the FLP comprises the backbone of a FLP, and it includes:

- FLP3-FLP with fuzzy constraints and fuzzy objective coefficients $\tilde{c}_i$;
- FLP4-FLP with fuzzy goal and fuzzy technical coefficients $\tilde{A}_i$ and/or the fuzzy resources/thresholds $\tilde{b}_i$;
- FLP5-FLP with fuzzy objective coefficients, *i.e.*, the benefit coefficients $\tilde{c}_i$ in the objective function are fuzzy numbers;
- FLP6-FLP with fuzzy technical coefficients and fuzzy thresholds, *i.e.*, the technical coefficients $\tilde{A}_i$ and threshold $\tilde{b}_i$ are fuzzy numbers; and
- FLP7-FLP with fuzzy coefficients, *i.e.*, the benefit coefficients, technical coefficients and resources/thresholds, are all fuzzy numbers.

The detailed formulation of the above classes of the FLP is given in Lai and Hwang (1992a,b). In the sense that fuzzy constraints are defined as fuzzy inequalities with tolerances, they are equivalent to fuzzy resources/thresholds in FLP. If fuzzy coefficients are modeled by a possibility distribution, the corresponding FLP is a possibility linear programming (PLP) problems. Under this circumstance, corresponding to the classification by Inuiguchi and Ramik (2000), FLP1-FLP2 is called the flexible programming, FLP3-FLP4 are the robust programming, and FLP5-FLP7 are the positivistic programming.

Other methods of classification of FLP can be found in Inuiguchi and Ramik (2000), Luhandjula (1989) and Rommelfanger and Slowinski (1998). Here fuzzy linear programming is distinguished from the positivistic linear programming.

3.6 Brief Summary of Solution Methods for FOP

Since the concept of the fuzzy decision was first proposed by Bellman and Zadeh (1970), fuzzy optimization has received much attention, and various models and methods have been proposed by many researchers. A recent survey on fuzzy optimization techniques can be found in Delgado *et al.* (1994), Inuiguchi and Ramik (2000), Kacprzyk and Orlovski (1987), Luhandjula (1989), Rommelfanger and Slowinski (1998), and Dubois and Prade (1994), focusing on a special category of fuzzy mathematical programming. Owing to the increasing work on the fuzzy mathematical programming, it is impossible to embrace all of the techniques in one chapter; hence, we will just have a brief summary on the techniques for FMP with vagueness and FMP with the fuzzy coefficients characterized by the membership functions. The approaches to the positivistic programming problems POP5–POP7 have been summarized in Inuiguchi and Ramik (2000). This brief summary is made trying to emphasize the understanding and interpretation of the problem and the optimal solution in a fuzzy sense.

3.6.1 Symmetric Approaches Based on Fuzzy Decision

Symmetric approach is an important approach to the fuzzy optimization problems, especially for FMP1. The word "symmetric" used here comes originally from the symmetric model by Zimmermann (1976). The symmetric approaches here cited by many researchers (Luhandjula 1989), usually refer to the approaches proposed by Bellman and Zadeh (1970), Tanaka *et al.* (1974) and Zimmermann (1976) to FMP1 firstly. It is then extended to represent a type of approach to symmetric mathematical programming models in the sense that the goals and the system of constraints involved in the problem are dealt with in a symmetric way with regards to fuzziness. It means that the scope of the symmetric and the asymmetric approach is made from the perspective of the ways in which the goal and the system of constraints are treated, and not from the viewpoint of the problem itself. The symmetric/asymmetric way in which the goals and the system of constraints are treated is understood in the same way as with the symmetric/asymmetric model. In this sense, the symmetric or asymmetric approach is named according to the symmetric or asymmetric model, and not to the symmetric or asymmetric problem. Symmetric approaches based on fuzzy decision are summarized as follows.

These types of approaches were developed originally to deal with decision-making problems with fuzzy goals and fuzzy constraints, *i.e.*, FMP1, based on the concept of the fuzzy decision, as proposed by Bellman and Zadeh (1970). In viewpoint of Bellman and Zadeh, symmetry between the goals and the constraints is an important feature in decision making under fuzzy environment, and the fuzzy goals and the fuzzy constraints can be considered to play the same roles in the problem, and hence can be dealt with symmetrically. The fuzzy decision is defined as a fuzzy set of alternatives resulting from the intersection of the goals and the constraints. By introducing the fuzzy decision D, the solution to FMP1 can be interpreted as the intersection of the fuzzy goal C_0 and the fuzzy constraints C, *i.e.*, $D = C_0 \cap C_1$, where $\cap$ is a conjunctive operator, which have different alternatives and different meanings in practical situations. In terms of the membership function, the fuzzy decision can be formulated as:

$$\mu_D(x) = \mu_{C_0}(x) \cap \mu_C(x), \quad \forall x \in X, \tag{3.23}$$

where μ_{C_0} and μ_C are the membership functions of the fuzzy goals and the fuzzy constraints, respectively, and preferences are involved.

A maximizing decision x^* is then defined to be an alternative with the highest membership in the fuzzy decision D, *i.e.*, $\mu_D(x^*) = \max \mu_D(x) \forall x \in X$.

More generally, maximizing decision x^* can be determined by

$$\mu_D(x^*) = \bigcup_{x \in X} \mu_D(x). \tag{3.24}$$

The maximizing decision x^* is the optimal solution in a sense that it can be interpreted in different ways, depending on the definitions of the operators $\cup$ and $\cap$.

The operator $\cap$ may be extended to various forms of conjunctive operators, such as minimum operator, weighted sum of the goals and the constraints, multiplication operator, mean-value operator, bounded product, Hamacher's min operator, *etc.*, and $\cup$ can be substituted by algebraic sum, bounded sum, Yager's max operator (Yager 1979), *etc.*, which are summarized in Lai and Hwang (1992a,b) in detail. Among these operators, the max-min operator is commonly used in practice. The selection of the operators depends on the preference of the DM and the problem-context and semantic interpretation.

This approach provides a framework for solving fuzzy optimization problems with fuzzy goals and fuzzy constraints, and it is well known as the fundamental of decision making under a fuzzy environment. Since then, various forms of symmetric approaches (Zimmermann 1976, 1991; Wang 1997; Tang and Wang 1997a,b; Tang *et al.* 1998; Tanaka *et al.* 1974; Werners 1987) have been developed by applying different combinations of operators. Among them, Tanaka *et al.* (1974) extended Bellman and Zadeh's approach to tackle multi-objective fuzzy mathematical programming problems. The tolerance approach proposed by Zimmermann (1976) is one of the most important and practical approaches. By using piecewise linear membership functions to represent fuzzy goal and fuzzy constraints, the original problem can then be translated into a linear programming model. A maximizing decision among the fuzzy decision set can be achieved by solving the linear programming. In addition, the DM may capture some essential features on other solutions in the neighborhood of the maximizing decision. Along this line, Verdegay (1984a,b) and Chanas (1983) proposed parametric programming techniques to obtain the whole fuzzy decision set and complete fuzzy decision set, respectively.

Apart from FMP1, this type of approach can also apply to the symmetric problems FMP3, FMP4 and FMP7, in which fuzzy coefficients are characterized by membership functions. These fuzzy coefficients are embedded into the objective function and/or the system of constraints, and their membership function $\mu_{\tilde{r}}$ and $\mu_{\tilde{s}}$ reflect the preference of the DM. When applying the symmetric approaches to these problems, the fuzzy objective function and the fuzzy system of constraints are treated as the fuzzy goal and the fuzzy constraints, respectively, in a symmetric way. Firstly, $\mu_{f(x,\tilde{r})}$ and $\mu_{C(x,\tilde{s})}$, the membership functions of the fuzzy objective function and the fuzzy system of constraints can be obtained via $\mu_{\tilde{r}}$ and $\mu_{\tilde{s}}$ using the extension principle, and then similar procedures can be applied by substituting μ_{C_0} and μ_C with $\mu_{f(x,\tilde{r})}$ and $\mu_{C(x,\tilde{s})}$, respectively. In addition, this approach can be applied to solve the asymmetric problem FMP2. Werners (1987) developed a symmetric approach to linear programming problems with fuzzy resources by treating the goal of the problem in the same way as the fuzzy constraints are treated.

This approach can be applied to the cases with single objective or multiple objectives, in the forms of linearity or nonlinearity. The types of optimal solutions to these approaches can be expressed in different forms, such as the fuzzy decision (Bellman *et al.* 1970), maximizing decision (Zimmermann 1976; Werners 1987), fuzzy optimal solution (Wang 1997; Wang and Fang 1997; Tang and Wang 1997a,b; Tang *et al.* 1998), depending on the operators and the interpretation applied.

3.6.2 Symmetric Approach Based on Non-dominated Alternatives

This approach is developed for solving FMP1, in which the fuzzy goal is expressed in a fuzzy utility function $\varphi(x, y) : X * Y \to [0, 1]$, and the fuzzy constraints are expressed in fuzzy preference relations, denoted by μ: $Y * Y \to [0, 1]$, where X and Y are a set of alternatives and a universal set of estimates, respectively, based on the concept of fuzzy strict preference relations and non-dominated alternatives (Orlovski 1977, 1980). In this case, given an alternative $x \in X$, the function φ gives the corresponding utility value $\varphi(x, y)$ in the form of a fuzzy set in Y. The basic rationale of this approach is as follows: Firstly, $\forall \tilde{x} \in X, \tilde{x} \neq x$, a fuzzy strict preference relation R^s in X is defined using the original fuzzy relation μ in Y. The membership function $\mu_R^s(\tilde{x}, x)$ of R^s, representing the degree that $\tilde{x}$ is strictly preferred to x, is defined as follows:

$$\mu_R(\bar{x}, x) = \sup_{y_1, y_2 \in Y} \min \{\varphi(\bar{x}, y_1), \varphi(x, y_2), \mu(y_1, y_2)\} , \tag{3.25}$$

$$\mu_R^s(\bar{x}, x) = \max \{0, \mu_R(\bar{x}, x) - \mu_R(x, \bar{x})\} , \tag{3.26}$$

where μ_R is a fuzzy preference relation induced in X. The elements of the general scheme are given as:

$$e(x, C_0, C) = \mu_R^s(\bar{x}, x) , \tag{3.27}$$

$$K(e(x, C_0, C)) = 1 - \sup_{\bar{x} \in X} \mu_R^s(\bar{x}, x) = \mu^{\mathrm{ND}}(x) , \tag{3.28}$$

$$TK(x) = \mu^{\mathrm{ND}}(x) , \tag{3.29}$$

where $\mu^{\mathrm{ND}}(x)$ is the degree to which x is non-dominated by any other elements $\bar{x}$, *e.g.*, for some x such that $\mu^{\mathrm{ND}}(x) = \alpha$, which means that this element is dominated by other elements to a degree not higher than α.

In this sense, the original FMP is stated as the following problem:

$$\max_{x \in X} \mu^{\mathrm{ND}}(x) , \tag{3.30}$$

which can be solved by transforming it into an equivalent semi-infinite programming model. The optimal solution is understood in the sense of non-dominated alternatives.

It can be seen from Equations 3.25–3.30 that the fuzzy goal and fuzzy constraints are treated in the same way as indicated in Bellman and Zadeh's approach, and hence it is a symmetric approach.

3.6.3 Asymmetric Approaches

In contrast to the symmetric approaches, the asymmetric approaches here refer to the type of approaches to the asymmetric mathematical model in the sense that the goals (or the objective functions) and the system of constraints are treated in an asymmetric way with regards to the fuzziness, *i.e.*, only one of the two constituents is treated as fuzziness and the counterpart as crispness no matter what fuzziness is involved in the problems. In this sense, the asymmetric approaches cannot only solve the asymmetric problems FMP2, FMP5 and FMP6, but also solve the symmetric problems FMP1, FMP3, FMP4 and FMP7, in which the goals and the system of constraints are treated in an asymmetric way. We first focus our attention to the approaches to FMP1. When solving the FMP1, the asymmetric approaches treat the fuzzy goal C_0 and the fuzzy constraints C in an asymmetric way, and usually via the following asymmetric form (Luhandjula 1989):

$$\max_{x \in C} \mu_{C_0}(x) , \tag{3.31}$$

where $\mu_{C_0}(x)$ is a crisply defined compatibility function. Similarly, the other symmetric problems FMP3, FMP4 and FMP7 can be treated in the same way using Equation 3.31 by substituting the μ_{C_0} and/or C with $\mu_{f(x,\tilde{r})}$ and $C(x,\tilde{s})$, respectively. On the other hand, the asymmetric problems FMP2 and FMP6 have the form as in Equation 3.31, and the fuzzy dual problem of FMP5 can also be expressed in this form. Hence, the approaches here for FMP1 can also be applied to the problems FMP2–FMP7.

Equation 3.31 is meaningless in mathematics, and hence the optimal solution of this problem should be understood in a fuzzy sense, which constitutes the fundamental part of the approaches. One possible interpretation is by the concept of maximizing set, which is a fuzzy set and reflects the compatibility of elements in the support of the feasible domain C with the fuzzy goal C_0. Other possible ways of interpretation and definition of the optimal solution includes fuzzy maximizing decision, maximum decision, fuzzy solution, α-optimal solution and fuzzy optimal solution set. Various approaches are available depending on the possible interpretation and the definition of the optimal solution, some of which are summarized as follows:

1. Fuzzy maximizing decision approach. According to the definition of the maximizing set, a maximizing set M is a fuzzy set, the membership function of which reflects the compatibility degree of the fuzzy goal C_0 and the support set S_C of the fuzzy feasible set C. The fuzzy maximizing decision M is a maximizing set and can be characterized by $\mu_M(x)$as follows:

$$\mu_M(x) = \frac{\mu_{C_0}(x) - \inf\limits_{x \in S_C} \mu_{C_0}(x)}{\sup\limits_{x \in S_C} \mu_{C_0}(x) - \inf\limits_{x \in S_C} \mu_{C_0}(x)} \tag{3.32}$$

   where S_C is a support set of the fuzzy set C. In comparison with the problems in Equations 3.31 and 3.11, one can see that Equation 3.31 is a special case of the fuzzy extreme problem. Hence, this approach can also be applied to the fuzzy extreme problem in Equation 3.11. The fuzzy maximizing decision M is regarded as the optimal solution in the sense of a fuzzy set, the membership function of which reflects the compatibility degree of the fuzzy goal and the support set of the fuzzy constraints. The fuzzy maximizing decision approaches are commonly applied to solve asymmetric problems like FMP2, FMP6 and fuzzy extreme problems. It can also be applied to FMP5 through transformation into its dual problem.

2. Crisp maximum decision approach. This approach originally was developed to solve asymmetric problems, and it comes from the idea that the objective should also be fuzziness owing to the fuzziness involved in the feasible domain. Hence, symmetric approach based on the fuzzy decision can be also applied to Equation 3.31 by regarding the fuzzy maximizing decision as the fuzzy decision. It aims to achieve the maximum degree of intersection between the fuzzy maximizing decision M and the fuzzy feasible set C. The alternatives with the highest degree in the fuzzy maximizing decision are interpreted as the optimal solutions, *i.e.*, $\mu_M(x^*) = \max\{\mu_M(x), |x \in S_C\}$. They are crisp solutions. When applying the fuzzy maximizing decision approach and the crisp maximum decision approach to solve the FMP2 and FMP6, the solution can be obtained by substituting μ_{C_0} and C with $f(x, r)$ and $C(x, \tilde{s})$, respectively.
3. Fuzzy solution approach (Tanaka and Asai 1984a,b; Orlovski 1977, 1980; Verdegay 1982). This approach is applied when one wants to know the extent to which the uncertain solution reflects the uncertainty of the problem's setting, especially to the asymmetric problems with respect to the fuzziness, *i.e.*, FMP2, FMP5 and FMP6. An important concept of the approach is the fuzzy solution which is a fuzzy set. The fuzzy solution can be expressed in various forms depending on the formulation of the membership function, which results in various forms of the fuzzy solution approaches. Among them, Orlovski (1977, 1980) firstly proposed the concept of fuzzy solution to the problems in Equation 3.22, and two methods are developed to formulate the fuzzy solutions denoted by $Sol1$ and $Sol2$ using an α-level cut set of the fuzzy feasible domain and the Praetor optimal solution, respectively. The concrete forms of the fuzzy solutions are defined by the membership functions in the form of Equation 3.24 and 3.27, respectively. Verdegay (1982, 1984a,b) investigated a fuzzy solution for fuzzy mathematical programming problems FLP2 and FLP5 based on the concept of an α-optimal solution. The fuzzy solution is understood as the optimal solution in the sense that it optimizes the objective function under a preferred level set of the fuzzy constraints, *i.e.*, α-optimal solution to the subproblem defined on the α-level cut set of the fuzzy domain C. Werners (1987) proposed a formulation for the fuzzy solution to FLP2 and FLP6, and named it the fuzzy set "decision" (Zimmermann 1991). It is interpreted as a fuzzy optimal solution set, which is a union of the set of α-optimal solution to the subproblem, and it has a different formulation from that of Verdegay. Tanaka and Asai (1984a,b) developed a fuzzy solution for FLP with fuzzy coefficients in the system of constraints using α-level cut set. The fuzzy solution with the widest spread is understood as the optimal solution in the sense that it satisfies the system of constraints to a given degree. In general, the fuzzy solutions can be obtained using parametric programming techniques or multi-objective programming. The possibility and necessity optimal solution sets (Inuiguchi and Ramik 2000) can take the form

of fuzzy solutions. The fuzzy solution expressed in various forms is regarded as the optimal solution in this approach.

$$\mu_{Sol1} = \begin{cases} \mu_{C_0}(x) & \text{if } x \in \bigcup_{k\in[0,1]} V(k)\,, \\ 0 & \text{else}\,, \end{cases} \tag{3.33}$$

where

$$V(k) = \{x \in X \,|\, \mu_{C_0}(x) = \max_{t\in D^k} \mu_{C_0}(t)\}\,, \tag{3.34}$$

$$D^k = \{x \in X \,|\, \mu_C(x) \geqslant k\}\,. \tag{3.35}$$

$$\mu_{Sol2} = \begin{cases} \mu_{C_0}(x) & \text{if } x \in E\,, \\ 0 & \text{else}\,, \end{cases} \tag{3.36}$$

where E is a set of efficient solutions of the multi-objective programming

$$\max_{x\in X} \{\mu_{C_0}(x), \mu_C(x)\}\ .$$

3.6.4 Possibility and Necessity Measure-based Approaches

The symmetric and asymmetric approaches are summarized mainly on the fuzzy optimization problems with vagueness and fuzzy coefficients characterized by membership functions. The approaches to solving the positivistic mathematical programming (PMP) problems can be found in the survey by Inuiguchi and Ramik (2000) and in Chapter 4 in Lai and Hwang (1992a,b). Among them, the approaches based on possibility and necessity measures are important approaches to PMP. They are briefly summarized below.

As indicated in the previous section, the fuzzy coefficients in the positivistic programming problems are viewed upon as the positivistic variables restricted by the possibility distributions. Under this circumstance, no matter if the objective or the system of constraints with the fuzzy coefficients is a positivistic function, the value of which are also ambiguous, and could not be determined uniquely. Hence, how to formulate and measure these values in an appropriate way are important constituents of the approaches to solving this category of problems. To do this, a specific interpretation should be introduced and developed based on the possibility theory (Zadeh 1978; Dubois and Prade 1988). The possible interpretation is summarized by Inuiguchi and Ramik in a recent survey (2000). Among the interpretations, two basic concepts are the possibility measure and the necessity measure.

Given a positivistic variable a restricted by a fuzzy set A with a possibility distribution μ_A, the possibility measure and the necessity measure, denoted by $\pi_A(B)$ and $N_A(B)$, respectively, represent the possibility degree and the necessity degree of the event that a is in the fuzzy set B, *i.e.*, the extent of possible and the extent of certain that a is in the fuzzy set B. They are defined as follows (Inuiguchi and Ramik 2000):

$$\pi_A(B) = \sup_a \min(\mu_A(a), \mu_B(a)), \tag{3.37}$$

$$N_A(B) = \inf_a \max(1 - \mu_A(a), \mu_B(a)). \tag{3.38}$$

Based on these two concepts, two positivistic function values, or a positivistic function value and a real value can be ranked with an index, *e.g.*, $Pos(a \leqslant g)$ or $Nes(a \leqslant g)$ in the sense of the possibility degree or the certainty degree. Here the indices $Pos(a \leqslant g)$ and $Nes(a \leqslant g)$ defined as follows indicate the degree of possibility and the degree of certainty to which the value a restricted by the possibility distribution μ_A is not greater than g. $Pos(a \geqslant g)$ and $Nes(a \geqslant g)$ can be defined and understood in the similar way. The selection of the indices depends on the form of the goal (*e.g.*, max or min), the inequality relations involved in the system of constraints and the DM's attitude.

$$Pos(a \leqslant g) = \pi_A((-\infty.g)) = \sup\{\mu_A(r), r \leqslant g\}, \tag{3.39}$$

$$Nes(a \leqslant g) = N_A((-\infty.g)) = 1 - \sup\{\mu_A(r)|r > g\}. \tag{3.40}$$

Using these indices, the positivistic objective function and the system of constraints can be formulated by an appropriate interpretation. This interpretation reflects the DM's preference on the degree of possibility and certainty, and the attitude to the treatment of the objective function and the system of constraints. From an attitude point of view, *i.e.*, the symmetric attitude or the asymmetric attitude to the treatment of the objective function and the system of constraints, the approaches to solve the PMP can be classified into asymmetric and symmetric approaches, which are introduced briefly as follows.

3.6.5 Asymmetric Approaches to PMP5 and PMP6

These approaches are developed to solve the PMP5 and the PMP6, in which the fuzzy objective function and the system of constraints are treated separately. When applying this type of approach, firstly define an appropriate index based on the possibility measure and the necessity measure. The succeeding procedure is to understand the problem and try to find an appropriate interpretation so as to transform the positivistic programming model into a crisp one using the concepts. Different interpretations result in various approaches to the problem.

1. Fractal approach to PMP5.

The fractal approach originates from the Kataoka's model (Stancu-Minasian 1984) for solving stochastic programming problems, and it can be applied to the treatment of the positivistic objective function and the system of constraints. The two important concepts of the fractal approach are p-possibility fractal and p-necessity fractal, which are defined as the smallest value of u satisfying $Pos(a \leqslant u) \geqslant p$ and $Nes(a \leqslant u) \geqslant p$, respectively.

If the DM has more interest in the objective function, *i.e.*, one pursues a maximum objective function with a high certainty, then the maximum objective function with a high certainty can be interpreted as a p-necessity fractal in the following equivalent form:

$$Nes(f(x, \tilde{r}) \geqslant u) \geqslant p\,, \tag{3.41}$$

where the p is a preferred value reflecting the DM's desire or preference on the certainty degree of the objective function.

In this case, PMP5 can be solved by transforming Equation 3.41 into the following equivalent model:

$$\begin{aligned} &\max u \\ &s.t.\ Nes(f(x, \tilde{r}) \geqslant u) \geqslant p\,, \\ &x \in C(x, s)\,. \end{aligned} \tag{3.42}$$

The solution to Equation 3.42 is an optimal solution to PMP5 in the sense of p-necessity fractal that the objective function is not less than u^* at a certainty degree with p.

2. Asymmetric approach to PMP6.

Similarly, in the case that the DM leans towards a higher degree of satisfaction of the constraints, it can be interpreted that the problem aims to pursue a maximum objective at a higher certainty degree of satisfying the constraints. This certainty degree is not the one in the sense of p-necessity fractal. With this interpretation, PMP6 can be solved by transforming it into an equivalent crisp model as follows:

$$\begin{aligned} &\max f(x.r) \\ &s.t.\ Nes(C(x, \tilde{s})) \geqslant p\,, \end{aligned} \tag{3.43}$$

where p is a preferred value reflecting the DM's preference on the certainty degree of the system of constraints.

The solution to Equation 3.43 is an optimal solution to PMP6 in the sense that the system of constraints are satisfied with a certainty degree not less than p.

Similarly, the objective function can be treated with p-possibility fractal, and the system of constraints can also be treated in terms of the possibility degree, when dealing with the PMP5 and PMP6, respectively.

Apart from the fractal approach to the FMP5, the modality approach (Inuiguchi and Ramik 2000) can also treat the objective function such that the DM puts more emphasis on a maximum certainty degree of which the objective function is not less than a preferred level.

3.6.6 Symmetric Approaches to the PMP7

The fractal approach can be applied to solve asymmetric problems, *i.e.*, the PMP5 and the PMP6, but it can also work with the symmetric problem PMP7 using an appropriate interpretation. In some cases, the DM not only pursues the objective, but is also concerned with the satisfaction with the system of constraints. It can be interpreted that the problem aims to pursue a maximum objective with a high possibility degree at a higher certainty degree of satisfying the constraints. The possibility degree and the certainty degree can be understood in the way of p-possibility (necessity) fractal. Owing to various combinations of the possibility measures and the necessity measures involved in the interpretation, various symmetric models and approaches can be developed to solve PMP7. For simplicity, the p-possibility fractal and the necessity degree are used to treat the objective function and the system of constraints, respectively, while PMP7 can be solved by transforming it into the following model:

$$
\begin{aligned}
&\max u \\
&s.t.\ pos(f(x,\tilde{r}) \geqslant u) \geqslant p_1\,, \\
&Nes(C(x,\tilde{s})) \geqslant p_2\,.
\end{aligned} \tag{3.44}
$$

p_1 and p_2 are the preferred levels of the DM.

If neither the possibility degree nor the necessity degree are the ones in the sense of p_1-fractal, and the objective function is treated as in the modality approach, PMP7 can be understood in terms of

$$
\begin{aligned}
&\max p \\
&s.t.\ pos(f(x,\tilde{r}) \geqslant u) \geqslant p\,, \\
&Nes(C(x,\tilde{s})) \geqslant p\,.
\end{aligned} \tag{3.45}
$$

u is the preferred level of the objective function.

3.6.7 Interactive Satisfying Solution Approach

The interactive satisfying solution approach is an important type of approach to the fuzzy optimization problems, especially to fuzzy multi-objective programming problems through an interactive fuzzy optimization procedure. The satisfying solution, compromise solution and Pareto optimal solution are understood as the optimal solutions to these problems. With this approach, the solution is determined step-by-step in an interactive process. Many procedures of this type of approach can be found in Sakawa and Yano (1989), Slowinski and Teghem (1990), and Rommelfanger (1990).

3.6.8 Generalized Approach by Angelov

The generalized approach, originally proposed by Angelov (1994), is viewed as a new approach to fuzzy optimization problems on the basis of the generalization of Bellman and Zadeh's concept (Bellman *et al.* 1970). It directly solves the fuzzy optimization problems through a parametric generalization of intersection of fuzzy sets and a generalized fuzzification procedure called BADD (Filev and Yager 1991) without the step of transforming the model into a crisp one. It can be outlined as follows:

Step 1 Specifying α and β, where $\alpha \in [0, +\infty)$ reflects the credibility of every fuzzy solution, and $\beta \in [0, +\infty)$ is the degree of strength of the flexible conjunction.

Step 2 Construction of fuzzy decision D as

$$\mu_D(x) = \frac{\mu_{C_0}(x)\mu_C(x)}{\beta + (1-\beta)(\mu_{C_0}(x) + \mu_C(x) - \mu_{C_0}(x)\mu_C(x))}. \tag{3.46}$$

Step 3 Determination of a crisp solution x_0 as

$$x_0 = \sum_{j=1}^{N} \frac{\mu_{D_j}^{\alpha}(x_j)}{\sum_{i=1}^{N} \mu_{D_i}^{\alpha}(x_i)} x_j, \quad N = Card(x). \tag{3.47}$$

With these procedures, a family of parametric crisp solutions of the FOP can be obtained, via the variations of α and β; whereas in Bellman and Zadeh's method, the decision with maximal degree of membership is taken. In this sense, Bellman and Zadeh's approach can be considered as a special case of this approach. The fuzzy solutions and the crisp solutions can be understood as the optimal solution to the FOP.

3.6.9 Fuzzy Genetic Algorithm

Buckley and Hayashi (1994) proposes a fuzzy genetic algorithm to solve the following type of fuzzy maximum problems approximately:

$$\max F(\tilde{X}), \tag{3.48}$$

where $\tilde{X}$ is any type of fuzzy subset in $[0, M]$, $M > 0$, and F is a crisply defined map.

The fundamental of the fuzzy genetic algorithm is that, first of all, define a measurement function $m(F(\tilde{y})) = \theta$; and then discredit $\tilde{X}$, *i.e.*,

$$\begin{aligned} \tilde{X} &= (x_0, x_1, x_2 \cdots, x_N), x_i = \mu_{\tilde{X}}(z_i) \\ z_i &= i * M/N, i = 0, 1, 2, \cdots N. \end{aligned}$$

Under this circumstance, the original fuzzy optimization problem in Equation 3.48 may be stated as how to determine $x_i, i = 0, 1, 2, \cdots N$ such that $m(f(\tilde{x})) = \theta \rightarrow \max$, to which a genetic algorithm can be applied, and an approximate optimal solution can be achieved.

3.6.10 Genetic-based Fuzzy Optimal Solution Method

Based on the method proposed by Zimmermann (1976), a genetic-based fuzzy optimal solution (Wang 1997; Wang and Fang 1997; Tang and Wang 1997a,b; Tang *et al.* 1998) is interpreted as the neighboring domain of an optimal solution, in which every solution is acceptable, *i.e.*, it is an optimal solution in a fuzzy sense. Using this method a family of solutions with acceptable degree of membership can be found through a genetic search, and the solutions preferred by the DM under different criteria can be achieved by means of the human-computer interactions. This method has been applied to fuzzy linear, quadratic and nonlinear programming problems of the types FMP1 and FMP4. Recently some other intelligent-based fuzzy optimization approaches (Jivenez and Verdegay 1999) have become popular.

3.6.11 Penalty Function-based Approach

This approach was first proposed by Lodwick and Jamison (1998) to solve fuzzy constrained optimization problems. The penalty functions are imposed to the objective as a penalty when the fuzzy constraints are "violated." It is useful in computation and reflects the practical scene. The authors in Tang and Wang (1997a,b) consider the penalty in the fuzzy nonlinear programming problems with fuzzy resources, and suggested some properties of a fuzzy optimal solution set of this model. A genetic-based approach for finding the maximum decision is developed.

Apart from the above approaches to the fuzzy optimization problems, parametric techniques (Carlsson and Korhonen 1986), dual approach (Verdegay 1984a,b), fuzzy dual decompose approach (Sakawa and Yano 1994) and differential equation approach (Ali 1998) are also proposed by many researchers. In addition, convergence analysis, stability analysis and sensitivity analysis of the algorithm for the FOP are also applied for fuzzy optimization.

References

Ali FM (1998) A differential equation approach to fuzzy non-linear programming problems. Fuzzy Sets Syst 93(1):57–61

Angelov P (1994) A generalized approach to fuzzy optimization. Int J Intell Syst 9(3):261–268

Bellman RE, Zadeh LA (1970) Decision making in a fuzzy environment. Manage Sci 17B:141–164

Bortolan G, Degani R (1985) A review of some methods for ranking fuzzy subsets. Fuzzy Sets Syst 15:1–19

Buckley JJ, Hayashi Y (1994) Fuzzy genetic algorithm and applications. Fuzzy Sets Syst 61(2): 129–136

Carlsson C, Korhonen P (1986) A parametric approach to fuzzy linear programming. Fuzzy Sets Syst 20(1):17–30

Chanas S (1983) Using parametric programming in fuzzy linear programming. Fuzzy Sets Syst 11(3):243–251

Chanas S, Kuchta D (1998) Discrete fuzzy optimization. In: Slowinski R (ed) Fuzzy sets in decision analysis operations research and statistics: the handbook of fuzzy sets series. Kluwer Academic, Dordrecht, pp 249–276

Delgado M, Kacprzyk S, Verdegay JL, Vila MA (1994) Fuzzy optimization: recent advances. Physica, Berlin

Dubois D, Prade H (1980) Fuzzy sets and systems: theory and applications. Academic, New York

Dubois D (1987) Linear programming with fuzzy data. In: Bezdek JC (ed) Analysis of fuzzy information. CRC, Boca Raton, FL, pp 241–263

Dubois D, Prade H (1988) Possibility theory: an approach to computerized processing of uncertainty. Plenum, New York

Dubois D, Prade H (1994) Decision making under fuzzy constraints and fuzzy criteria mathematical programming vs rule-based system approach. In: Delgado M, Kacprzyk S, Verdegay JL, Vila MA (eds) Fuzzy optimization: recent advances. Physica, Heidelberg, pp 21–32

Fang S, Hu CF *et al.* (1999) Linear programming with fuzzy coefficients in constraints. Comput Math Appl 37:63–76

Fang SC, Li G (1999) Solving fuzzy relation equations with a linear objective function. Fuzzy Sets Syst 103:107–113

Fedrizzi M, Kacprzyk J, Verdegar JL (1991a) A survey of fuzzy optimization and mathematical programming. In: Fedrizzi M, Kacprzyk J *et al.* (eds) Interactive fuzzy optimization. Lecture notes in economics and mathematical systems. Springer, Berlin Heidelberg New York, pp 15–28

Fedrizzi M, Kacprzyk J, Roubens M (eds) (1991b) Interactive fuzzy optimization. Springer, Berlin Heidelberg New York

Filev DP, Yager RR (1991) A generalized defuzzification method via BAD distribution. Int J Intell Syst 6:687–697

Hamacher H, Leberling H, Zimmermann HJ (1978) Sensitivity analysis in fuzzy linear programming. Fuzzy Sets Syst 1(1):269–281

Han S, Ishii H, Fujii S (1994) One machine scheduling problem with fuzzy due dates. Eur J Oper Res 79:1–12

Inuiguchi M (1997) Fuzzy linear programming: what, why and how? Tatra Mountains Math Publ 13:123–167

Inuiguchi M, Ramik J (2000) Possibility linear programming: a brief review of fuzzy mathematical programming and a comparison with stochastic programming in portfolio selection problem. Fuzzy Sets Syst 111:3–28

Ishibuchi H, Yamamoto N, Murata T, Tanaka H (1994) Genetic algorithms and neighborhood search algorithm for fuzzy flow shop scheduling problems. Fuzzy Sets Syst 67(1):81–100

Jivenez F, Verdegay JL (1999) Solving fuzzy solid transportation problems by an evolutionary algorithm based parametric approach. Eur J Oper Res 117:485–510

Kacprzyk J, Esogbue AO (1996) Fuzzy dynamic programming: Main developments and applications. Fuzzy Sets Syst 81(1):31–46

Kacprzyk J, Orlovski SA (1987) Fuzzy optimization and mathematical programming: a brief introduction and survey. In: Optimization models using fuzzy sets and possibility theory. Reidel, Dordrecht, pp 50–72

Lai YJ, Hwang CL (1992a) A new approach to some possibilistic linear programming problems. Fuzzy Sets Syst 49:121–133

Lai YJ, Hwang CL (1992b) Fuzzy mathematical programming. Lecture notes in economics and mathematical systems 394. Springer, Berlin Heidelberg New York
Liu B (1999) Uncertain programming. Wiley, New York
Liu J, Fang SC (2001) Solving nonlinear optimization problems with fuzzy relation equation constraints. Fuzzy Sets Syst 119:1–20
Liu B (2002) Theory and practice of uncertain programming. Physica, Heidelberg
Lodwick WA, Jamison KD (1998) A computational method for fuzzy optimization. In: Ayyub BM, Gupta MM (eds) Uncertainty analysis in engineering and sciences: fuzzy logic, statistics and neural network approach. Kluwer Academic, Boston, pp 291–300
Luhandjula MK (1989) Fuzzy optimization: an appraisal. Fuzzy Sets Syst 30(3):257–282
Orlovski SA (1980) On formulation of general fuzzy mathematical problems. Fuzzy Sets Syst 3(1):311–321
Orlovski SA (1977) On programming with fuzzy constraints sets. Kybernetes 6:197–201
Ostermark R (1987) Sensitivity analysis of fuzzy linear programs: an approach to parametric interdependence. Kybernetes 16:113–120
Ramik J, Rimanek J (1985) Inequality relation between fuzzy numbers and it's use in fuzzy optimization. Fuzzy Sets Syst 16:123–138
Ramik J, Rommelfanger H (1996) Fuzzy mathematical programming based on some new inequality relations. Fuzzy Sets Syst 81(1):77–87
Rommelfanger H (1990) FULPAL: an interactive method for solving multi-objective fuzzy linear programming. In: Slowinski R, Teghem J (eds) Stochastic vs fuzzy approaches to multiobjective mathematical programming under uncertainty. Kluwer Academic, Dordrecht, pp 279–299
Rommelfanger H (1996) Fuzzy linear programming and applications. Eur J Oper Res 92(3):512–527
Rommelfanger H, Hanuschek R, Wolf J (1989) Linear programming with fuzzy objective. Fuzzy Sets Syst 29:31–48
Rommelfanger H, Slowinski R (1998) Fuzzy linear programming with single and multiple objective functions. In: Slowinski R (ed) Fuzzy sets in decision analysis operations research and statistics. The handbook of fuzzy sets series. Kluwer Academic, Dordrecht, pp 179–207
Sakawa M, Yano H (1989) An interactive fuzzy satisfaction method for multi-objective nonlinear programming problems with fuzzy parameters. Fuzzy Sets Syst 30(10):221–238
Sakawa M, Yano H (1994) Fuzzy dual decomposition method for large-scale multi-objective nonlinear programming problem. Fuzzy Sets Syst 67:19–27
Simon F (1995) Uncertainty and imprecision: modeling and analysis. J Oper Res Soc 46(1):70–79
Slowinski R, Teghem J (1990) Stochastic vs fuzzy approaches to multi-objective mathematical programming under uncertainty. Kluwer Academic, Dordrecht, pp 249–262
Stancu-Minasian IM (1984) Stochastic programming with multiple objective functions. Reidel, Dordrecht
Stoica M *et al.* (1984) Fuzzy integer programming. In: Zimmermann HJ, Zadeh LA, Gaines BR (eds) Fuzzy sets and decision analysis. North Holland, Amsterdam, pp 123–132
Tanaka H, Asai K (1984a) Fuzzy linear programming problems with fuzzy numbers. Fuzzy Sets Syst 13:1–10
Tanaka H, Asai K (1984b) Fuzzy solution in fuzzy linear programming. IEEE Trans SMC 14(2):325–328
Tanaka H, Okudu T, Asai K (1974) On fuzzy mathematical programming. Cybernet 3:37–46
Tang J, Wang D (1997a) A non-symmetric model for fuzzy nonlinear programming problems with penalty coefficients. Comput Oper Res 24(8):717–725
Tang J, Wang D (1997b) An interactive approach based on GA for a type of quadratic programming problems with fuzzy objective and resources. Comput Oper Res 24(5):413–422
Tang J, Wang D, Fung RYK (1998) Model and method based on GA for non-linear programming problems with fuzzy objective and resources. Int J Syst Sci 29(8):907–913

Trappey JFC, Liu CR, Chang TC (1988) Fuzzy non-linear programming: theory and application in manufacturing. Int J Prod Res 26(5):957–985
Verdegay JL (1982) Fuzzy mathematical programming. In: Gupta M, Sanchez E (eds) Fuzzy information and decision processes. North Holland, Amsterdam, pp 231–237
Verdegay JL (1984a) A dual approach to solve the fuzzy linear programming problems. Fuzzy Sets Syst 14(1):131–141
Verdegay JL (1984b) Application of fuzzy optimization in operational research. Control Cybernet 13:229–239
Wang D, Fang SC (1997) A genetic-based approach for aggregate production planning in fuzzy environment. IEEE Trans SMC (A) 12(5):636–645
Wang D (1997) An inexact approach for linear programming with fuzzy objective and resource. Fuzzy Sets Syst 8(1):61–68
Werners B (1987) An interactive fuzzy programming systems. Fuzzy Sets Syst 23:131–147
Yager RR (1979) Mathematical programming with fuzzy constraints and a preference on the objectives. Kybernetics 9:109–114
Zadeh LA (1965) Fuzzy sets. Info Control 8:338–353
Zadeh LA (1978) Fuzzy sets as a basis for a theory of possibility. Fuzzy Sets Syst 1:3–28
Zimmermann HJ (1976) Description and optimization of fuzzy system. Int J General Syst 2:209–216
Zimmermann HJ (1985) Application of fuzzy set theory to mathematical programming, Info Sci 36:29–58
Zimmermann HJ (1991) Fuzzy set theory and its applications, 2nd edn. Kluwer-Nijhoff, Boston

Chapter 4
Genetic Algorithm-based Fuzzy Nonlinear Programming

A type of model of fuzzy quadratic programming problems (FQP) is proposed. It describes the fuzzy objective and resource constraints with different types of membership functions according to different types of fuzzy objective and fuzzy resource constraints in actual production problems. An inexact approach is developed to solve this type of model of quadratic programming problems with fuzzy objective and resource constraints. Instead of finding an exact optimal solution, we use a genetic algorithm (GA) with mutation along the weighted gradient direction to find a family of solutions with acceptable membership degrees. Then by means of the human-computer interaction, the solutions are preferred by the decision maker (DM). As an extension, a non-symmetric model for a type of fuzzy nonlinear programming problem with penalty coefficients (FNLP-PC) is proposed. Based on a fuzzy optimal solution set and optimal decision set, a satisfying solution method and a crisp optimal solution method with GA for FNLP-PC are developed. Finally, the analysis of simulation results of an example in actual production problems is also given.

4.1 GA-based Interactive Approach for QP Problems with Fuzzy Objective and Resources

4.1.1 Introduction

This section studies quadratic programming problems with a type of fuzzy objective and resource constraints and its solution method: an interactive approach. It describes the fuzzy objective and resource constraints with different types of membership functions according to different types of fuzzy objective and fuzzy resource constraints in actual production problems. This section develops an inexact approach to solve this type of model of quadratic programming problems with fuzzy objective and resource constraints. We use a genetic algorithm (GA) with mutation along the weighted gradient direction to find a family of solutions with acceptable membership degrees, and then by means of the human-computer interaction, the solutions preferred by the decision maker (DM) under different criteria can be achieved.

4.1.2 Quadratic Programming Problems with Fuzzy Objective/ Resource Constraints

In actual production problems, the available quantity of a resource during a given period is often uncertain and possesses different types of fuzziness. It must also be considered that the objective defined by the decision maker (DM) is an ill-defined goal, where the objective function has fuzzy parameters for various reasons, such as the extent of comprehension about the problem. The DM may also prefer to give some leeway rather than to actually maximize (minimize) the objective. This section discusses quadratic programming problems with the following types of fuzzy objective and fuzzy resource constraints.

1. The objective value which the DM desires is not an actual maximum, but a fuzzy value.
   The DM hopes to reach an aspiration level such as z_0, not less than the lowest level $z_0 - p_0$, and the DM's satisfaction degree increases as the objective value increases.
2. The available quantity of some type of resource $i (i = 1, 2, \cdots, m_1)$ has some increments which were accepted by DM by taking overtime work, using the inventory quantity, *etc*. Assume that the planned available quantity of this type of resource i is $b_i (i = 1, 2, \cdots, m_1)$, where the largest acceptable increment is p_i. The fuzzy available quantity is denoted by $\tilde{b}_i$, and is attached to a monotonic non-increasing membership function. This type of resource is only allowed to be used if it is smaller than the available quantity.
3. The available quantity of some other type of resource $i (i = m_1 + 1, m_1 + 2, \cdots, m)$ is imprecise. Assume that the available quantity $\tilde{b}_i (i = m_1 + 1, m_1 + 2, \cdots, m)$ of this type of resource i is an estimated value b_i and the estimated error is p_i^- and p_i^+, respectively, and $\tilde{b}_i$ has a pair-wise linear membership function. For this type of resource, the DM hopes to utilize them as fully as possible.

Further, assume that the objective function $f(x)$ and the resource constraints function $g_i(x)$ are quadratic and linear, respectively, and both are continuous and derivable in $(R^n)^+$.

The problem is how to make a reasonable plan such that the objective is optimal or to make the DM *most* satisfied with his preferred criteria in an environment of the type of fuzzy objective/resource constraints described above. This type of problems belongs to the class of fuzzy optimization problems.

The quadratic programming problems with fuzzy objective/resource constraints (FQP) has the following general form:

$$\begin{cases} \tilde{\max}\ f(x) = x^T Qx + c^T x \\ s.t.\quad g_1(x) = A_i^T x \le \tilde{b}_i, i = 1, 2, \cdots, m_1 \\ \qquad g_1(x) = A_i^T x = \tilde{b}_i, i = m_1 + 1, m_1 + 2, \cdots, m \\ \qquad x \ge 0 \end{cases} \tag{4.1}$$

where x is the n-dimensional decision variable vector, $x = (x_1, x_2, \cdots, x_n)^T$ A is the resource consumption coefficient matrix $A = (A_1, A_2, \cdots, A_m)^T$, Q is the symmetric objective matrix, c is the objective vector, and $\tilde{b}$ is the fuzzy available resource vector, $\tilde{b} = (\tilde{b}_1, \tilde{b}_2, \cdots, \tilde{b}_m)^T$.

We introduce the following types of membership functions to describe the fuzzy number $\tilde{b}_i$, fuzzy objective and fuzzy constraints.

For resource $i\,(i = 1, 2, \cdots, m_1)$ let $\mu_{\tilde{b}_i}(x)$ be the membership function, which indicates the attainability of fuzzy available resource $\tilde{b}_i$ and define

$$\mu_{\tilde{b}_i}(x) = \begin{cases} 1, & x \le b_i \\ 1 - \dfrac{(x - b_i)}{p_i}, & b_i \le x \le b_i + p_i \\ 0, & x > b_i + p_i \end{cases} \tag{4.2}$$

where $\mu_{\tilde{b}_i}(x)$ is a monotonic non-increasing linear function, which denotes the attainable degree of fuzzy available resource $\tilde{b}_i$.

For resource $i\,(i = m_1 + 1, m_1 + 2, \cdots, m)$, let $\mu_{\tilde{b}_i}(x)$ describe the estimation for fuzzy available resource $\tilde{b}_i$ and define

$$\mu_{\tilde{b}_i}(x) = \begin{cases} 1, & x \le b_i - p_i^- \\ 1 - \dfrac{b_i - x}{p_i^-}, & b_i - p_i^- \le x \le b_i \\ 1 - \dfrac{x - b_i}{p_i^+}, & b_i \le x \le b_i + p_i^+ \\ 0, & x > b_i + p_i^+ \end{cases} \tag{4.3}$$

where $\mu_{\tilde{b}_i}(x)$ is a pair-wise linear function, denoting the accurate level of estimation for fuzzy available resource $\tilde{b}_i$. Similarly, $g_i(x) \le \tilde{b}_i$ is a fuzzy subset of R^n, and so let $\mu_i(x)$ be the membership function of the ith fuzzy constraint. This represents the situation of subjection to the ith fuzzy constraint, according to the extension principle.

For $i = 1, 2, \cdots, m_1$

$$\mu_i(x) = \bigvee_{y \ge g_i(x)} \mu_{\tilde{b}_i}(y) = \mu_{\tilde{b}_i}(g_i(x)) = \begin{cases} 1, & g_i(x) \le b_i \\ 1 - \dfrac{g_i(x) - b_i}{p_i}, & 0 < g_i(x) - b_i \le p_i \\ 0, & else \end{cases} \tag{4.4}$$

$$\mu_i(x) = \bigvee_{y \geq g_i(x)} \mu_{\tilde{b}_i}(y) = \mu_{\tilde{b}_i}(g_i(x)) = \begin{cases} 1, & g_i(x) \leq b_i - p_i^- \\ 1 - \dfrac{b_i - g_i(x)}{p_i^-}, & b_i - p_i^- \leq g_i(x) \leq b_i \\ 1 - \dfrac{g_i(x) - b_i}{p_i^+}, & b_i \leq g_i(x) \leq b_i + p_i^+ \\ 0, & g_i(x) > b_i + p_i^+ . \end{cases} \tag{4.5}$$

Obviously, $\mu_i(x)$ has the same formula as $\mu_{\tilde{b}_i}(y)$ for $i = 1, 2, \cdots, m_1$ and $i = m_1 + 1, m_1 + 2, \cdots, m$ respectively. However, there are some conceptual differences. Moreover, it may not be the same for other types of fuzzy numbers $\tilde{b}_i$. The term $\mu_i(x)$ denotes the degree of DM's satisfaction with the ith fuzzy constraint at the point x.

Let $\mu_0(x)$ describe the DM's fuzzy objective $\tilde{\max}\, f(x)$. It is defined as follows:

$$\mu_0(x) = \begin{cases} 0, & f(x) \leq z_0 - p_0 \\ 1 - \dfrac{x_0 - f(x)}{p_0}, & z_0 - p_0 \leq f(x) \leq z_0 \\ 1, & f(x) \geq z_0 \end{cases} \tag{4.6}$$

$\mu_0(x)$ is a monotonic non-decreasing continuous linear function. It denotes the degree of the DM's satisfaction with the fuzzy objective function at the point x.

Certainly, the type of membership function which describes the DM's fuzzy objective and fuzzy resource constraints may be determined by the DM to be of another type, such as quadratic or exponential, and so on, as long as it satisfies the assumption.

FQP may be described as how to make a reasonable plan such that the DM is most satisfied with the fuzzy objective and fuzzy constraints, namely there is a highest intersection degree between fuzzy objective and fuzzy constraints. Or it may be described as how to find a plan such that there is a maximum objective or other types of targets under the condition that the DM is satisfied with both the fuzzy objective and fuzzy constraints. The former uses the highest satisfaction degree as criteria and the latter may use different types of targets as criteria under the consideration of a certain satisfaction degree. The former may be described as FQP-1, and the latter may be written as FQP-2:

FQP-1:

$$\begin{cases} \max \alpha \\ s.t. \quad \mu_0(x) \geq \alpha \\ \qquad \mu_i(x) \geq \alpha, \ i = 1, 2, \cdots, m \\ \qquad x \geq 0, \alpha \geq 0 \end{cases} \tag{4.7}$$

FQP-2:

$$\begin{cases} \max \text{ Target} \\ s.t. \quad \mu_0(x) \geq \alpha_0 \\ \qquad \mu_i(x) \geq \alpha_0, \ i = 1, 2, \cdots, m \\ \qquad x \geq 0 \end{cases} \tag{4.8}$$

where *Target* may be the objective function, the decision variable or the resource constraints, *etc.*, and $\alpha_0 (0 \le \alpha_0 \le 1)$ is an acceptable satisfaction degree preferred by the DM. Then the solution to FQP may be transformed into the solution for FQP-1 or FQP-2 according to different types of criteria.

4.1.3 Fuzzy Optimal Solution and Best Balance Degree

Generally, a unique optimal solution α^* can be found from FQP-1, the corresponding solution x^* is the solution with the highest membership degree to the FQP. It means that the best balance of objective and constraints might be achieved at the point x^*. However; the solution x^* probably is not desired by the DM. Furthermore, the exact optimal solution is meaningless to the DM under a fuzzy environment; the solution needed by the DM is multiple different solutions subject to both the objective and resource constraints under different criteria preferred by the DM.

Based on the above idea, we introduce the concept of a fuzzy optimal solution hitch, which is a neighbor domain including the optimal solution.

Definition 4.1. Fuzzy optimal solution of FQP is a fuzzy set $\tilde{S}$ defined by

$$\tilde{S} = \left\{ (X, \mu_{\tilde{S}}(x)) | x \in (R^n)^+ \right\} \tag{4.9}$$

with

$$\mu_{\tilde{S}}(x) = \min \{\mu_0(x), \mu_1(x), \cdots, i = 1, 2, \cdots, m\} \tag{4.10}$$

Let

$$S_\alpha = \left\{ x \in (R^n)^+ | \mu_{\tilde{S}}(x) \ge \alpha \right\} \tag{4.11}$$

$$\alpha \in [0, 1] \, . \tag{4.12}$$

Then, S_α is an α-level cut set of $\tilde{S}$, and is a general set.

Property 4.1. If Q is seminegative definite, then $\forall \alpha \in [0, 1]$, S_α is a convex set and $\tilde{S}$ is a convex fuzzy set.

Property 4.2. If $\alpha_k \le \alpha_{k+1}$, then $S_{\alpha_k} \supseteq S_{\alpha_{k+1}}$.

Property 4.3. The sufficient condition of $S_1 (S_\alpha, \alpha = 1)$ being non-empty is that there exists x_0 such that

$$\begin{cases} f(x_0) \ge z_0 \\ p_i(x_0) \le b_i, \ i = 1, 2, \cdots, m_1 \\ p_i(x_0) = b_i, \ i = m_1 + 1, m_1 + 2, \cdots, m \\ x_0 \ge 0 \end{cases} \tag{4.13}$$

Definition 4.2. α^* is the best balance degree if there exists an α^* such that $\forall 0 \le \alpha \le \alpha^*$, S_α is non-empty and $\forall \alpha > \alpha^*$, S_α is empty.

Definition 4.3. S_{α^*} is empty, if for $\forall 0 \geq \alpha \geq 1$, S_α is empty. Define α-QP as

$$\begin{cases} \max\ F(x) = x^T Qx + c^T x \\ s.t.\ \ \mu_i(x) \geq \alpha,\ i = 1, 2, \cdots, m \\ \qquad x \geq 0 \end{cases} \tag{4.14}$$

Let $F_0, F_{1/2}, F_1$ denote the optimal objective function values of α-QP as $\alpha = 0, 1/2, 1$, respectively.

Property 4.4. If $z_0 - p_0 = F_0$, then $\alpha^* = 0$.

Property 4.5. If $z_0 - 1/2p_0 > F_{1/2}$, then $\alpha^* < 0.5$.

Property 4.6. If $z_0 - 1/2p_0 = F_{1/2}$, then $\alpha^* = 0.5$.

Property 4.7. If $z_0 - 1/2p_0 < F_{1/2}$, then $\alpha^* > 0.5$.

Property 4.8. The sufficient condition of S_α being empty is that $z_0 - p_0 > F_0$.

Property 4.9. If $z_0 \leq F_1$, then $\alpha^* = 1.0$.

Property 4.10. If $m_1 = m$ and Q is a negative definite matrix, then S_{α^*} is a single point set.

Property 4.11. If Q is a negative definite matrix, and S_{α^*} is non-empty, then, for $0 \geq \alpha \geq \alpha^*$, the optimal solution to α-QP is equivalent to that of model FQP-2 when *Target* equals the objective function.

From the above properties, the "area" S_α becomes smaller and smaller as α increases, until there exists a δ^* such that $\forall \delta > 0, \alpha = \alpha^* + \delta$, where S_{α^*} is an empty set. α^* is the best balance degree, namely, the unique optimal solution to FQP-1; however, the corresponding x^* might not be the solution desired by the DM under some criteria preferred by the DM. The inexact approach proposed here is to find a neighbor domain of optimal solution such that every solution x in the domain is the "optimal" solution desired by the DM under the fuzzy environment.

4.1.4 A Genetic Algorithm with Mutation Along the Weighted Gradient Direction

According to the basic idea of fuzzy optimal solution, the aim of this approach is not to get an exact optimal solution, but to find a neighbor domain that includes the optimal solution such that the satisfaction degree of every solution in the neighbor domain is acceptable; namely, it is an optimal solution under the fuzzy environment. Therefore, the solution to FQP may be transformed into the inexact solution to FQP-1.

FQP-1 (Equation 4.7) may be rewritten as follows according to the definition of the membership function:

$$\begin{cases} \max \mu_{\tilde{S}}(x) = \max\{\min\{\mu_0(x), \mu_1(x), \mu_2(x), \cdots, \mu_m(x)\}\} \\ x \in (R_n)^+ . \end{cases} \tag{4.15}$$

The model P is an unconstrained optimization problem, but its objective function is not continuous and derivable, and thus cannot be solved by traditional optimization methods. In this section, we develop a genetic algorithm with mutation along the weighted gradient direction to solve it. The basic idea is described as follows. First, randomly produce an initial population with the size of *NP* (population size) individuals, where each individual is selected to reproduce children along the increment direction of membership degree of both fuzzy objective and constraints according to the selection probability depending on the fitness function value. For an individual with membership degree less than α_0 (acceptable membership degree), give it a smaller membership degree by means of penalty, so that it may have less of a chance than the others to be selected to reproduce children in the later generation. As the generation increases, the individuals with membership degree less than α_0 die out gradually. After a number of generations, the individual membership degrees are all greater than α_0, namely $S_{\alpha_k} \subseteq S_{\alpha_0}$ and most individuals will be close to the optimal solution.

For individual x, let $\mu_{\min}(x) = \min\{\mu_0(x), \mu_1(x), \mu_2(x), \cdots, \mu_m(x)$. If $\mu_{\min}(x) \le \mu_0(x) < 1$, then move along the $\mu_0(x)$ gradient direction, and the value of may be improved. The smaller $\mu_0(x)$ is, the greater the improvement of $\mu_0(x)$ that may be achieved.

Similarly, if $\mu_{\min}(x) \le \mu_i(x) < 1$, then move along the $\mu_i(x)$ gradient direction, and the value of $\mu_i(x)$ may be improved. The smaller $\mu_i(x)$ is, the greater the improvement of $\mu_i(x)$ that may be made.

Based on the above idea, we can construct

$$D(x) = w_0 \nabla \mu_0(x) + \sum_{i=1}^{m} w_i \nabla \mu_i(x) . \tag{4.16}$$

According to Equations 4.4, 4.5 and 4.6, for $0 < \mu_i(x) \le 1, i = 0, 1, 2, \cdots, m$

$$D(x) = w_0 \frac{\nabla f(x)}{p_0} - \sum_{i=1}^{m_1} w_i \frac{\nabla g_i(x)}{p_i} - \sum_{i=m_1+1}^{m} w_1 \operatorname{sgn}(g_i(x) - b_i) \frac{\nabla g_i(x)}{q_i} \tag{4.17}$$

where $q_i = \begin{cases} p_i^- & g_i(x)-b_i \le 0 \\ p_i^+ & g_i(x)-b_i > 0 \end{cases}$, and $\operatorname{sgn}(x)$ is a sign function.

For $\mu_0(x) = 0$ let $\nabla \mu_0(x) = \nabla f(x)$, and for $\mu_i(x) = 0$, let $\nabla \mu_i(x) = \nabla g_i(x)$.

$D(x)$ is called the weighted gradient direction of $\mu_{\tilde{S}}(x) w_i$, which is the gradient direction weight defined as the following:

$$w_i = \begin{cases} 0 & \mu_i = 1 \\ \dfrac{1}{\mu_i - \mu_{\min} + e} & \mu_{\min} \le \mu_i < 1 \end{cases} \tag{4.18}$$

where e is a sufficiently small positive number.

The child x_j^{k+1} is generated from x_i^k by mutation along the weighted gradient direction. $D(x)$ can be described as follows:

$$x_j^{k+1} = x_i^k + \beta^k D\left(x_i^k\right) . \tag{4.19}$$

In order to avoid local convergence, we select β^k as a step-length of Erlanger distribution random number with decline means, generated by a random number generator. Certainly, we may also select β^k as a step-length of other kinds of random numbers, such as a normal distribution.

Then, calculate the membership degree $\mu_{\tilde{S}}(x_j)$ as follows:

Let $\mu_{\min} = \min\{\mu_0(x_j), \mu_1(x_j), \mu_2(x_j), \cdots, \mu_m(x_j)\}$

$$\mu_{\tilde{S}}(x_j) = \begin{cases} \mu_{\min} & \text{if } \mu_{\min} \geq \alpha_0 \\ \varepsilon\, \mu_{\min} & \text{else} \end{cases} \tag{4.20}$$

where α_0 is an acceptable satisfaction degree preferred by DM, and $\varepsilon \in (0, 1)$.

From the formulas of weighted gradient direction and mutation (Equations 4.16–4.19), it can be seen that the constraints which cannot be met have the minimum membership degree 0. They will then get the largest weight $1/e$, and the mutation will lead the individual to the feasible area. When $\mu_{\tilde{S}}$ is greater than 0, the objective or constraints with minimum membership degree will get the largest weight. The weighted gradient direction will improve the minimum and $\mu_{\tilde{S}}(x)$ will be improved. From Equation 4.20, we may find that $x_j \notin S_{\alpha_0}$ may received a smaller membership degree that is non-zero so that it has little chance of being selected as parents to reproduce children. Therefore the weighted gradient direction will lead all individuals to be close to the exact optimal solution. The individuals will form a neighbor domain including the exact optimal solution.

4.1.5 Human–Computer Interactive Procedure

In the following paragraph, we design a human–computer interactive approach to help the DM select the solution desired from the fuzzy optimal solution under different criteria. It can be described as follows.

Firstly, the approach asks the DM for the acceptable membership degree of the fuzzy optimal solution, α_0. Secondly, by means of interaction with the computer, the prepared membership degree described fuzzy objective and resource constraints are displayed for the DM to select the type of membership degree. Then, it elicits the DM to find his preference structure and point out which criteria are most important to him. The criteria may be the objective, decision variables or some kinds of resources. The solutions are in the α-level cut set of fuzzy optimal solutions by GA with mutation along the weighted gradient direction and the highest and lowest bounds of these criteria are updated in each generation. When the iteration terminates these solutions with their criteria, values will be shown for the DM to select by human–computer interface.

Combining the genetic algorithm of mutation along the weighted gradient direction with the human–computer interaction to select a preferred solution is a new approach called an inexact approach for FQP is proposed. The step by step procedure of the inexact approach can be described as follows:

Step 1 Initialize.

Step 1.1 Input acceptable satisfaction degree α_0 and the largest generation number *NG*, population size *NP*.

Step 1.2 Input DM's most critical criteria index set $CS = \{0, 1, 2, \cdots, n, n+1, \cdots, n+m\}$ where 0 stands for objective function, $j = 1, 2, \cdots n$ for decision variables and $j = n+1, n+2, \cdots, n+m$ for the constraints, respectively. Give the initial values and the upper and lower values of criteria r, $r \in CS$.

Step 1.3 Input the type of membership function which describes fuzzy objective and fuzzy constraints.

Step 2 Produce the initial population randomly and calculate their membership degrees. $x_i(j) = \varepsilon UP_i$, $\varepsilon \in \mathrm{U}(0,1)$, $i = 1, 2, \cdots, n$, $j = 1, 2, \cdots, NP$, where UP_i is the upper bound of the ith element of variable x. Membership function $\mu_{\tilde{S}}(x_j)$ is calculated as Equation 4.20.

Step 3 Set iteration index $k = 1$.

Step 4 For individual $j (j = 1, 2, \cdots, NP)$, calculate their fitness function $F(j)$ and selection probability $P(j))$ as:

$$F(j) = \mu_{\tilde{S}}(x_j) + e\,, \quad P(j) = \frac{F(j)}{\sum_{i=1}^{NP} F(j)}\,. \tag{4.21}$$

Step 5 Produce new individuals $x_j = (j = 1, 2, \cdots, NP)$. x_i is selected to produce x_j as follows:

$$x_j^k = x_i^{k-1} + \beta^k D\left(x_i^{k-1}\right)$$

where β^k is a step-length of Erlang distribution random number with declining means, generated by random number generator, and $D(x)$ is defined as in Equation 4.16.

Step 6 For individual $j = (j = 1, 2, \cdots, NP)$, calculate the membership function $\mu_{\tilde{S}}(x_j)$ as Equation 4.20, update optimal membership degree $\mu_{\max}$ and the upper and lower bounds of criteria r.

Step 7 $k + 1 \to k$, *if* $k \le NG$, go to step 4; otherwise, go to step 8.

Step 8 Output the optimal membership degree $\mu_{\max}$ and the upper and lower value of criteria preferred by DM, stop.

4.1.6 A Numerical Illustration and Simulation Results

To clarify the algorithm, this section supplies an example and shows two iterative processes of a genetic algorithm with mutation along the weighted gradient direction. The results of the example by way of the interactive approach are also discussed in this section.

Example 4.1. A manufacturing factory is going to produce two kinds of products A and B in a period (such as one month). The production of A and B requires three kinds of resources R_1, R_2 and R_3. The requirements of each kind of resource to produce each product A are 2, 4 and 3 units, respectively. To produce each product B, 3, 2 and 2 units are required respectively. The planned available resource of R_1 and R_2 are 50 and 44 units, respectively, but there are an additional 30 and 20 units safety store of material, which are administrated by the general manager. The estimated value of the available quantity of resource R_3 is 36 units, with an estimated error of 5 units. Assuming that the planned production quantity of A and B is x_1 and x_2, respectively. The experience of the past data shows that the unit cost and sale price of product A and B are $UC_1 = 15 - 0.5x_1$, $UC_2 = 17 - 0.6x_2$ and $US_1 = 32 - 0.8x_1$, $US_2 = 35 - 1.0x_2$, respectively. The DM hopes that the total profits reach an aspiration level z_0 and not less than a lower level $z_0 - p_0$. This is a typical FQP problem. It can be described as:

$$\begin{cases} \tilde{\max} \; f(x) = x_1(US_1 - UC_1) + x_2(US_2 - UC_2) = -0.3x_1^2 - 0.4X_2^2 + 17x_1 + 18x_2 \\ s.t. \quad 2x_1 + 3x_2 \leq \tilde{5}0 \\ \qquad 4x_1 + 2x_2 \leq \tilde{4}4 \\ \qquad 3x_1 + 2x_2 = \tilde{3}6 \\ \qquad x_1, x_2 \geq 0 \\ p_1 = 30,\, p_2 = 20,\, p_3^- = p_3^+ = 5.0\,. \end{cases} \tag{4.22}$$

For simplicity, we select parameters: $NP = 3$, $\alpha_0 = 0.25$, $e = 0.05$, $z_0 = 250$, $P_0 = 20$, and $\beta^{k-1} = 0.5$, assuming that the objective function, production quantity of product A and consumption of resource R_1 are of the utmost importance to the DM. For some iteration $k - 1$, the population is

$$x_1^{k-1} = (5.5, 11.5)\,,\; x_2^{k-1} = (4.85, 12.6)\,,\; x_3^{k-1} = (6.55, 10.00)\,.$$

Example 4.2.

Step 1 According to Equations 4.4 through 4.6 and 4.20, calculate $\mu_{\tilde{S}}(x)$ and the current upper and lower bounds of the criteria.

$$\begin{aligned} \mu_{\tilde{S}}\left(x_1^{k-1}\right) &= \min\{0.4263, 1.0, 0.95, 0.30\} = 0.30 > \alpha_0 \\ \mu_{\tilde{S}}\left(x_2^{k-1}\right) &= \min\{0.4345, 1.0, 0.997, 0.25\} = 0.25 \geq \alpha_0 \\ \mu_{\tilde{S}}\left(x_3^{k-1}\right) &= \min\{0.4240, 1.0, 0.890, 0.27\} = 0.27 > \alpha_0 \\ \mu_{\max} &= \max\{0.3, 0.25, 0.27\} = 0.30 \end{aligned}$$

$$Obj_{max} = 238.69, Obj_{min} = 238.48, A_{min} = 4.85$$
$$(R_1)_{max} = 47.5, (R_1)_{min} = 43.1 .$$

Step 2 Calculate fitness function $F(j)$ and selection probability $P(j)$ as in Equation 4.21.
$F(1) = 0.35$, $F(2) = 0.30$, $F(3) = 0.32$, $P(1) = 0.361$, $P(2) = 0.31$, $P(3) = 0.329$. Set individual number $j = 1$.

Step 3 Produce new individual x_j^k, $j = 1, 2, 3$.

Step 3.1 Randomly generate $\varepsilon = 0.251 \in (0, 1)$. According to the proportional selection strategy, $\varepsilon < p(2)$, so that individual x_2^{k-1} is selected as parent to reproduce x_j^k.

Step 3.2 For individual x_2^{k-1}, calculate w_i and $D(x_2^{k-1})$, as in Equation 4.18, 4.16 and 4.17, respectively.

$$w_0 = 4.26 ,\ w_1 = 0.00 ,\ w_2 = 1.225 ,\ w_3 = 20 ,\ D\left(x_2^{k-1}\right) = (-0.821, -0.571) .$$

Step 3.3 x_j^k is generated from x_2^{k-1} as in Equation 4.19.

$$x_j^k = (4.727, 12.5144) ,\ j = 1 .$$

Similarly, we can obtain $x_2^k = (6.423, 9.920)$, $x_3^k = (5.32, 11.42)$, the corresponding $\mu_{\tilde{S}}(x_j^k)$, $\mu_{\tilde{S}}(x_1^k) = 0.31365$, $\mu_{\tilde{S}}(x_2^k) = 0.3006$, $\mu_{\tilde{S}}(x_3^k) = 0.31$. The kth generation generated from the $(k-1)$th generation is x_1^k, x_2^k, x_3^k, and for every individual x_j^k, $\mu_{\tilde{S}}(x_j^k) \geq \alpha_0$.

Step 4 Calculate μ_{max} and the upper and lower bounds of criteria in the kth generation.

$$\mu_{max}^k = \max\{0.31356, 0.3006, 0.31\} = 0.31356$$
$$Obj_{max}^k = 236.27 ,\ Obj_{min}^k = 235.34, A_{max}^k = 6.432, A_{min}^k = 4.727$$
$$(R_1)_{max}^k = 46.997 ,\ (R_1)_{min}^k = 42.61 .$$

Step 5 Update μ_{max} and the upper and lower bounds of the criteria.

$$\mu_{max} = 0.31356 > 0.30$$
$$Obj_{max} = 238.69 ,\ Obj_{min} = 235.34, A_{max} = 6.55, A_{min} = 4.727$$
$$(R_1)_{max} = 47.5 ,\ (R_1)_{min} = 42.61 .$$

These are the basic processes of iteration of the algorithm. The following is the simulation results. The results of the above FQP are shown in Tables 4.1 and 4.2 by means of an interactive approach based on a genetic algorithm, coded in Microsoft Fortran.

The optimal solution of the above crisp quadratic programming problem is $x^* = (4.83333, 10.750)$, $f_{opt} = 222.4334$.

From Table 4.1, we may find that the DM may have more candidates than the exact optimal solution to choose from to make a decision in the fuzzy environment.

Table 4.1 The results under different criteria, $z_0 = 150$, $z_0 - p_0 = 120$, $\alpha_0 = 0.25$

Criteria	x_1	x_2	$f(x)$	$\mu_{\tilde{S}}(x)$	Recourse	Meaning
α^*	5.48027	11.38033	237.19580	0.35970		Opt. Mem. Deg.
0(1)	5.37358	11.30064	235.01790	0.25090		Min. Obj.
0(2)	5.54528	11.55561	239.63290	0.25059		Max. Obj.
1(1)	4.58771	12.71712	235.89510	0.25059		Min. prod. A
1(2)	7.02248	9.29991	237.39070	0.26655		Max. prod. A
2(1)	7.02248	9.29991	237.39070	0.26655		Min. prod. B
2(2)	4.58771	12.71712	235.89510	0.26655		Max. prod. B
3(1)	7.02248	9.29991	237.39070	0.26655	41.94469	Min. Reso. R_1
3(2)	4.58771	12.71712	235.89510	0.26655	47.32679	Max. Reso. R_1
4(1)	4.58771	12.71712	235.89510	0.29476	43.78510	Min. Reso. R_2
4(2)	7.02248	9.29991	237.39070	0.26655	46.68974	Max. Reso. R_2
5(1)	5.37358	11.30064	235.01790	0.25090	38.72201	Min. Reso. R_3
5(2)	5.16244	12.13077	239.25780	0.25023	39.74885	Max. Reso. R_3

Table 4.2 The results at each iteration as $z_0 = 150$, $z_0 - p_0 = 120$, $\alpha_0 = 0.25$

Iteration No.	Individual No.	$\mu_{\tilde{S}}(x)$	Iteration No.	Individual No.	$\mu_{\tilde{S}}(x)$	Iteration No.	Individual No.	$\mu_{\tilde{S}}(x)$
1	17	0.107582	13	11	0.331429	25	79	0.356789
2	2	0.113397	14	72	0.334537	26	68	0.358515
3	2	0.113397	15	51	0.347050	27	68	0.358515
4	52	0.295307	16	51	0.347050	28	68	0.358515
5	52	0.295307	17	51	0.347050	29	5	0.356930
6	52	0.295307	18	51	0.347050	30	14	0.358998
7	52	0.295307	19	79	0.332684	31	14	0.358998
8	18	0.333055	20	72	0.342522	32	14	0.358998
9	18	0.333055	21	44	0.352423	33	67	0.359408
10	18	0.333055	22	79	0.356789	34	67	0.359408
11	11	0.331429	23	79	0.356789	35	75	0.359704
12	11	0.331429	24	79	0.356789	36	75	0.359704

4.2 Nonlinear Programming Problems with Fuzzy Objective and Resources

4.2.1 Introduction

As the extension to nonlinear programming problems, this section focuses on a symmetric model for a kind of fuzzy nonlinear programming problem (FO/RNP) by way of a special GA with mutation along the weighted gradient direction. It uses an r-power type of membership function to formulate a kind of fuzzy objective and two kinds of fuzzy resource constraints, which are commonly used in actual production problems. The solution to FO/RNP may be transformed into the solution to three kinds of models according to different kinds of criteria preferred by the decision maker (DM). This chapter develops an inexact approach to solve this type of model of nonlinear programming problems. Instead of finding an exact optimal solution, by using a GA with mutation along the weighted gradient direction, this approach aims at finding a neighboring domain of optimal solutions such that every solution in the neighboring domain can be acceptable; namely, it is an "optimal solution" under a fuzzy environment. Then, by means of the human–computer interaction, the solutions preferred by the decision maker (DM) under different criteria can be achieved. The overall procedure for FO/RNP is also developed here; it may supply a preliminary framework for the practical application of the FO/RNP model.

4.2.2 Formulation of NLP Problems with Fuzzy Objective/Resource Constraints

Generally speaking, nonlinear programming problems with resource constraints have the following general form:

$$\begin{cases} \max f(x) = f(x_1, x_2, \cdots, x_n) \\ s.t.\ \ g_i(x) \le b_i,\ i = 1, 2 \cdots, m \\ x \ge 0 \end{cases} \tag{4.23}$$

where, b_i is a crisp available resource.

In actual production planning problems, however, the available quantity of a resource in a certain period is uncertain. This possesses different types of fuzziness, but it must also be known that the objective defined by the decision maker (DM) is an ill-defined goal and that there may be some fuzzy parameters in the objective function due to different reasons, such as the extent of comprehension of the problem, or the needlessness of actual maximization (minimization), and the necessity for giving some leeway. Here we discuss nonlinear programming problems with the following types of fuzzy objective and fuzzy resource constraints.

1. The objective value which the DM desired is not an actual maximum, but a fuzzy value, and the DM hopes to reach an aspirational level such as z_0, and not less than a lowest level $z_0 - p_0$. Moreover, the DM's satisfaction degree increases as the objective value increases.
2. The available quantity of a type of resource i $(i = 1, 2, \cdots, m_1)$ has some increments which were accepted by the DM by way of taking overtime work, putting to use the inventory quantity, *etc.*. Assuming that the planned available quantity of this type of resource i is b_i $(i = 1, 2, \cdots, m_1)$, the largest acceptable increment by the DM is p_i, and the fuzzy available quantity is denoted by $\tilde{b}_i$. Moreover, it was attached to a monotonic nonincreasing membership function. For the DM, this type of resource is utilized no more than what is available.
3. The available quantity of another type of resource i $(i = m_1 + 1, m_1 + 2, \cdots, m)$ is imprecise. Assume that the available quantity $\tilde{b}_i$ $(i = m_1 + 1, m_1 + 2, \cdots, m)$ of this type of resource i is an estimate with mean b_i and error p_i^- and p_i^+,

respectively, and has L-R-type membership function. For the DM, this type of resource is utilized as fully as possible.

Furthermore, assume that the objective function $f(x)$ and the resource constraints function $g_i(x)$ are nonlinear, continuous and derivable in $(R^n)^+$. The problem is how to make a reasonable plan such that the objective can be optimal or the DM will be "most" satisfied with his preferred criteria in the environment of the above fuzzy objective and resource constraints. This type of problem belongs to class of fuzzy optimization problems.

The nonlinear programming problem with the above types of fuzzy objective/resource constraints (FO/RNP) has the following general form:

$$\begin{cases} \tilde{\max} f(x) \\ s.t.\ \ g_i(x) \lesssim \tilde{b}_i,\ i = 1, 2, \cdots, m_1 \\ g_i(x) = \tilde{b}_i,\ i = m_1 + 1, m_1 + 2, \cdots, m \\ x \geq 0 \end{cases} \tag{4.24}$$

where x is the n-dimensional decision variable vector, $x = (x_1, x_2, \cdots, x_n)^T$. $\tilde{b}$ is the fuzzy available resource vector, $\tilde{b} = (\tilde{b}_1, \tilde{b}_2, \cdots \tilde{b}_m)$.

We introduce the following type of membership function to describe the fuzzy number $\tilde{b}_i$, fuzzy objective and fuzzy constraints (Zimmermann 1976). For resource $i (i = 1, 2, \cdots, m_1)$, let $\mu_{\tilde{b}_i}(x)$ indicate the attainability of the fuzzy available resource $\tilde{b}_i$ and define $\mu_{\tilde{b}_i}(x)$ as follows:

$$\mu_{\tilde{b}_i}(x) = \begin{cases} 1 & x \leq b_i \\ 1 - \left(\dfrac{(x - b_i)}{p_i}\right)^r & b_i \leq x \leq b_i + p_i \\ 0 & x > b_i + p_i \end{cases} \tag{4.25}$$

where $r > 0$. $\mu_{\tilde{b}_i}(x)$ is a monotonic non-increasing function, which denotes the attainable degree of fuzzy available resource $\tilde{b}_i$.

For resource $i (i = 1, 2, \cdots, m_1)$, let $\mu_{\tilde{b}_i}(x)$ describe the estimation for fuzzy available resource $\tilde{b}_i$ and define $\mu_{\tilde{b}_i}(x)$ as

$$\mu_{\tilde{b}_i}(x) = \begin{cases} 0 & x \leq b_i - p_i^- \\ 1 - \left(\dfrac{(b_i - x)}{p_i^-}\right)^r & b_i - p_i^- \leq x \leq b_i \\ 1 - \left(\dfrac{(b_i - x)}{p_i^+}\right)^r & b_i \leq x \leq b_i + p_i^+ \\ 0 & x > b_i + p_i^+ \end{cases} \tag{4.26}$$

$\mu_{\tilde{b}_i}(x)$ is an $L - R$-type function, which denotes the accurate level of estimation for fuzzy available resource $\tilde{b}_i$.

Similarly, let $\mu_i(x)$, which is defined below, be the membership function of the ith fuzzy constraint and reflect the situation of the DM's satisfaction with the ith fuzzy constraint.

For $i = 1, 2, \cdots, m_1$, $\mu_i(x) = \max_{y \geq g_i(x)} \mu_{\tilde{b}_i}(y) = \mu_{\tilde{b}_i}(g_i(x))$. For $i = m_1 + 1, \cdots, m$, $\mu_i(x) = \mu_{\tilde{b}_i}(g_i(x))$.

$\mu_i(x)$ denotes the degree of satisfaction with the ith fuzzy constraint by the point x.

Let $\mu_0(x)$, which is defined below, describe the DM's fuzzy objective $\tilde{\max} f(x)$.

$$\mu_0(x) = \begin{cases} 0 & f(x) \leq z_0 - p_0 \\ 1 - \left(\dfrac{z_0 - f(x)}{p_0} \right)^r & z_0 - p_0 < x \leq z_0 \\ 1 & f(x) \geq z_0 \ . \end{cases} \tag{4.27}$$

$\mu_0(x)$ is a monotonic non-decreasing continuous function and denotes the degree of satisfaction with the fuzzy objective function at the point x.

Certainly, the types of membership functions which describe the DM's fuzzy objective and fuzzy resource constraints may be determined by the DM to be exponential, logarithmic, *etc.*

FO/RNP may be described as how to make a reasonable plan such that the DM is most satisfied with the fuzzy objective and fuzzy constraints; namely, that there is the best balance degree between fuzzy objective and fuzzy constraints, and that there is a maximum objective or other types of targets under this "most satisfied" condition. The former regard the highest satisfaction degree as criteria and the latter may consider different types of targets as criteria under the consideration of a definite satisfaction degree. Then FO/RNP may be formulated as the following three kinds of models.

1. Maximize the minimum satisfaction degree (model FO/RNP-1):

$$\begin{cases} \max \alpha \\ s.t.\ \mu_0(x) \geq \alpha \\ \mu_i(x) \geq \alpha \quad i = 1, 2, 3, \cdots, m \\ x \geq 0 \ . \end{cases} \tag{4.28}$$

2. Maximize the sum of weighted satisfaction degree (model FO/RNP-2):

$$\begin{cases} \max \sum\limits_{i=0}^{m} \beta_i \mu_i(x) \\ s.t.\ \mu_0(x) \geq \alpha_0 \\ \mu_i(x) \geq \alpha_0 \quad i = 1, 2, 3, \cdots, m \\ x \geq 0 \end{cases} \tag{4.29}$$

where α_0 is an acceptable satisfaction degree preset by the DM, and β_i $(i = 1, 2, 3, \cdots, m)$ is the weight for the objective and resources constraints. It reflects the relative importance of the objective and resources in the eye of the DM and is subject to $\sum_{i=0}^{m} \beta_i = 1$.

3. Maximize the decision target (model FO/RNP-3)

$$\begin{cases} \max \mathit{Target} \\ s.t.\ \mu_0(x) \geq \alpha_0 \\ \mu_i(x) \geq \alpha_0 \quad i = 1, 2, 3, \cdots, m \\ x \geq 0 \end{cases} \tag{4.30}$$

where *Target* may be the objective function, the decision variable or the resource constraints, *etc.*, and α_0 is an acceptable satisfaction degree preferred by the DM.

Then the solution to FO/RNP may be transformed into the solution to FO/RNP-1, FO/RNP-2 or FO/RNP-3, according to different types of criteria.

Definition 4.4. Fuzzy optimal solution of FO/RNP is a fuzzy set $\tilde{S}$ defined by

$$\tilde{S} = \left\{ \left(x, \mu_{\tilde{S}}(x)\right) | x \in (R^n)^+, \mu_{\tilde{S}}(x) = \min\{\mu_0(x), \mu_i(x),\ i = 1, 2, \cdots, m\} \right\}. \tag{4.31}$$

Let $S_\alpha = \{x \in (R^n)^+, |\mu_{\tilde{S}}(x) \geq \alpha\}, \alpha \in [0, 1]$. Then, S_α is a general set which is an α-level cut set of $\tilde{S}$.

Definition 4.5. α^* is the best balance degree, such that $\forall 0 \leq \alpha \leq \alpha^*$, S_α is non-empty.

In general, a unique optimal solution α^* can be found from FO/RNP-1. The corresponding solution x^*, which is not usually unique, is the solution with the highest membership degree to the FO/RNP; its meaning is that the best balance of objective and constraints might be achieved at the point x^*. However the solution x^* is probably not the most desired by the DM under some types of criteria. In addition, the exact optimal solution is meaningless to the DM under the fuzzy environment; the solution desired by the DM is multiple different solutions which satisfy both the objective and resource constraints under different criteria preferred by the DM. The inexact approach is based on the concept of finding a neighboring domain of optimal solutions such that every solution x in the domain can be the "optimal" solution desired by the DM in a fuzzy environment.

Based on the above analysis, in the following sections, we will propose the inexact approach based on GA to solve the FO/RNP-1 model, and then develop an overall procedure for the solution to FO/RNP by means of the human–computer interaction, which may supply a preliminary framework for the practical application of the FO/RNP model.

4.2.3 Inexact Approach Based on GA to Solve FO/RNP-1

Obviously, FO/RNP-1(6) is equivalent to the following optimization problem P

$$\begin{cases} \max \mu_{\tilde{S}}(x) = \max\ \min\{\mu_0(x), \mu_1(x), \mu_2(x), \cdots, \mu_m(x)\} \\ x \in (R_n)^+ \end{cases} \tag{4.32}$$

P is an unconstrained optimization problem, but its objective function is not continuous and derivable; it cannot be solved by the traditional optimization method. A special GA with mutation along the weighted gradient direction is developed in this section. It uses the mutation as the main operator, while the arithmetic combinatorial cross-over operator is only used in the later generation. The basic idea is the following. First, randomly produce an initial population with the size of *pop size* individuals, where each individual is selected to reproduce children along the direction with an increment of membership degree of both fuzzy objective and constraints.

For individuals with membership degree less than α_0 (acceptable membership degree preferred by the DM), give it a smaller membership degree so that it may have a less chance than others of being selected to reproduce children in the later generation. As the generation increases, the individuals with membership degree less than α_0 die out gradually and others exist. After a number of generations, the individual membership degree is all greater than α_0, namely $S_{\alpha_k} \subseteq S_{\alpha_0}$ and most individuals will be close to the optimal solution.

For individual x, let $\mu_{\min}(x) = \min\{\mu_0(x), \mu_1(x), \cdots, \mu_m(x)\}$. If $\mu_{\min}(x) \leq \mu_0(x) < 1$, then move along the gradient direction of $\mu_0(x)$, the value of $\mu_0(x)$ may be improved, the smaller $\mu_0(x)$ is, the greater the improvement of $\mu_0(x)$ may be achieved.

Similarly, if $\mu_{\min}(x) \leq \mu_i(x) < 1$, then move along the gradient direction of $\mu_i(x)$, the value of $\mu_i(x)$ may be improved, the smaller $\mu_i(x)$ is, the greater the improvement of $\mu_i(x)$ may be achieved. Based on the above idea, construct

$$D(x) = w_0 \nabla \mu_0(x) + \sum_{i=1}^{m} w_i \nabla \mu_i(x) , \tag{4.33}$$

where, for $0 < \mu_0(x) \leq 1, 0 < \mu_i(x) \leq 1$,

$$\begin{cases} \nabla \mu_0(x) = r(z_0 - f(x))^{r-1} \dfrac{\nabla f(x)}{p_0^r} \\ \nabla \mu_i(x) = -r(g_i(x) - b_i)^{r-1} \dfrac{\nabla g_i(x)}{p_i^r} \\ \qquad i = 1, 2, \cdots, m_1 \\ \nabla \mu_i(x) = -r(g_i(x) - b_i)^{r-1} \operatorname{sgn}(g_i(x) - b_i) \dfrac{\nabla g_i(x)}{q_i^r} \\ \qquad i = m_1 + 1, \cdots, m \end{cases} \tag{4.34}$$

for $\mu_0(x) = 0$, $\mu_0(x) = \nabla f(x)$; for $\mu_i(x) = 0$, $\nabla \mu_i(x) = \nabla g_i(x)$.

$$q_i = \begin{cases} p_i^- & g_i(x) \leq b_i \\ p_i^+ & g_i(x) \geq b_i \end{cases} . \tag{4.35}$$

$D(x)$ is called the weighted gradient direction of $\mu_{\tilde{S}}(x)$. sgn(x) is the sign function, and w_i is the gradient direction weight defined as follows:

$$w_i = \begin{cases} 0 & \mu_i = 1 \\ \dfrac{1}{\mu_i - \mu_{\min} + e} & \mu_{\min} \leq \mu_i < 1 \end{cases} . \tag{4.36}$$

e is a sufficiently small positive number. $1/e$ is the largest weight.

The child x_j^{k+1} generated from x_i^k by mutation along the weighted gradient direction $D(x)$ can be described as follows:

$$x_j^{k+1} = x_i^k + \beta^k D\left(x_i^k\right) . \tag{4.37}$$

β^k is a random step-length of Erlanger distribution generated by a random number generator with decline means. The membership degree $\mu_{\tilde{S}}(x_j)$. is calculated as follows:

$$\mu_{\tilde{S}}(x_j) = \begin{cases} \mu_{\min}(x_j) & \text{if } \mu_{\min} \geq \alpha_0 \\ \varepsilon\, \mu_{\min}(x_j) & \text{else} \end{cases} \tag{4.38}$$

where α_0 is the acceptable satisfaction degree preferred by the DM, and $\varepsilon \in \cup(0, 1)$.

From the formulae of weight and mutation (Equations 4.11–4.15), we can see that the constraints which cannot be met have the minimum membership degree of 0, and will get the largest weight $1/e$, the mutation will lead the individual to the feasible area. When $\mu_{\tilde{S}}(x)$ is greater than 0, the objective or constraint with minimum membership degree will get the largest weight. The weighted gradient direction will improve the minimum, which results in the improvement of $\mu_{\tilde{S}}(x)$. From Equation 4.16, we may find that $x_j \notin S_{\alpha_0}$ gives a less membership degree but remains non-zero so that it still has a small chance of being selected as parents to reproduce children. Therefore, the weighted gradient direction will lead all individuals close to the exact optimal solution, and the individuals form a neighboring domain including the exact optimal solution.

In the special GA we proposed, the mutation along the weighted gradient direction is the main operator, and the arithmetic combinatorial cross-over operator is only used in the later generation. The real decision vector x is taken as the scheme of gene representation; the most commonly used proportional selection strategy is also used in our algorithm.

4.2.4 Overall Procedure for FO/RNP by Means of Human–Computer Interaction

In this section, we design a human–computer interactive procedure to help the DM select the desired solution from the fuzzy optimal solution, *i.e.*, the neighboring domain of the optimal solution, under different criteria. It can be described as follows.

Firstly, the approach asks the DM for the acceptable membership degree of the fuzzy optimal solution, α_0. Secondly, by means of interaction with the computer, the prepared membership degree, which describes the fuzzy objective and resource constraints, is displayed for the DM to select the type of membership degree. This elicits the DM to find his preference structure and point out which criteria are most important to him. The criteria may include the best balance degree of objective and resource constraints (FO/RNP-1), the sum of weighted satisfaction degree (FO/RNP-2) and the objective, decision variables or some other kinds of resources

(FO/RNP-3). The solutions in the α-level cut set of the fuzzy optimal solution by GA with mutation along the weighted gradient direction and with the highest and lowest values of these criteria are updated in each generation. When the iteration terminates, these solutions, with their criteria values, will be shown for the DM to select by the human–computer interface. The step by step procedure can be described as follows:

Procedure 4.1. Overall Procedure for FO/RNP

Step 1 Initialize.

Step 1.1 Input acceptable satisfaction degree α_0 and the largest generation number *NG*, population size *pop size*.

Step 1.2 Input the criteria index set the DM is most concerned with, $CS = \{0, 1, 2, \cdots, n, n+1, \cdots, n+m, n+m+1\}$, where 0 stands for objective function, $j = 1, 2 \cdots, n$ for decision variables, $j = n+1, n+2 \cdots, n+m$ for the constraints, and $j = n+m+1$ for the sum of weighted satisfaction degree, respectively. Give the initial values and the upper and lower values of criteria $cs, cs \in CS$.

Step 1.3 Input the type of membership function which describe the fuzzy objective and fuzzy constraints.

Step 1.4 Input the weight of objective and resource constraints β_i.

Step 2 Randomly produce the initial population and calculate their membership degrees by $x_i(j) = \varepsilon UP_i, \varepsilon \in \mathrm{U}(0, 1), i = 1, 2, \cdots, n, j = 1, 2, \cdots, pop\,size$ where, UP_i is the upper bound of the ith element of variable x. Membership function $\mu_{\tilde{S}}(x_j)$ is calculated as in Equation 4.38.

Step 3 Set iteration index $k = 1$.

Step 4 For individual j ($j = 1, 2, \cdots, pop\,size$), calculate the fitness function $F(j)$ and selection probability $P(j)$ as:

$$F(j) = \mu_{\tilde{S}}(x_j) + e\,, \quad P(j) = \frac{F(j)}{\sum_{i=1}^{pop\,size} F(j)}\,.$$

Step 5 Produce new individuals $x_j = (j = 1, 2, \cdots, popsize)$.
The term x_i is selected to produce x_j as $x_j^k = x_i^{k-1} + \beta^k D(x_i^{k-1})$ where β^k is a random step-length of Erlanger distribution with decline means generated by a random number generator, and $D(x)$ defined as in Equations 4.33 and 4.34.

Step 6 For individual $j = (j = 1, 2, \cdots, pop\,size)$, calculate the membership function $\mu_{\tilde{S}}(x_j)$ as in Equation 4.38, update optimal membership degree $\mu_{\max}$ and the upper and lower value of criteria r.

Step 7 $k + 1 \rightarrow k$, if $k \leq NG$, go to step 4; otherwise, go to step 8.

Step 8 Output the optimal membership degree $\mu_{\max}$ and the upper and lower value of criteria preferred by the DM, stop.

4.2.5 Numerical Results and Analysis

To clarify the algorithm, this section gives an example as follows. The results by the inexact approach of combining a GA with human–computer interaction are also given in this section.

Example 4.3. A manufacturing factory is going to produce two kinds of products A and B in a period (such as one month). The production of A and B requires three kinds of resources R_1, R_2 and R_3. The requirements for each kind of resource to produce each product A are 2, 4 and 3 units, respectively. To produce each product B, 3, 2 and 2 units are required, respectively. The planned available resource of R_1 and R_2 are 50 and 44 units, respectively, but there are an additional 30 and 20 units safety store of material, which are administrated by the general manager. The estimated value of the available amount of resource R_3 is 36 units, with an estimated error of 5 units. Assume that the planned production quantity of A and B is x_1 and x_2, respectively. Furthermore, assume that the unit cost and sale price of product A and B are $UC_1 = c_1$, $UC_2 = c_2$ and $US_1 = k_1/(x_1^{1/a_1})$, $US_2 = k_2/(x_2^{1/a_2})$, respectively. The DM hopes that the total profits reach an aspiration level z_0 and no less than a lower level $z_0 - p_0$. This is a typical FO/RNP problem. It can be described as:

$$\begin{cases} \max \tilde{f} = k_1 x_1^{1-1/a_1} - c_1 x_1 + k_2 x_2^{1-1/a_2} - c_2 x_2 \\ s.t.\ 2x_1 + 3x_2 \lesssim \tilde{5}0 \quad 4x_1 + 2x_2 \lesssim \tilde{4}4 \\ 3x_1 + 2x_2 = \tilde{3}6 \quad x_1, x_2 \geq 0 \\ p_1 = 30, p_2 = 20, p_3^- = p_3^+ = 5.0, r = 1 \\ k_1 = 50, c_1 = 8.0, k_2 = 45, c_2 = 10, a_1 = a_2 = 2\,. \end{cases}$$

For the above FO/RNP, criteria 0, 1, 2, 3, 4, 5 are chosen by the DM, which means that the DM cares about net profit (objective), production quantities of product A and B, and the consumption quantities of resource R_1, R_2, and R_3. Setting population size equal to 80 results in the different criteria after 52 generations, shown in Table 4.3. Additionally, the maximum satisfaction degree of generation is shown in Table 4.3 by means of the inexact approach combining GA with the human–computer interaction.

From Table 4.3, the DM may choose an appropriate solution, for example he may choose the solution corresponding to the maximum objective, which is just the best balance between objective and all kinds of resources. From Table 4.4 we may find that the individual of each iteration can quickly reach $\tilde{S}$, the fuzzy optimal solution, and only after 18 generations are they closer to the exact optimal solution $(9.76222, 5.06271)$ with the best balance degree $\alpha^* = 0.29167$.

Table 4.3 The results under different criteria as $z_0 = 150$, $z_0 - p_0 = 120$, $\alpha_0 = 0.25$

Criteria	x_1	x_2	$f(x)$	$\mu_{\tilde{S}}(x)$	Recourse	Meaning
α^*	9.76222	5.06271	128.75000	0.29167		Opt. Mem. Deg.
0(1)	7.88529	6.00671	127.54300	0.25143		Min. Obj.
0(2)	9.76222	5.06271	128.75000	0.29167		Max. Obj.
1(1)	7.69458	4.80628	127.73060	0.25769		Min. prod. A
1(2)	10.64834	3.80453	127.70040	0.25668		Max. prod. A
2(1)	8.89719	3.76743	127.63340	0.25445		Min. prod. B
2(2)	8.63742	6.40122	127.68870	0.25629		Max. prod. B
3(1)	8.89719	3.76743	127.63340	0.25445	29.09666	Min. Reso. R_1
3(2)	8.89301	6.31004	127.90030	0.26334	36.71612	Max. Reso. R_1
4(1)	7.69458	4.80628	127.73060	0.25769	40.39088	Min. Reso. R_2
4(2)	10.64834	3.80453	127.70040	0.25668	50.20244	Max. Reso. R_2
5(1)	7.69458	4.80628	127.73060	0.25769	32.69630	Min. Reso. R_3
5(2)	9.78006	5.20287	128.74040	0.25082	39.74592	Max. Reso. R_3

Table 4.4 The maximum satisfaction degree of each iteration as $z_0 = 150$, $z_0 - p_0 = 120$, $\alpha_0 = 0.25$

Iteration No.	Individual No.	$\mu_{\tilde{S}}(x)$	Iteration No.	Individual No.	$\mu_{\tilde{S}}(x)$	Iteration No.	Individual No.	$\mu_{\tilde{S}}(x)$
1	66	0.26351	19	51	0.29167	37	67	0.29167
2	66	0.26351	20	51	0.29167	38	80	0.29167
3	66	0.26351	21	51	0.29167	39	80	0.29167
4	28	0.26393	22	51	0.29167	40	80	0.29167
5	28	0.26393	23	68	0.29167	41	78	0.29167
6	76	0.28529	24	68	0.29167	42	78	0.29167
7	76	0.28529	25	68	0.29167	43	80	0.29167
8	17	0.28894	26	75	0.29167	44	80	0.29167
9	17	0.28894	27	76	0.29167	45	80	0.29167
10	17	0.28894	28	71	0.29167	46	80	0.29167
11	47	0.28955	29	71	0.29167	47	80	0.29167
12	30	0.29081	30	76	0.29167	48	80	0.29167
13	30	0.29081	31	72	0.29167	49	80	0.29167
14	48	0.29150	32	67	0.29167	50	80	0.29167
15	48	0.29150	33	68	0.29167	51	80	0.29167
16	7	0.29164	34	74	0.29167	52	80	0.29167
17	76	0.29164	35	78	0.29167			
18	62	0.29167	36	79	0.29167			

4.3 A Non-symmetric Model for Fuzzy NLP Problems with Penalty Coefficients

4.3.1 Introduction

A non-symmetric model for a type of fuzzy nonlinear programming problem with penalty coefficients (FNLP-PC) is proposed. It uses a kind of nonlinear membership function to describe the fuzzy available resources and fuzzy constraints. Based on a fuzzy optimal solution set and optimal decision set, a satisfying solution method and a crisp optimal solution method with GA for FNLP-PC are developed. Finally, the analysis of simulation results of an example in actual production problems is also given.

4.3.2 Formulation of Fuzzy Nonlinear Programming Problems with Penalty Coefficients

Crisp nonlinear programming problems with resource constraints have the following general form:

$$\begin{cases} \max\ f(x) = f(x_1, x_2, \cdots, x_n) \\ s.t.\ \ g_i(x) \le b_i\ ,\ i = 1, 2, \cdots, m \\ \qquad x \ge 0 \end{cases} \tag{4.39}$$

where, b_i is a crisp available resource.

In actual production problems, however, the available quantity of resources (manufacturing ability) during a period is often uncertain and possesses different types of fuzziness. We discuss the following types of fuzzy resources.

1. The available quantity of some type of resource i $(i = 1, 2, \cdots, m_1)$ has some increments which were accepted by the DM by taking overtime work, using the inventory quantity, *etc.* Assume that the planned available quantity of this type of resource i is b_i $(i = 1, 2, \cdots, m_1)$, and the largest acceptable increment is p_i. The fuzzy available quantity is denoted as $\tilde{b}_i$; it is attached to a monotonic nonincreasing membership function. For this type of resource, it is allowed to utilize no more than the available quantity.
2. The available quantity of some other type of resource i $(i = m_1 + 1, m_1 + 2, \cdots, m)$ is imprecise. Assume that the available quantity $\tilde{b}_i$ $(i = m_1 + 1, m_1 + 2, \cdots, m)$ of this type of resource i is an estimated value b_i and the estimated error is p_i^- and p_i^+, respectively, and also $\tilde{b}_i$ has a pair-wise linear membership function. For this type of resource, the DM hopes to utilize them as fully as possible.

Let penalty coefficient γ_i denote the penalty cost brought by per unit increment of resource i. The value of γ_i reflects the importance of resource i. Furthermore, we assume that the benefits function $f(x)$ and the constraints function $g_i(x)$ are all continuous and derivable in $(R^n)^+$. Then, NLP with the above types of fuzzy resources under the consideration of penalty coefficients may be formulated as a FNLP-PC model:

$$\begin{cases} \max\ F(x) = f(x) - \sum_{i=1}^{m_1} \gamma_i \max\{0, g_i(x) - b_i\} \\ s.t.\quad g_i(x) \lesssim \tilde{b}_i \quad i = 1, 2, \cdots, m \\ \qquad x \geq 0 \end{cases} \tag{4.40}$$

where x is the n-dimensional decision variable vector $x = (x_1, x_2, \cdots, x_n)^T$, $g(x)$ is the constraint function vector, $g(x) = (g_1(x), g_2(x), \cdots, g_m(x))^T$ and $\tilde{b}$ is the fuzzy available resource vector, $\tilde{b} = (\tilde{b}_1, \tilde{b}_2, \cdots, \tilde{b}_m)^T$. If $f(x)$ denotes production cost function, then we may get the following minimization problems:

$$\begin{cases} \max\ F(x) = f(x) + \sum_{i=1}^{m_1} \gamma_i \max\{0, g_i(x) - b_i\} \\ s.t.\quad g_i(x) \lesssim \tilde{b}_i \quad i = 1, 2, \cdots, m \\ \qquad x \geq 0\,. \end{cases} \tag{4.41}$$

It may be transformed into the solution to maximization problems. In practical situations, assume that the largest acceptable penalty costs for the tolerance of resource i is cost (preferred by the DM). Then there exists the following relationship between the penalty coefficient γ_i and the largest acceptable tolerances p_i:

$$p_i^* \gamma_i = COST\,. \tag{4.42}$$

We introduce an r-power-type of membership function to describe the fuzzy number $\tilde{b}_i$ and fuzzy constraints.

For resource i $(i = 1, 2, \cdots, m_1)$, let $\mu_{\tilde{b}_i}(x)$ be the membership function, defined by

$$\mu_{\tilde{b}_i}(x) = \begin{cases} 1, x \leq b_i \\ 1 - \left[\dfrac{(x - b_i)}{p_i}\right]^r, b_i \leq x \leq b_i + p_i \\ 0,\ x > b_i + p_i\,. \end{cases} \tag{4.43}$$

It denotes the attainable degree of fuzzy available resource $\tilde{b}_i$, and $r > 0$.

For resource i $(i = m_1 + 1, m_1 + 2, \cdots, m)$, let

$$\mu_{\tilde{b}_i}(x) = \begin{cases} 1, x \leq b_i - p_i^- \\ 1 - \left[\dfrac{b_i - x}{p_i^-}\right]^r, b_i - p_i^- \leq x \leq b_i \\ 1 - \left[\dfrac{x - b_i}{p_i^+}\right]^r, b_i \leq x \leq b_i + p_i^+ \\ 0,\ x > b_i + p_i^+\,. \end{cases} \tag{4.44}$$

It denotes the accurate level of estimation for fuzzy available resource $\tilde{b}_i$.

Similarly, $g_i(x) \leq \tilde{b}_i$ is a fuzzy subset of R^n. Let $\mu_i(x)$ be the membership function of the ith fuzzy constraint. This represents the situation of subjection to the ith fuzzy constraint. According to the extension principle,

$$\mu_i(x) = \bigvee_{y \geq g_i(x)} \mu_{\tilde{b}_i}(y). \tag{4.45}$$

$\mu_i(x)$ denotes the degree of subjection to the ith constraint at the point x.

For $\tilde{b}_i$ with the type from Equation 4.5, $\mu_i(x) = \mu_{\tilde{b}_i}(g_i(x))$.

For $\tilde{b}_i$ with the type from Equation 4.6, $\mu_i(x) = \begin{cases} 1, & g_i(x) \leq b_i \\ \mu_{\tilde{b}_i}(g_i(x)), & \text{else} \end{cases}$.

For $\tilde{b}_i$ with general membership function $\mu_{\tilde{b}_i}(y)$, $\mu_i(x)$ have the following properties:

Let $b_i = \max\{r \in R | \mu_{\tilde{b}_i}(r) = \text{high} \geq \mu_{\tilde{b}_i}(y)\ \forall y \in \tilde{b}_i\}$, for the normal case, where high $= 1$.

Property 4.12. $\mu_i(x) = \mu_{\tilde{b}_i}(b_i)$, if $g_i(x) \leq b_i$.

Property 4.13. If $\mu_{\tilde{b}_i}(y)$ is a monotonic non-increasing continuous function for $y > b_i$, then, for $g_i(x) > b_i$, $\mu_i(x) = \mu_{\tilde{b}_i}(g_i(x))$.

Property 4.14. If $\mu_{\tilde{b}_i}(y)$ is a non-monotonic continuous function for $y > b_i$, then, for $g_i(x) > b_i$, $\mu_i(x)$ is a monotonic non-increasing continuous function.

Property 4.15. If the DM's degree of satisfaction with the ith constraint is μ_i, then the penalty cost, caused by the tolerance of resource i with the type from Equation 4.5, is $\cos t_i = COST(1 - \mu_i)^{1/r}$.

Proof. For resource i using the type as defined in Equation 4.5, according to the definition of $\mu_i(x)$ in Equation 4.7, $\mu_i(x) = \mu_{\tilde{b}_i}(g_i(x)) = \mu_i$.

Case 1 If $\mu_i = 1$, then $g_i(x) \leq b_i$, which means that there is no tolerance; namely, there is no penalty cost.

Case 2 If $\mu_i = 0$, then $g_i(x) \geq b_i + p_i$; however, in light of the assumption that the largest acceptable tolerance is p_i, the penalty cost is $\cos t_i = \gamma_i^* p_i = COST$.

Case 3 If $0 < \mu_i < 1$, then $g_i(x) - b_i = p_i^*(1 - \mu_i)^{1/r}$, and

$$\cos t_i = (g_i(x) - b_i)^* \gamma_i = \gamma_i^* p_i^* (1 - \mu_i)^{1/r} = COST(1 - \mu_i)^{1/r}.$$

In summary, for $0 \leq \mu_i \leq 1$, $\cos t_i = COST(1 - \mu_i)^{1/r}$.

From Property 4.4, it can been seen that $\mu_i = 1 - \left(\frac{\cos t_i}{COST}\right)^r$; namely, μ_i also reflects the situation of the DM's satisfaction with the penalty cost.

4.3.3 Fuzzy Feasible Domain and Fuzzy Optimal Solution Set

In this section we introduce the concept of fuzzy feasible domain and fuzzy optimal solution set.

Let $\tilde{Q} = \{(x, \mu_{\tilde{Q}}(x))|x \in (R^n)^+\}$, where $\mu_{\tilde{Q}}(x) = \min\{\mu_1(x), \mu_2(x), \cdots; \mu_m(x)\}$.

$\tilde{Q}$ is a fuzzy set, called a fuzzy feasible domain. $\mu_{\tilde{Q}}(x)$ denotes the feasible degree of point x belonging to the fuzzy feasible domain. Let

$$Q_\alpha = \left\{x \in (R^n)^+ | \mu_{\tilde{Q}}(x) \geq \alpha\right\}\ , \ \alpha \in [0,1]\ . \tag{4.46}$$

Q_α is an α-level cut set, which is a general set. Then, FNLP-PC(2) may be equivalent to the following extreme problem:

$$\max_{x \in Q} F(x)\ . \tag{4.47}$$

Definition 4.6. A solution $N(\alpha)$ is called an α-optimal solution to FNLP-PC, if $N(\alpha)$ is the optimal solution to α-NLP as follows:

$$\begin{cases} F_\alpha = \max F(x) \\ s.t.\ \ x \in Q_\alpha\ . \end{cases} \tag{4.48}$$

Then, $N(\alpha)$ may be rewritten as $N(\alpha) = \{x_\alpha | x_\alpha \in Q_\alpha, F(x_\alpha) = F_\alpha = \max F(x), x \in Q_\alpha\}$.

According to Property 4.4, α-NLP (Equation 4.48) may be transformed into the following parametric NLP, which may be solved by traditional nonlinear programming or parametric optimization methods.

$$\begin{cases} \max\ F(x) = f(x) - COST^* \sum_{i=1}^{m_1} [1 - \mu_i(x)]^{1/r} \\ s.t.\ \ \mu_i(x) \geq \alpha, i = 1, 2, \cdots, m \\ \qquad x \geq 0,\ 0 \leq \alpha \leq 1\ . \end{cases} \tag{4.49}$$

Certainly, in the case that the largest acceptable penalty cost for the tolerances of resource i may be different, Equation 4.49 is replaced by

$$\begin{cases} \max\ F(x) = f(x) - \sum_{i=1}^{m_1} COST_i^*[1 - \mu_i(x)]^{1/r} \\ s.t.\ \ \mu_i(x) \geq \alpha, i = 1, 2, \cdots, m \\ \qquad x \geq 0,\ 0 \leq \alpha \leq 1\ . \end{cases} \tag{4.50}$$

Definition 4.7. Fuzzy optimal solution set, denoted by $\tilde{O}$, is a fuzzy set such that

$$\begin{cases} \tilde{O} = \bigvee_{\alpha > 0} N(\alpha) \\ \mu_{\tilde{O}}(x) = \max\{\alpha | x \in \bigvee_{\alpha > 0} N(\alpha)\}\ . \end{cases} \tag{4.51}$$

Property 4.16. If $g_i(x)(i = 1, 2, \cdots, m)$ is a convex function in $(R^n)^+$, then $Q_\alpha(0 \le \alpha \le 1)$ is a convex set, and $\tilde{Q}$ is a convex fuzzy set.

Property 4.17. If $g_i(x)(i = 1, 2, \cdots, m)$ are concave and convex functions in $(R^n)^+$, respectively, then $N(\alpha)$ is a convex set as $r = 1$.

Proof. $\forall x_1, x_2 \in N(\alpha)$, $F(x_1) = F(x_2) = F_\alpha, 0 \le \alpha \le 1$; $\forall \lambda \in [0, 1]$, $x = \lambda x_1 + (1 - \lambda)x_2 \in Q_\alpha$, $F(x) \le F(\alpha)$.

Due to the assumption that $f(x)$ and $g_i(x)$ are concave and convex, respectively, it is easy to prove that $F(x)$ is also a concave function as $r = 1$. Hence, $F(x) \ge \lambda F(x_1) + (1 - \lambda)F(x_2) = \lambda F_\alpha + (1 - \lambda)F_\alpha = F_\alpha$.

Therefore, $x = \lambda x_1 + (1 - \lambda)x_2 \in N(\alpha)$, namely $N(\alpha)$ is convex set.

Property 4.18. If $f(x)$, are strictly concave and convex in $(R^n)^+$, respectively, then $N(\alpha)$ is a single point set as $r = 1$. Moreover, it is one of the extreme points of Q_α.

Property 4.19. If $f(x)$, $g_i(x)$ $(i = 1, 2, \cdots, m)$ are strictly concave and convex in $(R^n)^+$, respectively, then $\tilde{O}$ is a convex fuzzy set as $r = 1$. Moreover, $N(\alpha)$ is the α-level cut set of $\tilde{O}$. Thus, the solution to FNLP-PC(2) may be transformed into extreme problem (Equation 4.47). In theory, $\tilde{O}$ is the optimal solution to FNLP-PC, and has the following property.

Let $f_\alpha = \max f(x)|s.t.\ x \in Q_\alpha, 0 \le \alpha \le 1$.

Property 4.20. If $COST \ge f_0 - f_1$, then FNLP-PC has the same optimal solutions with the crisp NLP model; namely, for $0 \le \alpha \le 1$, the FNLP-PC model (Equation 4.40) has the same optimal solution $N(1)$.

Property 4.21. If $COST < f_0 - f_1$, then the optimal solution to FNLP-PC satisfies the following: (1) It would be beneficial to the objective to increase the tolerances of resources. (2) The more complete the utilization of the key resource (the resource with the largest penalty coefficient γ_i), the better the solution.

4.3.4 Satisfying Solution and Crisp Optimal Solution

In theory, $\tilde{O}$ is the optimal solution to FNLP-PC, which can be obtained by parametric optimization methods. In practice, however it cannot be accepted by the DM. In this section, we suggest a satisfying solution method and a crisp optimal solution method based on GA for FNLP-PC.

4.3.4.1 Satisfying Solution to FNLP-PC

Due to the fuzziness of systems, the unique optimal solution sometimes might not be desired by the DM. In that case, a satisfying solution is more acceptable and potential under the fuzzy environment. The following method is considered to be useful.

In this approach, make a suitable discrimination on [0, 1] in light of actual problems. Let $\alpha_i = i/K$, $i = 1, 2, \cdots, K$ (K is the total discrete point number). The solution to FNLP-PC may be considered as the solution to a series of α-NLP (10) problems, which may be solved by traditional optimization methods. Let their optimal solution be (α, N (α), F_α). Then, the DM may select a satisfying solution by means of the human–computer interaction from the relationship diagram of $\alpha - F_\alpha$. Also, he may obtain the satisfying solution at any level by means of convex combination.

4.3.4.2 The Crisp Optimal Solution Based on GA

Define the membership function of objective function value Z^* as follows:

$$\mu_{Z^*}(x) = \begin{cases} 0, \ F(x) \leq F_1 \\ \left[\dfrac{F(x) - F_1}{F_0 - F_1}\right]^r, \ F_1 \leq F(x) \leq F_0 \\ 1, \ F(x) \geq F_0 \end{cases} \tag{4.52}$$

where F_1, F_0 is the optimal objective function value of α-NLP(10) as $\alpha = 1$ and 0, respectively.

Definition 4.8. The optimal decision set $\tilde{D}$ is a fuzzy set, where the membership function $\mu_{\tilde{D}}(x)$ is defined as

$$\mu_{\tilde{D}}(x) = \min\left\{\mu_{Z^*}(x), \mu_{\tilde{Q}}(x)\right\} . \tag{4.53}$$

Then, the maximizing decision

$$x^* = \arg\left\{\max \mu_{\tilde{D}}(x)\right\} \tag{4.54}$$

is the crisp optimal solution to FNLP-PC.

Property 4.22. If $f(x)$ is strictly concave and $g_i(x)$, $i = 1, 2, \cdots, m$ are convex, then x^* is the unique optimal solution to Equation 4.54.

In the following, we discuss the algorithm to solve Equation 4.54. Equation 4.54 is equivalent to

$$\begin{cases} \max \ \min\{\mu_{Z^*}(x), \mu_i(x), i = 1, 2, \cdots, m\} \\ x \in (R_n)^+ \ . \end{cases} \tag{4.55}$$

Equation 4.55 is an unconstrained optimization problem, but its objective function is not continuous and derivable, so therefore it cannot be solved by traditional optimization methods. In this section we suggest a recommended GA with mutation along the weighted gradient direction to solve it. The basic idea is described as follows: first, select any small degree α_0 (generally, $\alpha_0 = 0.05$ or 0.1), randomly produce an initial population with the size of *NP* individuals, each individual is selected to reproduce children along the increment direction of membership degree of optimal objective and fuzzy constraints, according to selection probability, depending on its fitness function value. For an individual with membership degree less than α_{k-1} (the smallest membership degree of the individual in $(k-1)$th generation), give it a smaller membership degree by means of penalty so that it may have a smaller chance than others to be selected as parents to reproduce children in the later generation. As the generation increases, the individuals with a smaller membership degree die out gradually. After a number of generations, the individual's membership degree reaches the optimum or near optimum. For individual x, let $\mu_{\min}(x) = \min\{\mu_{Z^*}(x), \mu_1(x), \cdots, \mu_m(x)\}$.

If $\mu_{\min}(x) \le \mu_{Z^*}(x) < 1$, then move along the gradient direction of $\mu_{Z^*}(x)$, where the value of $\mu_{Z^*}(x)$ may be improved. The smaller $\mu_{Z^*}(x)$ is, the greater the improvement of $\mu_{Z^*}(x)$. Similarly, if $\mu_{\min}(x) \le \mu_{Z^*}(x) < 1$, then move along the gradient direction of $\mu_i(x)$, and the value of $\mu_i(x)$ may be improved. The smaller $\mu_i(x)$ is, the greater the improvement of $\mu_i(x)$.

Based on the above idea, construct

$$d(x) = w_0 \nabla \mu_{Z^*}(x) + \sum_{i=1}^{m} w_1 \mu_i(x) \,. \tag{4.56}$$

For $0 < \mu_{Z^*}(x) \le 1, 0 < \mu_i(x) \le 1, i = 1, 2, \cdots, m$,

$$\begin{cases} \nabla \mu_{Z^*}(x) = r(F_0 - F(x))^{r-1} \dfrac{\nabla F(x)}{(F_0 - F_1)^r} \\ \nabla \mu_i(x) = -r(g_i(x) - b_i)^{r-1} \dfrac{\nabla g_i(x)}{p_i^r}\,, \quad i = 1, 2, \cdots, m_1 \\ \nabla \mu_i(x) = -r(g_i(x) - b_i)^{r-1} \dfrac{\nabla g_i(x)}{(p_i^+)^r}\,, \quad i = m_1 + 1, \cdots, m\,. \end{cases} \tag{4.57}$$

For $\mu_{Z^*}(x) = 0$, let $\nabla \mu_{Z^*}(x) = \nabla F(x)$, for $\mu_i(x) = 0$, $\nabla \mu_i(x) = \nabla g_i(x)$. $d(x)$ is called the weighted gradient direction, where w_i is the gradient direction weight defined as follows:

$$w_i = \begin{cases} 0 & \mu_i = 1 \\ \dfrac{1}{\mu_i - \mu_{\min} + e}\,, & \mu_{\min} \le \mu_i < 1 \end{cases} \tag{4.58}$$

where e is a sufficiently small positive number. Then, x_i^{k+1} is generated from x_i^k by mutation along the weighted gradient direction $d(x)$ and can be described as

$$x_j^{k+1} = x_i^k + \beta^k d\left(x_i^k\right) \tag{4.59}$$

where β^k is a step-length of Erlanger distribution random number with declining means, generated by a random number generator. Calculate $\mu_{\tilde{D}}(x_j)$ as follows:

let $\mu_{\min} = \min\{\mu_{Z^*}(x_j), \mu_1(x_j), \cdots, \mu_m(x_j)\}$ and

$$\mu_{\tilde{D}}(x_j) = \begin{cases} \mu_{\min} & \text{if } \mu_{\min} \geq \alpha_{k-1} \\ \varepsilon^* \mu_{\min} & \text{else} \end{cases},$$

where, $\varepsilon \in \cup(0,1), \alpha_{k-1}$ satisfies:

$$\alpha_{k-1} = \min\left\{\mu_{\tilde{D}}\left(x_j^{k-1}\right) | x_j^{k-1} \text{ satisfy } \mu_{\tilde{D}}\left(x_j^{k-1}\right) \geq \alpha_{k-2}\right\}.$$

It can be seen that for an individual x_j with the membership degree of $\mu_{\tilde{D}}(x_j) < \alpha_{k-1}$, given a smaller membership degree, but not zero, it will have a small chance of being selected to reproduce children.

4.3.5 General Scheme to Implement the FNLP-PC Model

Based on the analysis of satisfying solution and crisp optimal solution to FNLP-PC, we may get the following general scheme to implement the FNLP-PC model.

Step 1 Formulate the flexibilization or fuzziness of resources by means of membership function, such as Equations 4.43 and 4.44, which is determined depending on the actual problems.

Step 2 Determine some initial parameters, such as $b_i, p_i, \alpha_0, COST, r, etc.$ The parameters *COST* and r are always determined by means of interaction with the DM.

Step 3 Ask the DM to prefer satisfying solution (mark $= 0$) or crisp optimal solution (mark $= 1$). If mark $= 0$, go to step 4; else go to step 5.

Step 4 Applying the traditional optimization method to α-NLP (10) problem. Generally, select $\alpha_k = \alpha_{k-1} + 0.05, \alpha_0 = 0.5, k = 1, 2, \cdots, 19$. The DM may get the preferred satisfying solution by means of the human–computer interaction from the relationship of $\alpha - F_\alpha$ or by convex combination.

Step 5 Find the crisp optimal strategy. Applying traditional optimization method to

$$f_\alpha = \max\{f(x) | s.t.\ x \in Q_\alpha\},\ \alpha = 0, 1$$

we may get f_0 and f_1; set the membership function of the objective function as shown in Equation 4.52. Then the optimal strategy is different as in the following cases.

Case 1 If $COST < f_0 - f_1$, the optimal strategy will satisfy Property 4.10.

Case 2 If $COST < f_0 - f_1$, then the DM hopes to reach the best balance between the resources and profits, by applying the recommended GA with mutation along the weighted gradient direction (Tang and Wang 1997a,b). We may get the best balance degree and the optimal total profits.

Case 3 If $COST \geq f_0 - f_1$, the optimal strategy is the optimal solution to the corresponding crisp NLP model.

4.3.6 Numerical Illustration and Analysis

To clarify the model and methods for FNLP-PC, in this section we give the illustration by means of an example.

Example 4.4. A manufacturing factory is going to produce two kinds of products A and B during a period (such as one month).The production of A and B requires three kinds of resources R_1, R_2 and R_3. The requirements of each kind of resource to produce each product A are 2, 4 and 3 units, respectively. To produce each product B, 3, 2 and 2 units are required, respectively. The planned available resources of R_1, R_2 and R_3 are 50, 44 and 36 units, respectively. However, there are an additional 30, 20 and 5.0 units safe store of material which are administrated by the general manager. Assume that the planned production quantity of A and B is x_1 and x_2, respectively. The experience of the past data shows that the unit cost and sale price of product A and B are $UC_1 = 15 - 0.1x_1$, $UC_2 = 17 - 0.1x_2$ and $US_1 = 32 - 0.4x_1$ $US_2 = 35 - 0.5x_2$, respectively. The DM hopes that the total profits reach a maximum level.

For simplicity, we select $\gamma_1 = 0.5$, $\gamma_2 = 0.75$ and $\gamma_3 = 3.0$.

$$\begin{cases} \max\ f(x) = x_1(US_1 - UC_1) + x_2(US_2 - UC_2) = -0.3x_1^2 - 0.4x_2^2 + 17x_1 + 18x_2 \\ s.t.\quad 2x_1 + 3x_2 \le \tilde{50} \\ \qquad\ 4x_1 + 2x_2 \le \tilde{44} \\ \qquad\ 3x_1 + 2x_2 \le \tilde{36} \\ \qquad\ x_1, x_2 \ge 0 \end{cases}$$

$$p_1 = 30\,,\ p_2 = 20\,,\ p_3 = 5.0\,,\ \gamma_1 = 0.5\,,\ \gamma_2 = 0.75\,,\ \gamma_3 = 3.0\,. \tag{4.60}$$

For resources R_1, R_2 and R_3, we select the membership function with Equation 4.5 as $r = 1$ to describe the flexibility of the resources.

The comparison results of the above example under the environment of no fuzzy and fuzzy with penalty coefficients are shown in Table 4.5 and Figure 4.1 by means of the satisfying solution method and crisp optimal solution method based on GA.

From Table 4.5 and Figure 4.1, one can obtain the optimal production strategy. Under the consideration of penalty cost, the optimal strategy is different for the following cases.

Case 1 If the penalty coefficients γ_i satisfy $COST = \gamma_i^* p_i < f_0 - f_1 = 22.8754$, then the optimal strategy satisfies Property 4.10.

Case 2 If the penalty coefficient γ_i satisfies $COST = \gamma_i^* p_i < 22.8754$, the DM hopes to reach the best balance between the consumptions of resources and the total profits, then the optimal solution corresponding to $COST = 15.0$ is $(x_1, x_2) = (5.38146, 10.94204)$ with the best balance degree of 0.59423 and the total profit of 225.7768.

Case 3 If the penalty coefficient γ_i satisfies $COST = \gamma_i^* p_i \ge 22.8754$, then the optimal solution is $(x_1, x_2) = (4.8333, 10.25)$ with the total profits of 222.43286.

Table 4.5 The comparison results under the environment of no fuzzy and fuzzy with penalty coefficients

Content	Crisp NLP Model	FNLP-PC Model
Type of optimal	Crisp x^*	Uncertain, fuzzy optimal solution set maximizing decision x^*
Optimal objective	222.43286	225.7768
x_1^*	4.83333	5.38146
x_2^*	10.75	10.94204
Consumptions of R_1	41.9166	43.58904
Consumptions of R_2	40.8332	43.40992
Consumptions of R_3	35.9999	38.0285
$\mu_{\tilde{D}}(x^*)$		0.59423

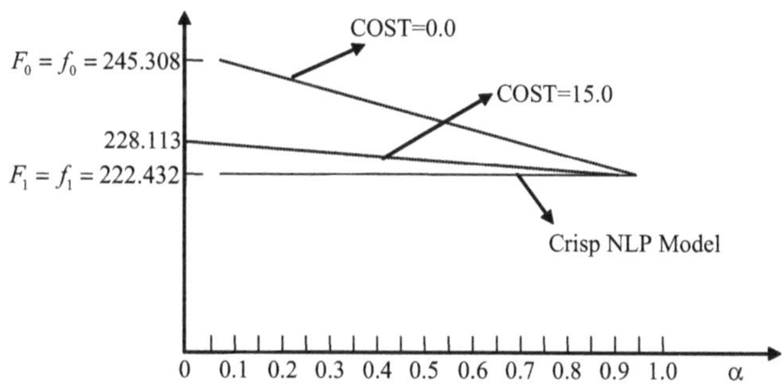

Figure 4.1 The comparison of α-optimal solution under fuzzy environment with penalty coefficients

4.4 Concluding Remarks

A type of model of quadratic programming problems with fuzzy objective and resource constraints was proposed based on the description of fuzzy objective and resource constraints with different types of membership functions in accordance with the different types of fuzzy objective and fuzzy resource constraints in actual production problems. An interactive approach is developed to solve this type of model of quadratic programming problems with fuzzy objective and resource constraints. It cannot only find a family of solutions with acceptable membership degrees, but also the solutions preferred by DM under different criteria can be achieved by means of the human–computer interaction.

In Section 4.2, a type of model of nonlinear programming problems with fuzzy objective and resource constraints (FO/RNP) was proposed and an inexact approach based on a special genetic algorithm for one kind of model FO/RNP, as well as the overall procedure for FO/RNP by means of the human–computer interaction were developed. The simulation of some examples show that: (1) this approach cannot only find an exact near optimal solution with best balance degree, but also a family of solutions with acceptable membership degree, especially, by means of the human–computer interaction. The preferred solutions that can give more candidates than only the exact optimal solution as a choice under different criteria can be achieved. (2) This approach is appropriate for any type of membership function such as linear and nonlinear, exponential, *etc.* The overall procedure supplies a preliminary framework for the practical application of the model FO/RNP in the area of production planning, *etc.*

Finally, a non-symmetric model for a type of fuzzy nonlinear programming problem with penalty coefficients (FNLP-PC) is proposed. It uses a kind of nonlinear membership function to describe the fuzzy available resources and fuzzy constraints. Based on a fuzzy optimal solution set and optimal decision set, a satisfying solution method and a crisp optimal solution method with GA for a FNLP-PC were developed. This also provides a satisfying solution and a crisp optimal solution based on GA for a FNLP-PC model. It is "flexible" and can give the DM some decision support.

References

Tang JF, Wang D (1997a) An interactive approach based on GA for a type of quadratic programming problems with fuzzy objective and resources. Comput Oper Res 24(5):413–422

Tang JF, Wang D (1997b) A non-symmetric model for fuzzy nonlinear programming problems with penalty coefficients. Comput Oper Res 24(8):717–725

Zimmermann HJ (1976) Description and optimization of fuzzy systems. Int J General Syst 2:209–216

Chapter 5
Neural Network and Self-organizing Maps

Self-organizing map (SOM) is a famous type of artificial neural network, which was first developed by Kohonen (1997). The SOM algorithm is vary practical and has many useful applications, such as semantic map, diagnosis of speech voicing, solving combinatorial optimization problem, and so on. However, its theoretical and mathematical structure is not clear. In this chapter, we discuss a special property, *i.e.*, monotonicity, of model functions in fundamental SOM with one-dimensional array of nodes and real-valued nodes. Firstly, so-called quasiconvexity and quasiconcavity for model functions have been suggested. Then it has been shown that the renewed model function of a quasiconvex (quasiconcave) model function is also quasiconvex (quasiconcave), and quasiconvex states or quasiconcave states of a model function appear in the previous stage of the monotonic states.

This chapter is organized as follows. Section 5.1 gives a simple review of neural network and self-organizing map and introduces our motivation for the research. Section 5.2 presents the basic concept and algorithm of the SOM. Section 5.3 gives the main theoretical results and detail proof. In Section 5.4 numerical examples are given to illustrate the properties of the SOM. Concluding remarks are given in the final section.

5.1 Introduction

A neural network is an information processing paradigm that is inspired by the way biological nervous systems, such as the brain, process information. The key element of this paradigm is the novel structure of the information processing system. It is composed of a large number of highly interconnected processing elements (neurons) working in unison to solve specific problems. A neural network is configured for a specific application, such as pattern recognition or data classification, through a learning process. Learning in biological systems involves adjustments to the synaptic connections that exist between the neurons. Neural networks, with their remarkable ability to derive meaning from complicated or imprecise data, can be used to extract patterns and detect trends that are too complex to be noticed by either humans or other computer techniques. A trained neural network can be thought of as an "expert" in the category of information it has been given to analyze. This expert can then be used to provide projections given new situations of interest, and answer "what if" questions. Several advantages include:

1. Adaptive learning: An ability to learn how to do tasks based on the data given for training or initial experience.
2. Self-organization: A neural network can create its own organization or representation of the information it receives during learning time.
3. Real-time operation: Neural network computations may be carried out in parallel, and special hardware devices are being designed and manufactured which take advantage of this capability.
4. Fault tolerance via redundant information coding: Partial destruction of a network leads to the corresponding degradation of performance. However, some network capabilities may be retained even with major network damage.

Neural networks learn by example. They cannot be programmed to perform a specific task. The examples must be selected carefully otherwise useful time is wasted or even worse the network might be functioning incorrectly. The disadvantage is that because the network finds out how to solve the problem by itself, its operation can be unpredictable.

Self-organizing map (SOM) is a famous type of neural network. With self-organizing map, clustering is performed by having several unit computers for the current unit. The unit whose weight vector is closest to the current unit becomes the winning or active unit. So, to move even close to the input unit, the weights of the winning unit are adjusted, as well as those of its nearest neighbors. SOM assume that there is some topology or ordering among the input units and that the units will eventually take on this structure in space. The organization of units is said to form a feature map, where SOM are believed to resemble processing that can occur in the brain and are useful for visualizing high-dimensional data in two- or three-dimensional space. Several views of SOM can be introduced to understand the nature of SOM (Kohonen 1997) as described below.

1. The SOM can be understood as an artificial neural network model of the brain, especially of the experimentally found ordered "maps" in the cortex. There exists a large amount of neurophysiological evidence to support the idea that the SOM captures some of the fundamental processing principles of the brain. However, the SOM principle has turned out to produce the brain-like maps most efficiently.
2. The SOM can be viewed as a model of unsupervised (machine) learning, and as an adaptive knowledge representation scheme. The relationship between the SOM (especially the word category map) and the semantic networks is considered. The traditional knowledge representation formalisms such as semantic networks, frame systems, predicate logic, to provide some examples, are static and the reference relations of the elements are determined by a human. Moreover, those formalisms are based on the tacit assumption that the relationship

between natural language and world is one-to-one: the world consists of objects and the relationships between the objects, and these objects and relationships have a straightforward correspondence to the elements of language. Related to knowledge representation and learning, the cognitive and philosophical aspects are highly relevant.

3. The SOM is nowadays often used as a statistical tool for multivariate analysis. The SOM is both a projection method, which maps high-dimensional data space into low-dimensional space, and a clustering method so that similar data samples tend to be mapped to nearby neurons. From the methodological and computational point of view the mathematical and statistical properties of the algorithm can be considered (for instance, the time and space complexity, the convergence properties), as well as the nature of the input data (signals, continuous statistical indicators, symbol strings) and their preprocessing. There exist a number of variants of the SOM in which, *e.g.*, the adaptation rules are different, various distance measures can be used, and the structure of the map interconnections is variable.
4. The SOM is widely used as a data mining and visualization method for complex data sets. Application areas include, for instance, image processing and speech recognition, process control, economical analysis, and diagnostics in industry and in medicine. An example of the engineering applications of the SOM is given in Chapter 14.

However, as Cottrell and Fort (1997) pointed out, the SOM algorithm is very astonishing. On the one hand, it is very simple to write down and to simulate; its practical properties are clear and easy to observe. But, on the other hand, its theoretical properties still remain without proof in the general case, despite the great effort of several authors. This chapter describes a contribution of the theoretical properties of SOM. In particular, we discuss the monotonicity of model functions in SOM with one-dimensional array of nodes and real-valued nodes. We suggest quasiconvexity and quasiconcavity for model functions. Based on that, we show that the renewed model function of a quasiconvex (quasiconcave) model function is also quasiconvex (quasiconcave), and quasiconvex states or quasiconcave states of a model function appear in the previous stage of the monotonic states. Further research efforts will be devoted to extend the model to higher-dimensional cases, which have more practical applications.

5.2 The Basic Concept of Self-organizing Map

SOM can be visualized as a sheet-like neural network array, the cells (or nodes) of which become specifically tuned to various input signal patterns or classes of patterns in an orderly fashion. The learning process is competitive and unsupervised, meaning that no teacher is needed to define the correct output (or actually the cell into which the input is mapped) for an input. In the basic version, only one map

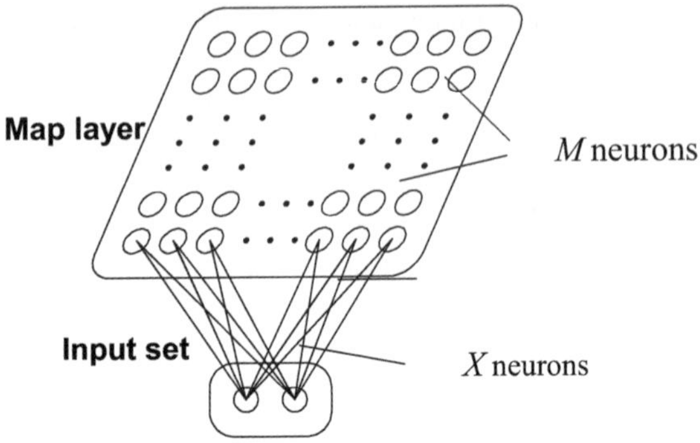

Figure 5.1 The self-organizing maps (SOM)

node (winner) at a time is activated corresponding to each input. The locations of the responses in the array tend to become ordered in the learning process as if some meaningful nonlinear coordinate system for the different input features were being created over the network.

Assume that some sample datasets have to be mapped onto the array depicted in Figure 5.1. The set of input samples described by a real vector X is the *input set*: $x_0, x_1, x_2, \ldots \in X \subset R$. Each node has its *value*, respectively. A mapping $m : I \to R$ transforming each node i to its value $m(i)$ is called a *model function* or a *reference function*. R is the set of all real numbers. Let M be the set of all model functions. $m_0 : I \to R$ is the initial model function (model vector).

The stochastic SOM algorithm performs a regression process. Thereby, the initial values of the components of the model vector, m_0, may even be selected at random. In practical applications, however, the model vectors are more profitably initialized in some orderly fashion, *e.g.*, along a two-dimensional subspace spanned by the two principal eigenvectors of the input data vectors. Any input sample is thought to be mapped into the location, the $m(i)$ of which matches best with X in some metric. The self-organizing algorithm creates the ordered mapping as a repetition of the following basic tasks:

1. An input vector X is compared with all the model vectors $m(i)$. The best-matching unit (node) on the map, *i.e.*, the node where the model vector is most similar to the input vector in some metric (*e.g.*, Euclidean) is identified. This best matching unit is often called the winner.
2. The model vectors of the winner and a number of its neighboring nodes in the array are changed towards the input vector according to the learning principle specified below.

The basic idea in the SOM learning process is that, for each sample input vector X, the winner and the nodes in its neighborhood are changed closer to X in the input data space. During the learning process, individual changes may be contradictory, but the net outcome in the process is that ordered values for the $m(i)$ emerge over the array. If the number of available input samples is restricted, the samples must be presented reiteratively to the SOM algorithm.

Adaptation of the model vectors in the learning process may take place according to the following equations: we assume two learning processes defined below. These learning processes are essential in theoretical study and application of self-organizing map models.

1. Learning Process L_A

Winner and its neighborhood:

$$I(m_k, x_k) = \{i^* \in I \mid |m_k(i^*) - x_k| = \inf_{i \in I} |m_k(i) - x_k|\} \ (m_k \in M, x_k \in X),$$
$$N_1(i) = \{j \in I \mid |j - i| \le 1\} \quad (i \in I).$$

Learning-rate factor: $0 \le \alpha \le 1$.
Learning:

$$m_{k+1}(i) = \begin{cases} (1-\alpha)m_k(i) + \alpha x_k & \text{if } i \in \bigcup_{i^* \in I(m_k, x_k)} (i^*), \\ m_k(i) & \text{if } i \notin \bigcup_{i^* \in I(m_k, x_k)} (i^*), \ k = 0, 1, 2, \ldots \end{cases}$$

2. Learning Process L_m

Winner and its neighborhood:

$$J(m_k, x_k) = \min \{i^* \in I \mid |m_k(i^*) - x_k| = \inf_{i \in I} |m_k(i) - x_k|\} \ (m_k \in M, x_k \in X),$$
$$N_1(i) = \{j \in I \mid |j - i| \le 1\} \quad (i \in I).$$

Learning-rate factor: $0 \le \alpha \le 1$.
Learning:

$$m_{k+1}(i) = \begin{cases} (1-\alpha)m_k(i) + \alpha x_k & \text{if } i \in N_1(J(m_k, x_k)), \\ m_k(i) & \text{if } i \notin N_1(J(m_k, x_k)), \quad k = 0, 1, 2, \ldots. \end{cases}$$

where k is the discrete index of the variables, the factor $0 \le \alpha \le 1$ is a scalar that defines the relative size of the learning step, and $N_1(i)$ specifies the *neighborhood* around the winner in the map array. Note the learning process L_m is the same as L_A except there is only one node $J(m_k, x_k)$, which was selected from $I(m_k, x_k)$ by a given rule.

For simplicity, we use a one dimensional example to illustrate the learning process. Denote $m_0 = [m_0(1), m_0(2), \ldots, m_0(n)]$ where n nodes exist and each node has its value. That is, nodes $1, 2, \ldots, n$ have values $m_0(1), m_0(2), \ldots, m_0(n)$, respectively. At the beginning of the learning process, if an input $x_0 \in X$ is given, then we choose node i^*, which has the most similar value to x_0 within

$m_0(1), m_0(2), \ldots, m_0(n)$. Node i^* and the nodes that are in the neighborhood of i^* learn x_0 and their values change to new values $m_1(i) = (1-\alpha)m_0(i) + \alpha x_0$. However the nodes not in the neighborhood of i^* have not been learned and their values were not changed. Repeating these, updating for the inputs $x_1, x_2, x_3, \ldots$, the value of each node is renewed sequentially. Simultaneously, model functions $m_1, m_2, m_3, \ldots$ are also generated sequentially. We denote $m_k = [m_k(1), m_k(2), \ldots, m_k(n)]$. It can be observed that the radius of the neighborhood is fairly large at the beginning of the learning process, but it is made to shrink during learning. This ensures that the global order is obtained already at the beginning, whereas towards the end, as the radius gets smaller, the local corrections of the model vectors in the map will be more specific. The factor α also decreases during learning. Following properties of the learning process are well known. In learning process L_A, for model functions $m_1, m_2, m_3, \ldots$, the following hold:

1. If m_k is increasing on I, then m_{k+1} is increasing on I.
2. If m_k is decreasing on I, then m_{k+1} is decreasing on I.
3. If m_k is strictly increasing on I, then m_{k+1} is strictly increasing on I.
4. If m_k is strictly decreasing on I, then m_{k+1} is strictly decreasing on I.

Such property is called monotonicity, which represents the absorbing states of SOM models in the sense that once model function leads to an increasing state: it never leads to other states for the learning by any input data. By repeating the learning process, some model functions have properties such as monotonicity and certain regularity may appear in the relation between the array of nodes, and the values of nodes. Such a phenomenon of appearing in the above process is called *organization*. This organization often appears in various sets of nodes, various spaces of the values of nodes, and various learning processes. Moreover, many applications of SOM in many practical problems have been accomplished by using these properties. Chapter 13 gives a real application example of automatic exterior inspection in electronic device industry.

5.3 The Trial Discussion on Convergence of SOM

The mathematical analysis of the SOM algorithm has turned out to be very difficult. The proof of the convergence of the SOM learning process in the one-dimensional case was first given by Cottrell and Fort (1987). Convergence properties are more generally studied, *e.g.*, in the following references (Erwin *et al.* 1991, 1992a,b; Horowitz and Alvarez 1996; Flanagan 1997). We introduce the certain regularity like as quasiconvexity and quasiconcavity of model function in SOM (Hoshino *et al.* 2004, 2006). Generally, we use convexity, concavity, quasiconvexity, and quasiconcavity for functions on convex sets. However, model functions in SOM are not defined on a linear space, and therefore, are not defined on a convex set in the usual sense. Now, we define quasiconvexity and quasiconcavity for a function on a partially ordered set instead of a convex set. A definition and properties of quasiconvex functions and quasiconcave functions on convex sets are described in Hoshino *et al.* (2004, 2006) in detail.

Definition 5.1. Let $(Y, \leq)$ be a partially ordered set and let f be a real-valued function on Y. Then f is said to be quasiconvex if for any $y_1, y_2, y_3 \in Y$ with $y_1 \leq y_2 \leq y_3$,

$$f(y_2) \leq \max\{f(y_1), f(y_3)\}.$$

Also f is said to be quasiconcave if for any $y_1, y_2, y_3 \in Y$ with $y_1 \leq y_2 \leq y_3$, $f(y_2) \geq \min\{f(y_1), f(y_3)\}$.

Figure 5.2 shows the examples of quasiconvex and quasiconcave functions. Additionally, f is said to be strongly quasiconvex if for any $y_1, y_2, y_3 \in Y$ with $y_1 < y_2 < y_3$,

$$f(y_2) < \max\{f(y_1), f(y_3)\}.$$

f is said to be strongly quasiconcave if for any $y_1, y_2, y_3 \in Y$ with $y_1 < y_2 < y_3$,

$$f(y_2) > \min\{f(y_1), f(y_3)\}.$$

For instance, the function in Figure 5.2 (a) is quasiconvex, not strongly quasiconvex. The function in Figure 5.2 (b) is strongly quasiconcave.

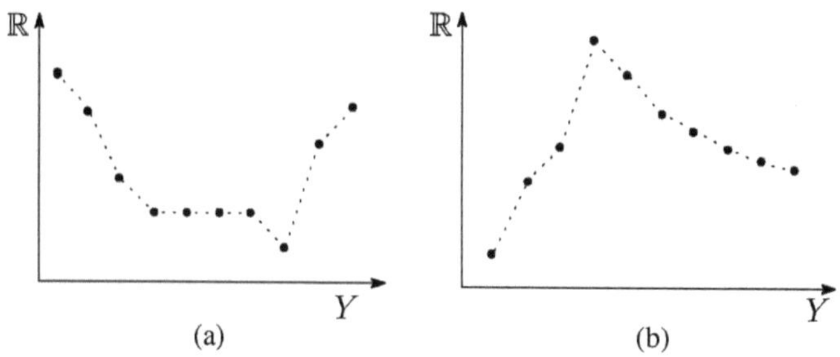

Figure 5.2 (a) A quasiconvex function and (b) a quasiconcave function

We give a necessary and sufficient condition for the quasiconvexity as follows.

Property 5.1. Let Y be a partially ordered set. Let f be a real-valued function on Y. For each $a \in R$, we put $L_a(f) = \{y \in Y \mid f(y) \leq a\}$, which is a level sets of f. Then the following are equivalent.

f is a quasiconvex function;

for all $a \in R$, if $y_1, y_3 \in L_a(f)$ and $y_1 \leq y_2 \leq y_3$, then $y_2 \in L_a(f)$.

Proof. (i) $\Rightarrow$ (ii) if suppose $y_1, y_3 \in L_a(f)$ and $y_1 \leq y_2 \leq y_3$. Then, we have $f(y_1) \leq a$ and $f(y_3) \leq a$.

By the quasiconvexity of f, $f(y_2) \leq \max\{f(y_1), f(y_3)\} \leq a$.

This implies $y_2 \in L_a(f)$.

(ii) $\Rightarrow$ (i) Let $y_1 \leq y_2 \leq y_3$. Putting $a = \max\{f(y_1), f(y_3)\}$, we have $f(y_1) \leq a$ and $f(y_3) \leq a$.

This implies $y_1, y_3 \in L_a(f)$. By condition (ii), we have $y_2 \in L_a(f)$.

Therefore, $f(y_2) \leq a = \max\{f(y_1), f(y_3)\}$.

Thus f is a quasiconvex function. □

According to Property 5.1 above, the following property is easily understood.

Property 5.2. Let Y be a partially ordered set. Let f be a real-valued function on Y. Then the following hold:

1. f is a quasiconvex function if and only if $-f$ is a quasiconcave function.
2. f is monotonic if and only if f is quasiconvex and quasiconcave.

Using these properties we can prove a SOM with learning process L_A has the following properties.

Theorem 5.1. *Take the learning process L_A. For model functions $m_1, m_2, m_3, \ldots$, the following statements hold:*

1. *If m_k is quasiconvex on I, then m_{k+1} is quasiconvex on I.*
2. *If m_k is quasiconcave on I, then m_{k+1} is quasiconcave on I.*

Proof. (i) Suppose that m_k is quasiconvex on I. Take any $i_1, i_2, i_3 \in I$ with $i_1 < i_2 < i_3$. Let x_k be the current input. We put

$$Q = \max\{m_{k+1}(i_1), m_{k+1}(i_3)\} - m_{k+1}(i_2)\,.$$

In order to prove that m_{k+1} is quasiconvex, we show $Q \geq 0$ in the following eight cases (A–H).

Case A $i_1, i_2, i_3 \in \bigcup_{i^* \in I(m_k, x_k)} N_1(i^*)$. We have

$$\begin{aligned} Q &= \max\{(1-\alpha)m_k(i_1) + \alpha x_k, (1-\alpha)m_k(i_3) + \alpha x_k\} - ((1-\alpha)m_k(i_2) + \alpha x_k) \\ &= \max\{(1-\alpha)(m_k(i_1) - m_k(i_2), (1-\alpha)(m_k(i_3) - m_k(i_2))\} \\ &= (1-\alpha)\,(\max\{m_k(i_1), m_k(i_3)\} - m_k(i_2)) \geq 0\,. \end{aligned}$$

Case B $i_1, i_2 \in \bigcup_{i^* \in I(m_k, x_k)} N_1(i^*)$ and $i_3 \notin \bigcup_{i^* \in I(m_k, x_k)} N_1(i^*)$. We have

$$\begin{aligned} Q &= \max\{(1-\alpha)m_k(i_1) + \alpha x_k, m_k(i_3)\} - ((1-\alpha)m_k(i_2) + \alpha x_k) \\ &= \max\{(1-\alpha)(m_k(i_1) - m_k(i_2)), (1-\alpha)(m_k(i_3) - m_k(i_2)) + \alpha(m_k(i_3) - x_k)\}. \end{aligned}$$

B1: If $m_k(i_1) \geq m_k(i_2)$, then the left term of max in Q is nonnegative. Hence we have $Q \geq 0$.

B2: If $m_k(i_1) < m_k(i_2)$, then $m_k(i_3) \geq m_k(i_2)$. We show $m_k(i_2 - 1) \leq m_k(i_2)$. Suppose $m_k(i_2 - 1) > m_k(i_2)$. Then we have $m_k(i_2 - 1) > \max\{m_k(i_1), m_k(i_2)\}$. This inequality contradicts that m_k is quasiconvex. Similarly, we have

$$m_k(i_2) \leq m_k(i_2 + 1) \leq m_k(i_3)$$

by the quasiconvexity of m_k. Now, we show $m_k(i_3) \geq x_k$. Suppose $m_k(i_3) < x_k$. Since

$$m_k(i_2 - 1) \leq m_k(i_2) \leq m_k(i_2 + 1) \leq m_k(i_3) < x_k$$

and $i_2 \in \bigcup_{i^* \in I(m_k, x_k)} N_1(i^*)$, we have $i_3 \in I(m_k, x_k)$. This contradicts the condition of Case B. Therefore, $m_k(i_3) \geq m_k(i_2)$ and $m_k(i_3) \geq x_k$ imply that the right term of max of Q is nonnegative. Thus, we have $Q \geq 0$.

Case C $i_1, i_3 \in \bigcup_{i^* \in I(m_k, x_k)} N_1(i^*)$ and $i_2 \notin \bigcup_{i^* \in I(m_k, x_k)} N_1(i^*)$. We have

$$\begin{aligned} Q &= \max\{(1-\alpha)m_k(i_1) + \alpha x_k, (1-\alpha)m_k(i_3) + \alpha x_k\} - m_k(i_2) \\ &= \max\{(1-\alpha)(m_k(i_1) - m_k(i_2)) + \alpha(x_k - a_k(i_2)), \\ &\qquad (1-\alpha)(m_k(i_3) - m_k(i_2)) + \alpha(x_k - a_k(i_2))\}. \end{aligned}$$

C1: If $m_k(i_1) \geq m_k(i_2)$ and $m_k(i_3) \geq m_k(i_2)$, then it follows from the quasiconvexity of m_k that $m_k(i_1 - 1) \geq m_k(i_1)$ and $m_k(i_3 + 1) \geq m_k(i_3)$. Moreover, by the quasiconvexity of m_k, we have $m_k(i_2) \leq m_k(i_1 + 1) \leq m_k(i_1)$ or $m_k(i_2) \leq m_k(i_3 - 1) \leq m_k(i_3)$. Now, we show $x_k \geq m_k(i_2)$. Suppose $x_k < m_k(i_2)$. Since

$$\begin{aligned} &x_k < m_k(i_2) \leq m_k(i_1 + 1) \leq m_k(i_1) \leq m_k(i_1 - 1) \quad \text{or} \\ &x_k < m_k(i_2) \leq m_k(i_3 - 1) \leq m_k(i_3) \leq m_k(i_3 + 1), \end{aligned}$$

$i_1, i_3 \in \bigcup_{i^* \in I(m_k, x_k)} N_1(i^*)$ implies $i_2 \in I(m_k, x_k)$. This contradicts the condition of Case C. Therefore, $Q \geq 0$ holds in case (C1).

C2: If $m_k(i_1) < m_k(i_2) \leq m_k(i_3)$, then it follows from the quasiconvexity of m_k that $m_k(i_2) \leq m_k(i_3 - 1) \leq m_k(i_3) \leq m_k(i_3 + 1)$. Now, suppose $x_k < m_k(i_2)$. Then $i_3 \in \bigcup_{i^* \in I(m_k, x_k)} N_1(i^*)$ implies $i_2 \in I(m_k, x_k)$. This contradicts the condition of Case C. Therefore, $x_k \geq m_k(i_2)$. Hence the right term of max of Q is nonnegative and $Q \geq 0$ holds in Case C1.

C3: If $m_k(i_3) < m_k(i_2) \leq m_k(i_1)$, then, from the proof of Case C2 and symmetry of i_1 and i_3, it follows that the left term of max of Q is nonnegative and $Q \geq 0$ holds in Case C3.

Case D $i_1 \in \bigcup_{i^* \in I(m_k, x_k)} N_1(i^*)$ and $i_2, i_3 \notin \bigcup_{i^* \in I(m_k, x_k)} N_1(i^*)$. We have

$$\begin{aligned} Q &= \max\{(1-\alpha)m_k(i_1) + \alpha x_k, m_k(i_3)\} - m_k(i_2) \\ &= \max\{(1-\alpha)(m_k(i_1) - m_k(i_2)) + \alpha(x_k - m_k(i_2)), m_k(i_3) - m_k(i_2)\}. \end{aligned}$$

D1: If $m_k(i_3) \geq m_k(i_2)$, then the right term of max in Q is nonnegative. Hence, we have $Q \geq 0$.

D2: If $m_k(i_3) < m_k(i_2)$, then $m_k(i_2) \geq m_k(i_1)$. Moreover, by the quasiconvexity of m_k, we have $m_k(i_1 - 1) \geq m_k(i_1) \geq m_k(i_1 + 1) \geq m_k(i_2)$. Now, suppose $x_k < m_k(i_2)$. Then, $i_1 \in \bigcup_{i^* \in I(m_k, x_k)} N_1(i^*)$ implies $i_2 \in I(m_k, x_k)$. This contradicts the condition of Case D. Therefore, $x_k \geq m_k(i_2)$. Hence, the left term of max of Q is nonnegative and $Q \geq 0$ holds in Case D2.

Case E $i_2, i_3 \in \bigcup_{\cup} i^* \in I(m_k, x_k)(i^*)$ and $i_1 \notin \bigcup_{\cup} i^* \in I(m_k, x_k)(i^*)$. By the symmetry of i_1 and i_3, it follows from the proof of Case B that $Q \geq 0$.

Case F $i_2 \in \bigcup_{i^* \in I(m_k, x_k)} N_1(i^*)$ and $i_1, i_3 \notin \bigcup_{i^* \in I(m_k, x_k)} N_1(i^*)$. We have

$$\begin{aligned} Q &= \max\{m_k(i_1), m_k(i_3)\} - ((1-\alpha)m_k(i_2) + \alpha x_k) \\ &= \max\{(1-\alpha)(m_k(i_1) - m_k(i_2)) + \alpha(a_k(i_1) - x_k), \\ &\qquad (1-\alpha)(m_k(i_3) - m_k(i_2)) + \alpha(a_k(i_3) - x_k)\} . \end{aligned}$$

F1: If $m_k(i_1) \geq m_k(i_2)$ and $m_k(i_3) \geq m_k(i_2)$, then, from the quasiconvexity of m_k, it follows that $m_k(i_2 - 1) \leq m_k(i_1)$ and $m_k(i_2 + 1) \leq m_k(i_3)$. Now, suppose $m_k(i_1) < x_k$ and $m_k(i_3) < x_k$. Then $i_2 \in \bigcup_{i^* \in I(m_k, x_k)} N_1(i^*)$ implies $i_1, i_3 \in I(m_k, x_k)$. This contradicts the condition of Case F. Therefore, we have $m_k(i_1) \geq x_k$ or $m_k(i_3) \geq x_k$. Hence, $Q \geq 0$ holds in Case F1.

F2: If $m_k(i_1) \geq m_k(i_2) > m_k(i_3)$, then, by using the quasiconvexity of m_k, we have $m_k(i_1) \geq m_k(i_2 - 1) \geq m_k(i_2) \geq m_k(i_2 + 1)$. Now, suppose $m_k(i_1) < x_k$. Then, by $i_2 \in \bigcup_{i^* \in I(m_k, x_k)} N_1(i^*)$, we have $i_1 \in I(m_k, x_k)$, which contradicts the condition of Case F. Therefore, $m_k(i_1) \geq x_k$. It follows that the left term of max of Q is nonnegative and $Q \geq 0$ holds in Case F2.

F3: If $m_k(i_3) \geq m_k(i_2) > m_k(i_1)$, then, by the proof of Case F2 and symmetry of i_1 and i_3, the left term of max of Q is nonnegative and $Q \geq 0$ holds in Case F3.

Case G $i_3 \in \bigcup_{i^* \in I(m_k, x_k)} N_1(i^*)$ and $i_1, i_2 \notin \bigcup_{i^* \in I(m_k, x_k)} N_1(i^*)$. By the symmetry of i_1 and i_3, it follows from the proof of Case D that $Q \geq 0$.

Thus, m_{k+1} is quasiconvex. Similarly, (ii) is also proved. □

Note according to Theorem 5.1, quasiconvex states or quasiconcave states of a model function appear in the previous stage of the monotonic states.

Theorem 5.2. *Take the learning process L_m. For model functions $m_1, m_2, m_3, \ldots$, the following statements hold:*

1. *If m_k is strictly increasing on I, then m_{k+1} is strictly increasing on I.*
2. *If m_k is strictly decreasing on I, then m_{k+1} is strictly decreasing on I.*

5.4 Numerical Example

We give a simple numerical example of the case of the one-dimensional array of nodes and R-valued nodes.

Example 5.1. Consider the following six nodes model with $I = \{1, 2, 3, 4, 5, 6\}$. The initial model function is $m_0 = [2, 4, 2, 2, 5, 0]$. Now, assume that we observe sequentially $x_0 = 5$, $x_1 = 4$, $x_2 = 2$, $x_3 = 1$, $x_4 = 2$, $x_5 = 4$, $x_6 = 0$, $x_7 = 2$, $x_8 = 1$, $x_9 = 1$, $x_{10} = 1$, $x_{11} = 4$, $x_{12} = 3$, $x_{13} = 3$, $x_{14} = 1$, $x_{15} = 1, \ldots$

as inputs. Consider learning process L_A. Repeating updates, we sequentially obtain the following model functions:

$$
\begin{aligned}
m_0 &= [2, & 4, && 2, && 2, && 5, && 0] \\
m_1 &= [2, & 4, && 2, && 3.5, && 5, && 2.5] \\
m_2 &= [3, & 4, && 3, && 3.5, && 5, && 2.5] \\
m_3 &= [3, & 4, && 3, && 3.5, && 3.5, && 2.25] \\
m_4 &= [3, & 4, && 3, && 3.5, && 2.25, && 1.625] \\
m_5 &= [3, & 4, && 3, && 2.75, && 2.125, && 1.8125] \\
m_6 &= [3.5, & 4, && 3.5, && 2.75, && 2.125, && 1.8125] \\
m_7 &= [3.5, & 4, && 3.5, && 2.75, && 1.0625, && 0.90625] \\
m_8 &= [3.5, & 4, && 2.75, && 2.375, && 1.53125, && 0.90625] \\
m_9 &= [3.5, & 4, && 2.75, && 2.375, && 1.26563, && 0.953125] \\
m_{10} &= [3.5, & 4, && 2.75, && 2.375, && 1.13281, && 0.976563] \\
m_{11} &= [3.5, & 4, && 2.75, && 2.375, && 1.06641, && 0.988281] \\
m_{12} &= [3.75, & 4, && 3.375, && 2.375, && 1.06641, && 0.988281] \\
m_{13} &= [3.75, & 3.5, && 3.1875, && 2.6875, && 1.06641, && 0.988281] \\
m_{14} &= [3.75, & 3.25, && 3.09375, && 2.84375, && 1.06641, && 0.988281] \\
m_{15} &= [3.75, & 3.25, && 3.09375, && 2.84375, && 1.0332, && 0.994141] \\
\cdots &\ .
\end{aligned}
$$

We notice that the model function m_k is quasiconcave on I for $k \geq 5$ and decreasing on I for $k \geq 13$.

We give the numerical example for the more nodes case.

Example 5.2. There are 100 nodes in the system, that is, $I = \{1, 2, 3, \ldots, 100\}$. Suppose that initial values of nodes are given by the following initial model function.

$$
\begin{aligned}
m_0 = [&5, 1, 6, 6, 3, 1, 0, 3, 0, 7, 9, 2, 2, 10, 5, 7, 9, 5, 6, 1, 7, 6, 8, 5, 9, 3, 9, 1, 9, 2, 4, 9, \\
&9, 10, 3, 9, 1, 9, 8, 10, 0, 7, 2, 1, 3, 0, 9, 6, 4, 10, 4, 1, 8, 0, 0, 9, 6, 8, 0, 10, 3, 6, \\
&4, 8, 0, 10, 3, 9, 9, 0, 4, 10, 6, 9, 1, 7, 8, 5, 9, 5, 1, 9, 6, 3, 7, 5, 2, 2, 3, 5, 0, 7, 0, \\
&2, 2, 4, 3, 1, 10, 3].
\end{aligned}
$$

100 000 inputs are generated by random variable with uniform distribution over interval [0, 10]. The sequence of inputs is the following.

$$
\begin{aligned}
x = &6.17655, 5.74143, 3.09101, 8.82768, 0.419905, 5.44219, 2.87489, 9.34485, \\
&2.83286, 8.54906, 4.73626, 0.181078, 2.97653, 4.9316, 5.73355, 2.63117, \\
&4.64547, 5.61251, 5.69556, 0.192715, 5.92268, 5.77079, 5.84419, 0.160254, \\
&\ldots .
\end{aligned}
$$

Assume the learning process L_A with learning-rate factor $\alpha = \frac{1}{2}$.

Then, after 2000 renewals, the model function is the following. We observe that the values of nodes become a little smooth.

$m_{2000} =$

[4.39191, 4.43309, 4.44767, 5.69379, 6.90303, 7.12243, 7.63443, 7.74315, 7.78238, 7.84754, 8.70004, 9.15484, 9.1742, 9.58232, 9.73811, 9.74484, 9.1116, 8.77991, 8.68443, 8.29606, 8.25261, 7.8229, 6.75517, 6.72703, 6.77499, 6.829, 6.96068, 7.20724, 7.31213, 7.44808, 7.47451, 7.55659, 6.75899, 6.37561, 6.33933, 6.21124, 5.4436, 5.41351, 5.0501, 4.72463, 4.71753, 4.63983, 4.01396, 3.93112, 3.91836, 3.86842, 3.85154, 4.72429, 4.94823, 4.91625, 3.77401, 3.06773, 2.96493, 2.84378, 2.41838, 1.47959, 1.5463, 1.5893, 1.59581, 1.95646, 2.05274, 2.18739, 2.34186, 2.3952, 2.53792, 2.65806, 3.38887, 5.52287, 5.52756, 6.2061, 6.25335, 6.17359, 6.0252, 5.86021, 5.23624, 5.13327, 4.05768, 3.50199, 3.48541, 3.4497, 3.59418, 3.68822, 3.69598, 3.72741, 2.15921, 1.57556, 1.22533, 1.07455, 0.716105, 0.539657, 0.317723, 0.482298, 0.712784, 0.74381, 0.994052, 1.05325, 1.28815, 1.63699, 1.83334, 1.98754].

After 20 000 renewals, the model function is the following. It is more smooth and more monotonic, however, not exactly monotonic.

$m_{20\,000} =$

[8.11775, 8.16118, 8.18684, 8.22328, 8.43226, 8.49329, 8.69656, 9.01596, 9.26446, 9.50196, 9.59347, 9.74019, 9.81579, 9.89013, 9.45529, 9.3072, 9.28937, 9.095, 8.66878, 8.58777, 8.47044, 7.90849, 7.73593, 7.7157, 7.62052, 7.34094, 7.27112, 7.18642, 7.1584, 6.97865, 6.77412, 6.67021, 6.57016, 6.54738, 6.23037, 6.15797, 6.10431, 5.96406, 5.78678, 5.76907, 5.59959, 5.44722, 5.39526, 5.28729, 5.1264, 5.0137, 4.83816, 4.76619, 4.71392, 4.60424, 4.58023, 4.37331, 4.29064, 4.21043, 4.1229, 4.00474, 3.91723, 3.89741, 3.77574, 3.75713, 3.39992, 3.25678, 3.28632, 3.3876, 3.44338, 3.46383, 3.50154, 3.61255, 3.29348, 3.1122, 3.0908, 3.03899, 2.9463, 2.85865, 2.79937, 2.64571, 2.5486, 2.44539, 2.40141, 2.29236, 2.1278, 1.97047, 1.82176, 1.78269, 1.73525, 1.62233, 1.55548, 1.39198, 1.23384, 1.10855, 1.08451, 1.03314, 0.853397, 0.816412, 0.723927, 0.60732, 0.472897, 0.271546, 0.182418, 0.149336].

After 40 000 renewals, the model function is the following. It is completely monotonic.

$m_{40\,000} =$

[9.97285, 9.86483, 9.79325, 9.77096, 9.65799, 9.51986, 9.43672, 9.42295, 9.33566, 9.1873, 9.14053, 9.02399, 8.99165, 8.93791, 8.77375, 8.6733, 8.56, 8.54632, 8.48743, 8.21424, 8.19628, 8.01563, 7.89914, 7.88513, 7.77079, 7.71722, 7.68404, 7.51271, 7.27206, 7.14383, 6.99086, 6.91394, 6.8272, 6.6745, 6.56568, 6.42838, 6.23068, 6.15585, 6.132, 6.082, 5.96179, 5.90954, 5.84668, 5.76472, 5.76023, 5.58424, 5.42572, 5.41527, 5.37392, 5.25088, 4.96241, 4.82772, 4.77318, 4.64659, 4.46483, 4.3582, 4.21838, 3.94902, 3.90375, 3.83041, 3.77986, 3.74044, 3.64656, 3.4752, 3.37017, 3.32334, 3.21827, 2.99897, 2.97323, 2.73626, 2.6555, 2.64423, 2.46857, 2.38732, 2.27121, 2.25192, 2.18371, 2.08563, 2.06315, 1.93825, 1.8184, 1.76525, 1.64021, 1.60615, 1.58154, 1.32032, 1.30627, 1.29105, 1.09399, 0.959558, 0.842863, 0.694888, 0.681571, 0.642036, 0.444499, 0.412211, 0.3483, 0.275458, 0.0813817, 0.0201232].

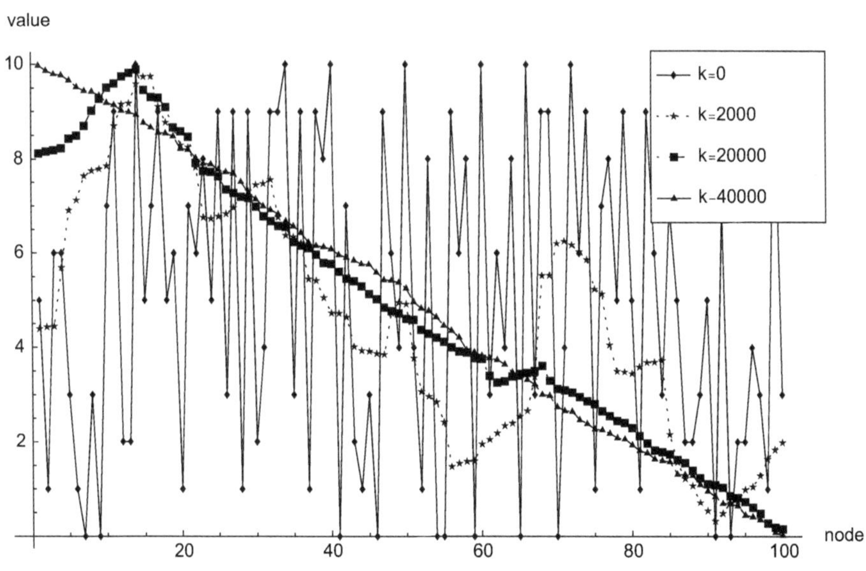

Figure 5.3 Updated model functions ($k = 0, 2000, 20\,000, 40\,000$)

It can be observed from Figure 5.3 that if k is greater than approximately 20 000 (20 295), then the model function m_k is quasiconcave on I; moreover, if k is greater than approximately 38 000 (38 276), then the model function m_k is decreasing on I.

5.5 Conclusion

In this chapter, we have discussed the monotonicity of model functions in fundamental self-organization maps with a one-dimensional array of nodes and real-valued nodes. We suggested quasiconvexity and quasiconcavity for model functions. Moreover, we have shown that the renewed model function of a quasiconvex (quasiconcave) model function is also quasiconvex (quasiconcave), and quasiconvex states or quasiconcave states of a model function appear in the previous stage of the monotonic states. Further research efforts will be devoted to extend the model to higher-dimensional cases, which have more practical applications. Finally, we hope these results will be useful for many problems.

Acknowledgements This work was a cooperation between researchers including Dr. Mituhiro Hoshino, Dr. Yutaka Kimura and Dr. Kaku. Their contribution is very much appreciated.

References

Cottrell M, Fort JC (1987) Étude d'un processus d'auto-organisation. Annales de l'Institut Henri Poincaré 23(1):1–20 (in French)

Erwin E, Obermeyer K, Schulten K (1991) Convergence properties of self-organizing maps. In: Kohonen T, Mäkisara K, Simula O, and Kangas J (eds) Artificial Neural Networks, pp 409–414. Elsevier, Amsterdam

Erwin E, Obermayer K, Schulten K (1992a) Self-organizing maps: ordering, convergence properties and energy functions. Bio Cybernetics 67(1):47–55

Erwin E, Obermayer K, Schulten K (1992b) Self-organizing maps: stationary states, metastability and convergence rate. Bio Cybernetics 67(1):35–45

Flanagan JA (1997) Self-organisation in the one-dimensional SOM with a reduced width neighbourhood. In: Proceedings of WSOM'97, Workshop on Self-Organizing Maps, pp 268–273. Helsinki University of Technology, Neural Networks Research Centre, Espoo, Finland

Horowitz R, Alvarez L (1996) Self-organizing neural networks: convergence properties. In: Proceedings of ICNN'96, 1996 IEEE International Conference on Neural Networks, vol 1, pp 7–12. IEEE, New York

Hoshino M, Kimura Y, Kaku I (2004) On the stationarity in some self-organizing maps. In: Proceedings of the 7th International Conference on Industrial Management, pp 754–760

Hoshino M, Kimura Y, Kaku I (2006) Quasi-convexity and monotonicity of model function of node in self-organizing maps. In: Proceedings of the 8th International Conference on Industrial Management, pp 1173–1178

Kohonen T (1997) Self-organizing maps. Springer, Berlin Heidelberg New York

Chapter 6
Privacy-preserving Data Mining

Privacy-preserving data mining (PPDM) is one of the newest trends in privacy and security research. It is driven by one of the major policy issues of the information era: the right to privacy.

Data mining is the process of automatically discovering high-level data and trends in large amounts of data that would otherwise remain hidden. The data-mining process assumes that all the data is easily accessible at a central location or through centralized access mechanisms such as federated databases and virtual warehouses. However, sometimes the data are distributed among various parties. Privacy in terms of legal and commercial concerns may prevent the parties from directly sharing some sensitive data. Sensitive data usually includes information regarding individuals' physical or mental health, financial privacy, *etc.* Privacy advocates and data mining are frequently at odds with each other, and bringing the data together in one place for analysis is not possible due to the privacy laws or policies. How parties collaboratively conduct data mining without breaching data privacy presents a major challenge. The problem is not data mining itself, but the way data mining is done. In this chapter, some techniques for PPDM are introduced.

This chapter is organized as follows. Section 6.1 introduces the issues about privacy and data mining. Section 6.2 discusses the relationship between security, privacy and data mining. Section 6.3 introduces the foundation for PPDM. Section 6.4 discusses the collusion behaviors in PPDM. Concluding remarks are given in the Section 6.5.

6.1 Introduction

Today, with the development of e-commerce and e-government and more and more personal data exchanged online, data privacy has become one of the most important issues in the information era. Protection of privacy from unauthorized access is one of the primary concerns in data use, from national security to business transactions.

Data mining and knowledge discovery in databases are important areas that investigate the automatic extraction of previously unknown patterns from large amounts of data. The power of data-mining tools to extract hidden information from large collections of data leads to an increase in data collection efforts by companies and government agencies. Naturally, this raises privacy concerns about collected data. Due to the increasing ability to trace, collect and analyze large amounts of personal or sensitive data, privacy has become an important issue in various domains. This is a challenge to the security and privacy community. In response to that, data-mining researchers started to address privacy concerns by developing special data-mining techniques under the framework of "privacy-preserving data mining." The aim of privacy-preserving data mining (PPDM) is to develop data-mining techniques that could be applied on databases without violating the privacy of individuals. In recent years, PPDM has become an important research issue.

Example 6.1. Consider the following problems: When researchers try to predicate diseases in a certain area, do they release or infer individual medical records? When the Department of Homeland Security tries to track down terrorists, they may need to access the travel or financial information of an individual. It is very difficult to define the line of privacy. There are two different kinds of privacy violations in data mining, one is the data access phase; another is in the release of data-mining results. What is the sensitive data and what should not be accessed? Privacy does not necessarily mean individuals, but can mean a group of individuals. What is the sensitive information in the data-mining results? Some sensitive relations cannot be revealed. What can be inferred from different sets of data-mining results? Privacy issues are indeed very complicated (Liu 2008).

On one side, the huge amount of data collection is available almost everywhere, and on the other side the data-mining tools are very powerful. People would worry about the risk of privacy violations.

Example 6.2. Imagine the following scenario. A law enforcement agency wants to cluster individuals based on their financial transactions, and study the differences between the clusters and known money laundering operations. Knowing the differences and similarities between normal individuals and known money launderers would enable a better direction of investigations. Currently, an individual's financial transactions may be divided between banks, credit card companies, tax collection agencies, *etc.* Each of these has effective controls governing the release of the information. These controls are not perfect, but violating them reveals only a subset of an individual's financial records. The law enforcement agency could promise to provide effective controls, but now overcoming those gives access to an individual's entire financial history. This raises justifiable concerns among privacy advocates.

Can we solve the above challenging problems? Can privacy and data mining coexist? These problems are not the data-mining results, but how they are obtained. If the results could be obtained without sharing information between the data sources, and the results were truly a summary and could not be used to deduce private information, there would be no loss of privacy through data mining. Can we perform data mining even without access to the original data? There are some questions we must ask. What data is used for data mining? And what is in the data-mining result? The data-mining tool itself does not need to invade people's privacy. Can we still do data mining while protecting privacy? PPDM is the right answer to this. The goal of this chapter is to introduce PPDM (Vaidya 2004).

PPDM typically concerns itself with one of two problems. The first is preparing the data for release. That is, the privacy data will be released to the data miner, but the data must first be altered in such a way so as to prevent any privacy breaches. The second angle of attack for PPDM research is to modify the data-mining algorithm itself to allow it to be run in a distributed way, such that no private information is released to the other participating parties (Shaneck 2007).

The research of PPDM is aimed at bridging the gap between collaborative data mining and data confidentiality (Guo 2007). It involves many areas such as statistics, computer sciences, and social sciences. It is of fundamental importance to homeland security, modern science, and to our society in general.

Table 6.1 provides an example of n customers' original personal information, which includes various attributes (Guo 2007). Disclosures that can occur as a result of inferences by snoopers include two classes: identity disclosure and value disclosure. Identity disclosure relates to the disclosure of identities of individuals in the database while value disclosure relates to the disclosure of the value of a certain confidential attribute of those individuals. There is no doubt that identity attributes, such as *SSN* and *Name*, should be masked to protect privacy before the data is released. However, some categorical attributes, such as *Zip*, *Race*, and *Gender*, can also be used to identify individuals by linking them to some public available data set. Those attributes hence are called quasi-identifiers. There has been much research on how to prevent identity disclosure, such as the well-known statistical disclosure control (SDC) method, k-anonymity. To prevent value disclosures, various randomization-based approaches have been investigated.

The privacy issues in data mining started to be investigated in the late 1990s. Over the past several years, a growing number of successful techniques were proposed in the literature to obtain valid data-mining results while preserving privacy at different levels. This chapter reviews the existing PPDM techniques and outlines the important research issues addressed in this book.

Table 6.1 Personal information of n customers

ID	SSN	Name	Zip	Race	...	Age	Gender	Balance ($1000)	Income ($1000)	...	Interest paid ($1000)
1	***	***	28223	Asian	...	20	M	10	85	...	2
2	***	***	28223	Asian	...	30	F	15	70	...	18
3	***	***	28262	Black	...	20	M	50	120	...	35
4	***	***	28261	White	...	26	M	45	23	...	134
...	...	...	...	...	...	...	...	...	...	...	...
N	***	***	28223	Asian	...	20	M	80	110		15

6.2 Security, Privacy and Data Mining

This section provides the background materials required to give an appropriate perspective for the work done in this chapter.

6.2.1 Security

Security is a very common concept. In Wikipedia Encyclopedia, information security means protecting information and information systems from unauthorized access, use, disclosure, disruption, modification or destruction. The goal of information security is to protect the confidentiality, integrity and availability of information.

- Confidentiality

  Confidentiality is preventing disclosure of information to unauthorized individuals or systems. For example, a credit card transaction on the Internet requires the credit card number to be transmitted from the buyer to the merchant and from the merchant to a transaction processing network. The system attempts to enforce confidentiality by encrypting the card number during transmission, by limiting the places where it might appear (in databases, log files, backups, printed receipts, and so on), and by restricting access to the places where it is stored. If an unauthorized party obtains the card number in any way, a breach of confidentiality has occurred.
  Confidentiality is necessary (but not sufficient) for maintaining the privacy of the people whose personal information a system holds.
- Integrity

  In information security, integrity means that data cannot be modified without authorization. This is not the same thing as referential integrity in databases. Integrity is violated when an employee accidentally or with malicious intent deletes important data files, when a computer virus infects a computer, when an employee is able to modify his own salary in a payroll database, when an unauthorized user vandalizes a website, when someone is able to cast a very large number of votes in an online poll, and so on.
  There are also many ways in which integrity could be violated without malicious intent. In the simplest case, a user on a system could mistype someone's address. On a larger scale, if an automated process is not written and tested correctly, bulk updates to a database could alter data in an incorrect way, leaving the integrity of the data compromised. Information security professionals are tasked with finding ways to implement controls that prevent errors of integrity.
- Availability

  For any information system to serve its purpose, the information must be available when it is needed. This means that the computing systems used to store and process the information, the security controls used to protect it, and the communication channels used to access it must be functioning correctly. High availability systems aim to remain available at all times, preventing service disruptions due to power outages, hardware failures, and system upgrades. Ensuring availability also involves preventing denial-of-service attacks.

Today, governments, military, corporations, financial institutions, hospitals, and private businesses amass a great deal of confidential information about their employees, customers, products, research, and financial status. Most of this information is now collected, processed and stored electronically on computers and transmitted across networks to other computers. Should confidential information about a business' customers or finances or new product line fall into the hands of a competitor, such a breach of security could lead to lost business, law suits or even bankruptcy of the business. Protecting confidential information is a business requirement, and in many cases also an ethical and legal requirement.

For the individual, information security has a significant effect on privacy, which is viewed very differently in different cultures. The field of information security has grown and evolved significantly in recent years. As a career choice there are many ways of gaining entry into the field. It offers many areas for specialization including: securing network(s) and allied infrastructure, securing applications and databases, security testing, information systems auditing, business continuity planning and digital forensics science, to name a few.

6.2.2 Privacy

Security and privacy are related but different. Usually, achieving privacy depends on security. Preserving privacy when data are shared for mining is a challenging problem. The traditional methods in database security, such as access control and authentication have been adopted to successfully manage access to data but present some limitations in the context of data mining. While access control and authentication protections can safeguard against direct disclosures, they do not address disclosures based on inference detection, which is beyond the reach of the existing methods. Therefore, we need to study new methods to solve the issues of privacy in data mining (Oliveira 2005).

In the paper by Zhan (2008), the author categorizes the protection into two layers. One is protection against the collaborative parties, the other is protection against network attackers. Without loss of generality, let us call attacks from collaborative parties *inside attacks*. These parties are called *inside attackers*; let us call attacks outside the collaborative parties *outside attacks*, and the attackers who conduct the attacks are called *outside attackers*.

To protect against outside attackers, we need to rely on secure channels. Prevention of inside attacks is different from prevention of outside attacks in that the inside attackers usually have more knowledge about private data than outside attackers. Furthermore, the goal of collaborative data mining is to obtain a valid data-mining result. However, the result itself may disclose the private data to inside attackers. Therefore, we cannot hope to achieve the same level of protection in privacy-preserving collaborative data mining as we do in general secure communications which protect against outside attacks. However, we would like to prevent the private data from being disclosed during the mining stage. In order to state more precisely how we understand privacy in the data-mining context, Zhan proposes the following definition (2008):

Definition 6.1. A privacy-oriented scheme S preserves data privacy if for any private data T, the following is held:

$$|\Pr(T|PPDMS) - \Pr(T)| \leq \varepsilon .$$

- *PPDMS*. Privacy-preserving data-mining scheme.
- ε. A probability parameter.
- $\Pr(T|PPDMS)$. The probability that the privacy data T is disclosed after *PPDMS* has been applied.
- $\Pr(T)$. The probability that the private data T is disclosed without any *PPDMS* being applied.
- $\Pr(T|PPDMS) - \Pr(T)$. The probability that private data T is disclosed with and without *PPDMS* being applied.

We call $1 - \varepsilon$ the privacy level that the privacy-oriented scheme S can achieve. The goal is to make ε as small as possible.

Besides the above definition, several definitions of privacy have been given, and they vary according to context, culture, and environment. Understanding privacy in data mining requires understanding how privacy can be violated and the possible means for preventing privacy violation.

In general, privacy preservation occurs in two major dimensions: users' personal information and information concerning their collective activity. We refer to the former as individual privacy preservation and the latter as collective privacy preservation (Oliveira 2005).

- Individual privacy preservation. The primary goal of data privacy is the protection of personally identifiable information. In general, information is considered personally identifiable if it can be linked, directly or indirectly, to an individual person. Thus, when personal data are subjected to mining, the attribute values associated with individuals are private and must be protected from disclosure. Miners are then able to learn from global models rather than from the characteristics of a particular individual.
- Collective privacy preservation. Protecting personal data may not be enough. Sometimes, we may need to protect against learning sensitive knowledge representing the activities of a group. We refer to the protection of sensitive knowledge as collective privacy preservation. The goal here is quite similar to that for

statistical databases, in which security control mechanisms provide aggregate information about groups and, at the same time, should prevent disclosure of confidential information about individuals. However, unlike as is the case for statistical databases, another objective of collective privacy preservation is to preserve strategic patterns that are paramount for strategic decisions, rather than minimizing the distortion of all statistics. In other words, the goal here is not only to protect personally identifiable information but also some patterns and trends that are not supposed to be discovered.

In the case of collective privacy preservation, organizations have to cope with some interesting conflicts. For instance, when personal information undergoes analysis, processes that produce new facts about users' shopping patterns, hobbies, or preferences, these facts could be used in recommender systems to predict or affect their future shopping patterns. In general, this scenario is beneficial to both users and organizations. However, when organizations share data in a collaborative project, the goal is not only to protect personally identifiable information but also to protect some strategic patterns. In the business world, such patterns are described as the knowledge that can provide the knowledge discovered from confidential information (*e.g.*, medical, financial, and crime information). The absence of privacy safeguards can equally compromise individuals' privacy. While violation of individual privacy is clear, violation of collective privacy can lead to violation of an individual's privacy.

6.2.3 *Data Mining*

The main purpose of the data mining is to take the large amounts of information, which would be impossible to analyze on an individual record by record basis, and extract some interesting trends or statistics. A simple approach to data mining over multiple sources that will not share data is to run existing data-mining tools at each site independently and combine the results. However, this will often fail to give globally valid results. Issues that cause a disparity between local and global results include the following (Vaidya 2004):

- Values for a single entity may be split across sources. Data mining at individual sites will be unable to detect cross-site correlations.
- The name item may be duplicated at different sites, and will be over-weighted in the results.
- Data at a single site is likely to be from a homogeneous population. Important geographic or demographic distinctions between that population and others cannot be seen on a single site.

With distributed data, the way the data is distributed also plays an important role in defining the problem. Data could be partitioned into many parts either vertically or horizontally. The different partitioning poses different problems and can lead to different algorithms for PPDM.

6.2.3.1 Vertical Partitioning

Vertical partitioning of data implies that though different sites gather information about the same set of entities, they collect different feature sets. For example, financial transaction information is collected by banks, while the IRS (Internal Revenue Service) collects tax information for everyone. An illustrative example of vertical partitioning and the kind of useful knowledge we can hope to extract is given in Figure 6.1 (Vaidya 2004). The figure describes two databases, one contains medical records of people while another contains cell phone information for the same set of people. Mining the joint global database might reveal information like "cell phones with lithium batteries lead to brain tumors in diabetics."

The model assumed is as follows: there are k parties, $P_0, \ldots, P_{k-1}$. There are a total of n transactions for which information is collected. Party P_i collects information about m_i attributes, such that $m = \sum_{i=0}^{k-1} m_i$ is the total number of attributes.

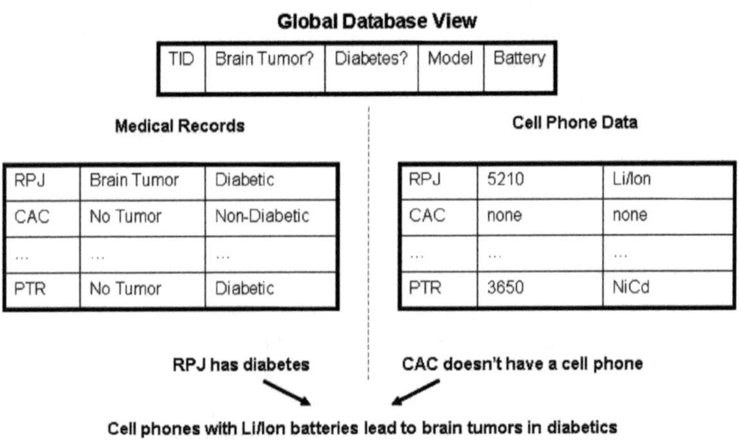

Figure 6.1 Vertically partitioned database

6.2.3.2 Horizontal Partitioning

In horizontal partitioning, different sites collect the same set of information, but about different entities. An example would be grocery shopping data collected by different supermarkets. Figure 6.2 illustrates horizontal partitioning and shows the credit card databases of two different credit unions (Vaidya 2004). Taken together, one may find that fraudulent customers often have similar transaction histories, *etc.*

The model assumed is as follows: there are k parties, $P_0, \ldots, P_{k-1}$. There are a total of N transactions for which information is collected. Party P_i collects information about n_i transactions, such that $N = \sum_{i=0}^{k-1} n_i$ is the total number of transactions.

Figure 6.2 Horizontal partitioned database

Global Database View

TID	Status	Credit	#Transactions	ZIP

Bank A (Credit Card)

RPJ	Active	<$1000	<20	47906
...	...	...	...	...
CAC	Passive	$5000	<5	98052

Bank B (Credit Card)

ABC	Passive	$1000	<20	85732
...	...	...	...	...
XYZ	Active	>$50000	>100	47907

6.3 Foundation of PPDM

PPDM has been studied extensively during the past several years. Various PPDM techniques have been developed to address different privacy issues. Several techniques ranging from perturbation to secure multiparty computation (SMC) have been explored. In this section, we will introduce the foundation and the main techniques of PPDM.

6.3.1 The Characters of PPDM

Before we describe the general parameters for characterizing scenarios in PPDM, let us consider two real-life motivation examples that pose different constraints (Oliveira 2005):

Example 6.3. A hospital shares some data for research purposes. The hospital's security administrator may suppress some identifier (*e.g.*, name, address, phone number, *etc.*) from patient records to meet privacy requirements. However, the released data may not be fully protected. A patient record may contain other information that can be linked with other datasets to re-identify individuals or entities (Samarati 2001). How can we identify groups of patients with a similar disease without revealing the values of the attributes associated with them?

Example 6.4. Two or more companies have a very large dataset of records on their customers' buying activities. These companies decide to cooperatively conduct association rule mining on their datasets for their mutual benefit since this collaboration brings them an advantage over other competitors. However, some of these companies may not want to share some strategic patterns hidden within their own data (also called sensitive association rules) with the other parties. They would like to transform their data in such a way that these sensitive association rules cannot be discovered but others can be. Is it possible for these companies to benefit from such collaboration by sharing their data while preserving some sensitive association rules?

Note that the above examples describe different privacy preservation problems. Each example poses a set of challenges. For instance, Example 6.3 is a typical example of an individual's privacy preservation, while Example 6.4 refers to collective privacy preservation.

In Clifton *et al.* (2002), some parameters are suggested as follows:

1. Outcome. Refers to the desired data-mining results. For instance, someone may look for association rules identifying relationships among attributes, or relationships among customers' buying behaviors as in Example 6.4, or may even want to cluster data as in Example 6.3.
2. Data distribution. How are the data available for mining: are they centralized or distributed across many sites? In the case of data distributed throughout many sites, are the entities described with scheme in all sites (horizontal partitions), or do different sites contain different attributes for one entity (vertical partitions)?
3. Privacy preservation. What are the privacy preservation requirements? If the concern is solely that values associated with an individual entity not be released (*e.g.*, personal information), techniques must focus on protecting such information. In other cases, the notion of what constitutes "sensitive knowledge" may not be known in advance. This would lead to human evaluation of the intermediate results before making the data available for mining.

6.3.2 Classification of PPDM Techniques

We classify representative PPDM techniques into several categories. Various approaches are based on different assumptions or domain knowledge. They are categorized in Liu (2007), and shown in Figure 6.3 (Liu 2008).

Previous work in data hiding is based on two approaches: data perturbation and secure multiparty computation (SMC). In data perturbation approaches, the aim is to preserve privacy by perturbing the data values. The main premise of this approach is that the perturbed data by adding noise does not reveal private information, and thus is "safe" to use for data mining. Based on the different noise addition techniques, this technique can be categorized as additive perturbation method, multiplicative perturbation, data microaggregation, data anonymization, data swapping and other randomization techniques.

The other approach uses cryptographic tools to build data-mining models. This approach treats PPDM as a special case of SMC and not only aims for preserving individual privacy but also tries to preserve leakage of any information other than

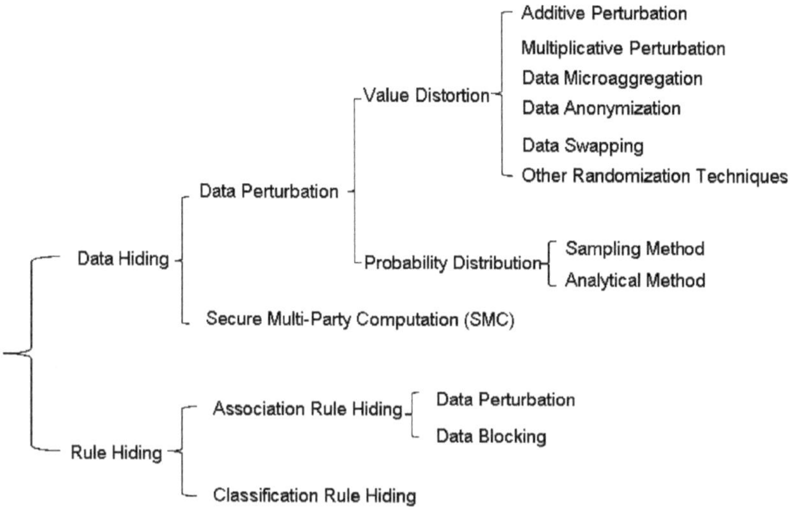

Figure 6.3 A brief overview of PPDM techniques

the final result. Unfortunately, in this approach, the communication and computation cost grows significantly as the number of parties increases.

Rule hiding is to transform the database so that only the sensitive rules are disguised, but other useful information can still be revealed.

6.3.2.1 Data Perturbation Technique

The basic ideal of data perturbation is to alter the data so that real individual data values cannot be recovered, while preserving the utility of the data for statistical summaries.

A primary perturbation technique used is data swapping: exchanging data values between records in ways that preserve certain statistics, but destroy real value. An alternative is randomization: adding noise to data to prevent discovery of the real values. Since the data no longer reflects real-world values, it cannot be used to violate individual privacy. The challenge is obtaining valid data-mining results from the perturbed data.

In 2000, Agrawal and Srikant firstly presented a solution to this problem (2000). Given the distribution of noise added to the data and the randomized data set, they were able to reconstruct the distribution of the data set.

6.3.2.2 Secure Multiparty Computation-based Solutions

In the context of PPDM over distributed data, cryptography-based techniques have been developed to solve problems of the following nature: two or more parties want to conduct a computation based on their private inputs. The issue here is how to conduct such a computation so that no party knows anything except its own input and the results. This problem is referred to as the SMC problem.

Consider a set of parties who do not trust each other, or the channels by which they communicate. Still, the parties wish to correctly compute some common function of their local inputs, while keeping their local data as private as possible. This, in a nutshell, is the problem of SMC. It is clear that the problem we wish to solve, PPDM, is a special case of the SMC problem (Vaidya 2004).

Generally speaking, SMC is the branch of cryptography that deals with the realization of distributed tasks in a secure manner; in this case, the definition of security can have different flavors, such as preserving the privacy of the data or protecting the computation against malicious attacks. Typically, SMC consists of computing some function $f(x, y)$, where x is in the hands of one participant and input y is in the hands of the other. For the computation to be secure, no more information is revealed to a participant than can be inferred from that participant's input and the output of the function itself (Lindell and Pinkas 2009).

Lindell and Pinkas first introduce a SMC technique for classification using the ID3 algorithm over horizontally partitioned data (Lindell and Pinkas 2000). From there, many researchers propose a cryptographic protocol for making the ID3 algorithm privacy preserving over vertically partitioned data and a secure way for clustering using the EM algorithm over horizontally partitioned data and so on.

There are four models of computation in SMC:

1. Trust third party model
   The goal standard for security is the assumption that we have a trusted third party to whom we can give all data. The third party performs the computation and delivers only the results – except for the third party, it is clear that no one learns anything that is not inferable from its own input and the results. The goal of secure protocols is to reach this same level of privacy preservation without the problem of finding a third party that everyone trusts.
2. Semihonest model
   In the semihonest model, every party follows the rules of the protocol using its correct input, but after the protocol is free to use whatever it sees during execution of the protocol to compromise security.
3. Malicious model
   In the malicious model, no restrictions are placed on any of the participants. Thus any party is completely free to indulge in whatever actions it pleases. In general, it is quite difficult to develop efficient protocols that are still valid under the malicious model. However, the semihonest model does not provide sufficient protection for many applications.
4. Other models: incentive compatibility
   While the semihonest and malicious models have been well researched in the cryptographic community, other models outside the purview of cryptography are possible. One example is the interesting economic notion of incentive compatibility. A protocol is incentive compatible if it can be shown that a cheating

party is either caught or else suffers an economic loss. Under the rational model of economics, this would serve to ensure that parties do not have any advantage by cheating. Of course, in an irrational model, this would not work.

There are two distinct problems that arise in the setting of PPDM. The first is to decide which functions can be safely computed, where safety means that the privacy of individuals is preserved. For example, is it safe to compute a decision tree on confidential medical data in a hospital, and publicize the resulting tree? For the most part, we will assume that the result of the data-mining algorithm is either safe or deemed essential. Thus, the question becomes how to compute the results while minimizing the damage to privacy. For example, it is always possible to pool all of the data in one place and run the data-mining algorithm on the pooled data. However, this is exactly what we do not want to do (hospitals are not allowed to hand their raw data out, security agencies cannot afford the risk, and governments risk citizen outcry if they do). Thus, the question we address is how to compute the results without pooling the data, and in a way that reveals nothing but the final results of the data-mining computation (Lindell and Pinkas 2009).

This question of PPDM is actually a special case of a long-studied problem in cryptography called SMC. This problem deals with a setting where a set of parties with private inputs wish to jointly compute some function of their inputs. Loosely speaking, this joint computation should have the property that the parties learn the correct output and nothing else, even if some of the parties maliciously collude to obtain more information. Clearly, a protocol that provides this guarantee can be used to solve PPDM problems of the type discussed above.

- Security in multiparty computation. As we have mentioned above, the model that we consider is one where an adversarial entity controls some subset of the parties and wishes to attack the protocol execution. The parties under the control of the adversary are called corrupted, and follow the adversary's instructions. Secure protocols should withstand any adversarial attack (where the exact power of the adversary will be discussed later). In order to formally claim and prove that a protocol is secure, a precise definition of security for multiparty computation is required. A number of different definitions have been proposed and these definitions aim to ensure a number of important security properties that are general enough to capture most (if not all) multiparty computation tasks. We now describe the most central of these properties.
- Privacy. No party should learn anything more than its prescribed output. In particular, the only information that should be learned about other parties' inputs is what can be derived from the output itself. For example, in an auction where the only bid revealed is that of the highest bidder, it is clearly possible to derive that all other bids were lower than the winning bid. However, this should be the only information revealed about the losing bids.
- Correctness. Each party is guaranteed that the output it receives is correct. To continue with the example of an auction, this implies that the party with the highest bid is guaranteed to win, and no party including the auctioneer can alter this.

- Independence of inputs. Corrupted parties must choose their inputs independently of the honest parties' inputs. This property is crucial in a sealed auction, where bids are kept secret and parties must fix their bids independently of others. We note that independence of inputs is not implied by privacy. For example, it may be possible to generate a higher bid, without knowing the value of the original one. Such an attack can actually be carried out on some encryption schemes (*i.e.*, given an encryption of \$100, it is possible to generate a valid encryption of \$101, without knowing the original encrypted value).
- Guaranteed output delivery. Corrupted parties should not be able to prevent honest parties from receiving their output. In other words, the adversary should not be able to disrupt the computation by carrying out a "denial of service" attack.
- Fairness. Corrupted parties should receive their outputs if and only if the honest parties also receive their outputs. The scenario where a corrupted party obtains output and an honest party does not should not be allowed to occur. This property can be crucial, for example, in the case of contract signing. Specifically, it would be very problematic if the corrupted party received the signed contract and the honest party did not.

We stress that the above list does not constitute a definition of security, but rather a set of requirements that should hold for any secure protocol. Indeed, one possible approach to defining security is to just generate a list of separate requirements (as above) and then say that a protocol is secure if all of these requirements are fulfilled. However, this approach is not satisfactory for the following reasons. First, it may be possible that an important requirement was missed. This is especially true because different applications have different requirements, and we would like a definition that is general enough to capture all applications. Second, the definition should be simple enough so that it is trivial to see that all possible adversarial attacks are prevented by the proposed definition.

6.4 The Collusion Behaviors in PPDM

Based on cryptographic techniques and secure multiparty computations (Yao 1982, 1986), privacy-preserving protocols or algorithms have been designed for PPDM. However, many of these algorithms make strong assumptions about the behavior of the participating entities, such as, they assume that the parties are semihonest, that is, they always follow the protocol and never try to collude or sabotage the process.

As mentioned in previous works on privacy-preserving distributed mining (Lindell and Pinkas 2002), the participants are assumed to be semihonest that is rational for distributed data mining, but these kinds of assumptions fall apart in real life and the collusion of parties happen easily to gain additional benefits. For example (Kargupta *et al.* 2007), the US Department of Homeland Security funded PURSUIT project involves privacy-preserving distributed data integration and analysis of network traffic data from different organizations. However, network traffic is usually privacy sensitive and no organization would be willing to share their network traffic with a third party. PPDM offers one possible solution, which would allow comparing and matching multiparty network traffic for detecting common attacks, stealth attacks and computing various statistics for a group of organizations without necessarily sharing the raw data. However, participating organizations in a consortium like PURSUIT may not all be ideal. Some may decide to behave like a "leach" exploiting the benefit of the system without contributing much. Some may intentionally try to sabotage the multiparty computation. Some may try to collude with other parties for exposing the private data of a party.

Applications of game theory in SMC and PPDM are relatively new (Abraham *et al.* 2006; Agrawal and Terzi 2006; Jiang and Clifton 2006). Kargupta *et al.* (2007) argue that large-scale multiparty PPDM can be thought of as a game where each participant tries to maximize its benefit by optimally choosing the strategies during the entire PPDM process. With a game theoretic framework for analyzing the rational behavior of each party, authors present detailed equilibrium analysis of the well known secure sum computation (Clifton *et al.* 2002) as an example. A new version of the secure sum is proposed as follows and interested readers can find a detailed analysis in the work by Kargupta *et al.* (2007).

Secure sum computation. Suppose there are n individual nodes organized in a ring topology, each with a value v_j, $j = 1, 2, \cdots, n$. It is known that the sum $v = \sum_{j=1}^{n} v_j$ (to be computed) takes an integer value in the range $[0, N-1]$.

The basic idea of secure sum is as follows. Assuming nodes do not collude, node 1 generates a random number R uniformly distributed in the range $[0, N-1]$, which is independent of its local value v_1. Then node 1 adds R to its local value v_1 and transmits $(R + v_1) \bmod N$ to node 2. In general, for $i = 2, \cdots, n$, node i performs the following operation: receive a value z_{i-1} from previous node $i-1$, add it to its own local value v_i and compute its modulus N. In other words, $z_i = (z_{i-1} + v_i) \bmod N = (R + \sum_{j=1}^{i} v_j) \bmod N$, where z_i is the perturbed version of local value v_i to be sent to the next node $i+1$. Node n performs the same step and sends the result z_n to node 1. Then node 1, which knows R, can subtract R from z_n to obtain the actual sum. This sum is further broadcasted to all other sites.

Collusion analysis. It can be shown that any z_i has a uniform distribution over the interval $[0, N-1]$ due to the modulus operation. Further, any z_i and v_i are statistically independent, and hence, a single malicious node may not be able to launch a successful privacy-breaching attack. Then how about collusion?

Assume that there are $k (k \geq 2)$ nodes acting together secretly to achieve a fraudulent purpose. Let v_i be an honest node, who is worried about her privacy. We also use v_i to denote the value in that node. Let v_{i-1} be the immediate predecessor of v_i and v_{i+1} be the immediate successor of v_i. The possible collusions that can arise are:

- If $k = n - 1$, then the exact value of v_i will be disclosed.
- If $k \geq 2$ and the colluding nodes include both v_{i-1} and v_{i+1}, then the exact value of v_i will be disclosed.

- If $n-1 > k \geq 2$ and the colluding nodes contain neither v_{i-1} nor v_{i+1}, or only one of them, then v_i is disguised by $n-k-1$ other nodes' values.

The first two cases need no explanation. Now let us investigate the third case. Without loss of generality, we can arrange the nodes in an order such that $v_1, v_2 \cdots v_{n-k-1}$ are the honest sites, v_i is the node whose privacy is at stake and $v_{i+1}, \cdots, v_{i+k}$ form the colluding group. We have

$$\underbrace{\sum_{j=1}^{n-k-1} v_j}_{\text{denoted by } X} + \underbrace{v_i}_{\text{denoted by } Y} = \underbrace{v - \sum_{j=i+1}^{i+k} v_j}_{\text{denoted by } W}$$

where W is a constant and is known to all the colluding nodes. Now, it is clear that the colluding nodes will know v_i is not greater than W, which is some extra information contributing to the utility of the collusions. To take a further look, the colluding nodes can compute the *a posteriori* probability of v_i and further use that to launch a maximum *a posteriori* probability (MAP) estimate-based attack. It can be shown that this *a posteriori* probability is:

$$f_{posteriori}(v_i) = \frac{1}{(m+1)(n-k-1)} \times \sum_{j=0}^{r} (-1)^j c_j^{(n-k-1)} \times c_{(n-k-1)+(r-j)(m+1)+t-1}^{(r-j)(m+1)+t}$$

where

$$v_i \leq W,\ r = \left\lfloor \frac{W - v_i}{m+1} \right\rfloor \quad \text{and} \quad t = W - v_i - \left\lfloor \frac{W - v_i}{m+1} \right\rfloor (m+1).$$

When $v_i > W$, $f_{posteriori}(v_i) = 0$. Due to space constraints, we have not included the proof of this result here.

Game analysis. In a multiparty PPDM environment, each node has certain responsibilities in terms of performing their part of the computations, communicating correct values to other nodes and protecting the privacy of the data. Depending on the characteristics of these nodes and their objectives, they either perform their duties or not, sometimes, they even collude with others to modify the protocol and reveal others' private information. Let M_i denote the overall sequence of computations node i has performed, which may or may not be the same as what it is supposed to do, defined by the PPDM protocol. Similarly, let R_i be the messages node i has received, and S_i the messages it has sent. Let G_i be a subgroup of the nodes that would collude with node i. The strategy of each node in the multiparty PPDM game prescribes the actions for such computations, communications, and collusions with other nodes, *i.e.*, $\sigma_i = (M_i, R_i, S_i, G_i)$. Further let $c_{i,m}(M_i)$ be the utility of performing M_i, and similarly we can define $c_{i,r}(R_i)$, $c_{i,s}(S_i)$, $c_{i,g}(G_i)$. Then the overall utility of node i will be a linear or nonlinear function of utilities obtained by the choice of strategies in the respective dimensions of computation, communication and collusion. Without loss of generality, we consider a utility function, which is a weighted linear combination of all of the above dimensions:

$$u_i\left(\{\sigma_i,\sigma_{-i}\}\right) = \omega_{i,m}c_{i,m}(M_i) + \omega_{i,s}c_{i,s}(S_i) + \omega_{i,r}c_{i,r}(R_i) + \omega_{i,g}c_{i,g}(G_i)$$

where $\omega_{i,m}, \omega_{i,s}, \omega_{i,r}, \omega_{i,g}$ represent the weights for the corresponding utility factors. Note that we omitted other node strategies in the above expression just for simplicity.

In secure sum computation, the derived *a posteriori* probability can be used to quantify the utility of collusion, *e.g.*,

$$g(v_i) = Posteriori - Prior = f_{posteriori}(v_i) - \frac{1}{m+1}.$$

We see here that this utility depends on $W - v_i$ and the size of the colluding group k. Now we can put together the overall utility function for the game of multiparty secure sum computation:

$$u_i\left(\{\sigma_i,\sigma_{-i}\}\right) = \omega_{i,m}c_{i,m}(M_i) + \omega_{i,s}c_{i,s}(S_i) + \omega_{i,r}c_{i,r}(R_i) + \omega_{i,g}\sum_{j\in P-G_i} g(v_j)$$

where P is the set of all nodes and G_i is the set of nodes colluding with node i.

Now considering a special instance of the overall utility where the node performs all the communication and computation related activities as required by the protocol. This results in a function: $u_i(\{\sigma_i,\sigma_{-i}\}) = \omega_{i,g}\sum_{j\in P-G_i} g(v_j)$, where the utilities due to communication and computation are constant and hence can be neglected for determining the nature of the function.

From the above analysis we can see, the collusion of parties happen easily to gain additional benefits in multiparty PPDM, because the strategies of following protocol are not always optimal. Based on the penalty mechanism without having to detect collusion, a cheap-talk protocol is proposed to offer a more robust process, and the optimal strategy is to following the protocol for secure computation with punishment strategy.

In addition to the work in Kargupta *et al.* (2007), Jiang and Clifton (2006) provide an alternative solution to the traditional semihonest adversary model by proposing an accountable computing framework in which malicious nodes can be detected in polynomial time. Ge and Zhu (2009) propose a collusion-resistant protocol of distributed association rules mining based on the threshold homomorphic encryption scheme, which can prevent effectively the collusion behaviors and conduct the computations across the parties without compromising their data privacy.

In a word, the semihonest assumption sometimes deviates from the real-life application of privacy-preserving distributed data mining. Therefore, a new trend for PPDM is to make the collusion resistant protocols or algorithms work well in real life.

6.5 Summary

In this chapter, we introduced issues with PPDM and discussed some problems concerning the privacy of data mining. We know that tools from PPDM and secure multiparty computation make it possible to process the data without disclosure, but do not address the privacy implication of the results.

In the general information security model, the threats and security fears come from the inside attackers and the outside attackers. In data mining, the inside attackers are the collaborative parties and the outside attackers are the other network attackers. Prevention of inside attackers is different from prevention of outside attackers in that the inside attackers usually have more knowledge about private data than outside attackers.

In PPDM, there are two methods to protect actual data from being disclosed, *i.e.*, data perturbation methods (randomization-based techniques), and the secure computation method (encryption technique). Many papers and published algorithms are based on those two methods. Data perturbation techniques are used to protect individual privacy for classification, by adding random values from a normal distribution of mean 0 to the actual data value. One problem with this approach is the existing tradeoff between the privacy and the accuracy of the results. While secure computation has an advantage over perturbation in that it provides accurate results and not approximation, it requires considerable computation and communication overhead for each secure computation step. At the same time, the collusion behaviors in privacy-preserving distributed data mining have been brought to our attention.

References

Abraham I, Dolev D, Gonen R, Halpern J (2006) Distributed computing meets game theory: robust mechanisms for rational secret sharing and multiparty computation. In: PODC 2006, pp 53–62, Denver, CO

Agrawal R, Srikant R (2000) Privacy-preserving data mining. In: Proceeding of the 2000 ACM SIGMOD Conference on Management of Data, pp 439–450, Dallas, TX, 14–19 May 2000

Agrawal R, Terzi E (2006) On honesty in sovereign information sharing. In: *EDBT'06*, pp 240–256, Munich, Germany, March 2006

Clifton C, Kantarcioglu M, Vaidya J, Lin X, Zhu MY (2002) Tools for privacy preserving distributed data mining. SIGKDD Explor 4(2):28–34

Ge X, Zhu J (2009) Collusion-resistant protocol for privacy-preserving distributed association rules mining. In: Information and Communications Security, 11th International Conference, ICICS 2009, Beijing, China

Guo S (2007) Analysis of and techniques for privacy preserving data mining. PhD thesis, University of North Carolina at Charlotte, 2007

Jiang W, Clifton C (2006) Transforming semi-honest protocols to ensure accountability. In: PADM'06, pp 524–529, Hong Kong, China, 2006

Kargupta H, Das K, Liu K (2007) Multi-party, privacy-preserving distributed data mining using a game theoretic framework. In: PKDD, vol 4702, Lecture Notes in Computer Science, pp 523–531. Springer, Berlin Heidelberg New York

Lindell Y, Pinkas B (2000) Privacy preserving data mining. In: Advances in Cryptology. CRYPTO 2000, pp 36–54, 20–24 August 2000. Springer, Berlin Heidelberg New York

Lindell Y, Pinkas B (2002) Privacy preserving data mining. J Cryptology 15(3):177–206

Lindell Y, Pinkas B (2009) Secure multiparty computation for privacy-preserving data mining. J Privacy Confident 1(1):59–98

Liu K (2007) Multiplicative data perturbation for privacy preserving data mining. PhD thesis, University of Maryland Baltimore County, Baltimore, MD, 2007

Liu L (2008) Perturbation based privacy preserving data mining techniques for real-world data. PhD thesis, University of Texas at Dallas, 2008

Oliveira SM (2005) Data transformation for privacy-preserving data mining. PhD thesis, University of Alberta, Edmonton, Alberta, 2005

Samarati P (2001) Protecting respondents' identities in microdata release. IEEE Trans Know Data Eng 13(6):1010–1027

Shaneck M (2007) Privacy preserving nearest neighbor search and its applications. PhD thesis, University of Minnesota, 2007

Vaidya JS (2004) Privacy preserving data mining over vertically partitioned data. PhD thesis, Purdue University, 2004

Yao AC (1982) Protocols for secure computations. In: Proceedings of the 23rd Annual IEEE Symposium on Foundations of Computer Science

Yao AC (1986) How to generate and exchange secrets. In: Proceedings of the 27th IEEE Symposium on Foundations of Computer Science, pp 162–167, Toronto, Ontario, Canada, 27–29 Oct 1986

Zhan J (2008) Privacy-preserving collaborative data mining. IEEE Comput Intell 3:31–41

Chapter 7
Supply Chain Design Using Decision Analysis

In this chapter, analytical models are developed to study the benefits from cooperation and leadership in a supply chain. In addition to providing analytical expressions for cooperation and leadership, we investigate conditions under which cooperation and leadership policies should be taken by the leader of the supply chain. A total of eight cooperation/leadership policies of the leader company are analyzed by using four models. We give optimal decisions for the leader company under different cost combinations. Some analytical implications obtained from this study show that if the benefit of leadership is high, then the leadership is always an optimal policy.

7.1 Introduction

A company in the supply chain usually holds a win-win or lose-lose relationship with its upstream players (suppliers) and downstream players (customers). They are partners and need to cooperate together to achieve win-win and avoid lose-lose. Brandenburger and Nalebuff (1996) describe the total value created by players of a supply chain as a pie. The pie will grow bigger (*i.e.*, win-win relationship) if the company, suppliers of the company and customers of the company cooperate together.

Among players in the supply chain, one often emerges as the leader that can control the performance of the whole supply chain. Majumder and Srinivasan (2006) summarize companies such as Wal-Mart, Ikea and Nike that hold contract leadership (*i.e.*, the ability to offer wholesale price and two-part tariff contracts). These companies have a strong influence on the supply chain.

Another good example of strong cooperation/leadership relationships is the structure of assembler-supplier relationships of the Japanese auto industry. This structure is called *keiretsu*, which enables Japanese auto assemblers to remain lean and flexible while enjoying a level of control over supply chain akin to that of vertical integration (Ahmadjian and Lincoln 2001; Schonberger 2007). Toyota and its partners (suppliers such as *Denso* and customers such as dealers) are a conspicuous example

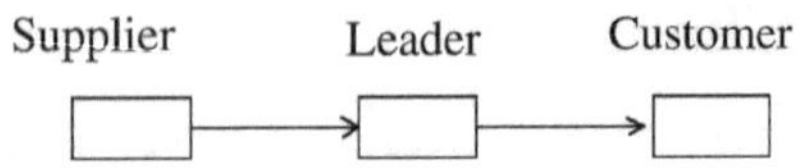

Figure 7.1 A supply chain with leader

of keiretsu. In a keiretsu, assembler (*e.g.*, Toyota) is often the leader of the supply chain. Many publications (Smitka 1991; Clark and Fujimoto 1991; Nishiguchi 1994; Ahmadjian and Lincoln 2001; Liker 2004) describe keiretsu as high-trust cooperation, strong leadership, long-term purchasing relations, intense collaboration, cross-shareholding, and the frequent exchange of personnel and technology.

The current literature on keiretsu or keiretsu-like supply chains contains many popular articles that are descriptive or provide qualitative studies. The above examples point to a need for developing quantitative models to analyze the performance of a supply chain which involves both cooperation and leadership. The goal of this chapter is to contribute to this objective.

We consider a supply chain which contains both cooperation and leadership relations. There are three players in the supply chain: the leader company (*e.g.*, Toyota), the supplier of the leader (*e.g.*, *Denso*) and the customer of the leader (*e.g.*, dealers), see Figure 7.1 for detail.

Two types of benefits are considered in this chapter: the benefit from the cooperation and the benefit from the leadership. We explain these two types of benefits as follows.

- The benefit from cooperation (Figure 7.2). We assume in this chapter that the cooperation only occurs between neighboring players. That is, benefit occurs if the supplier cooperates with the leader (we call it *upstream cooperation benefit*). Similarly, there is a benefit between the leader and customer if they cooperate with each other (we call it *downstream cooperation benefit*). However, because there is no direct cooperative relation between the supplier and the customer, no cooperative benefit will occur between them.
- The benefit from the leadership (Figure 7.3). It is our position that cooperation is a prerequisite of the leadership. This is reasonable because if the supplier or customers do not cooperate with the leader (*e.g.*, those short-term purchasing contracts or one-time players), how can they accept the leadership of the leader company? Therefore, the benefit from the leadership only occurs when both the supplier and customer cooperate with the leader. Accordingly, in this chapter, we assume that only both supplier and customer cooperative relations exist, and the supply chain can obtain a benefit from the leadership of the leader company.

The leader company is often a big player (*e.g.*, Toyota, Wal-Mart, Ikea, *etc.*), which consists of different divisions (*e.g.*, purchasing division, production division,

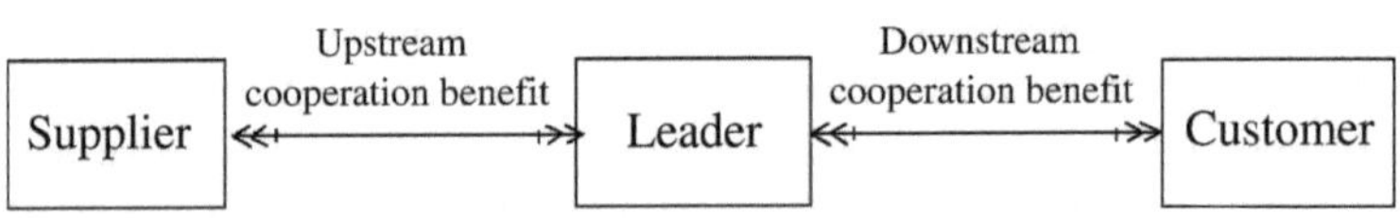

Figure 7.2 The benefit from cooperation

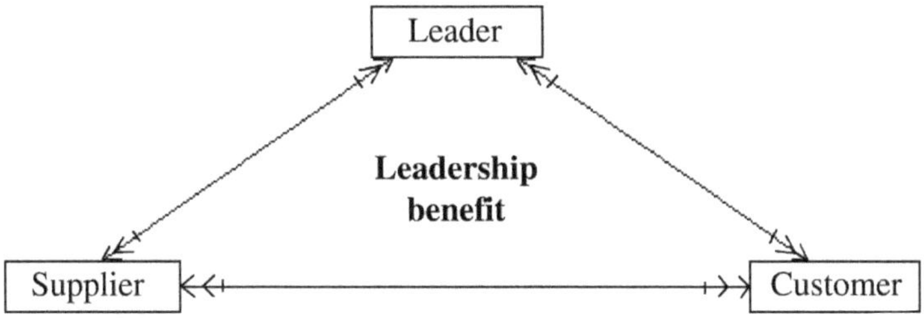

Figure 7.3 The benefit from leadership

marketing division, *etc.*). The upstream cooperation benefit often occurs between supplier and purchasing division of the leader. Similarly, the downstream cooperation benefit often occurs between customer and marketing division of the leader.

On the other hand, the leadership benefit occurs among all three players. In other words, the leadership benefit is a supply-chain-wide benefit. It is possible that there is benefit occurred between supplier and customer under the leadership of the leader company (see Figure 7.3). A good example of leadership is the process of developing new product models in Toyota. A special team that includes members from Toyota (leader company), parts manufacturer (supplier) and dealer (customer) is constructed. The team leader is often a senior product manager from Toyota. Clark and Fujimoto (1991) compile required skills and behaviors of a senior product manager. For example, coordination responsibility in wide areas, including production and sales as well as engineering; responsibility for specification, cost target and major part choices; possess market imagination and the ability to forecast future customer expectations based on ambiguous and equivocal clues in the present market; and other. The members of the special team collaborate with each other under the coordination of the senior product manager. From this example, to obtain leadership benefit, three sides (*i.e.*, supplier, leader company and customer) need to collaborate closely with each other under the leadership of the leader company.

In this chapter, we develop analytical models to study the two types of benefits. We investigate cooperation/leadership policies of the leader company. The chapter is organized as follows. We provide a literature review in Section 7.2. Analytical models are constructed in Section 7.3. We present the result from comparative statics in Section 7.4. Finally, we conclude the study in Section 7.5.

7.2 Literature Review

There are a large number of research papers related to supply chain. Chen and Paulraj (2004) analyze over 400 articles and synthesize the large, fragmented body of work dispersed across many disciplines such as purchasing and supply, logistics and transportation, marketing, organizational dynamics, information management, strategic management, and operations management literature.

Sarmah *et al.* (2006) review literature dealing with buyer vendor coordination models that have used quantity discount as coordination mechanism under deterministic environment and classified the various models.

Long-term contracting relationships between various partners generate a relatively cooperative environment. Karaesmen *et al.* (2002) study two issues: how should the additional information be used in order to improve performance and what is the potential benefit that can be expected. They investigated these problems by analyzing the structure of optimal control policies for a discrete-time make-to-stock queue with advance order information.

Dawande *et al.* (2006) study conflict and cooperation issues arising in a supply chain where a manufacturer makes products that are shipped to customers by a distributor. The manufacturer and the distributor each has an ideal schedule, determined by cost and capacity considerations. However, these two schedules are in general not well coordinated, which leads to poor overall performance. They then study two problems from literature and gave algorithms for each problem.

Majumder and Srinivasan (2006) consider a multistage supply chain with price dependent deterministic demand and increasing marginal costs. They analyze the effect of contract leadership (*i.e.* the ability to offer wholesale price and two-part tariff contracts) on the supply chain performance and use that as a basis to study coordination and cooperation. They also examined the implications of leader location in the supply chain.

Anderson (2002) discusses the question "sharing the wealth: when should firms treat customers as partners?". The author uses the example of a firm's choice of product configuration to demonstrate two effects. First, the author shows that a firm may configure a product in a manner that reduces total surplus but increases firm profits. Second, one might conjecture that increased competition would eliminate this effect, but the author shows that in a duopoly firm profits may be increasing in the cost of product completion. This second result suggests that firms may prefer to remain inefficient and/or stifle innovations. Both results violate a fundamental premise of partnering – that firms and consumers should work together to increase total surplus and reach Pareto-efficient agreements.

McCarter and Northcraft (2007) argue that a primary reason that inter-firm rivalry may abound in supply chain alliances is that a supply chain alliance represents a social dilemma. They consider how viewing supply chains as a social dilemma implicates trust as a key factor in supply chain success, and how different forms of interdependence structures within the supply chain might differentially influence the presence or growth of trust.

7.3 The Model

We use Figures 7.4 and 7.5 to interpret our models. As introduced in Section 7.1, there are two types of activities (*i.e.*, cooperation and leadership) and three benefits: the benefit from the upstream cooperation (we use circled 1 to represent it); the

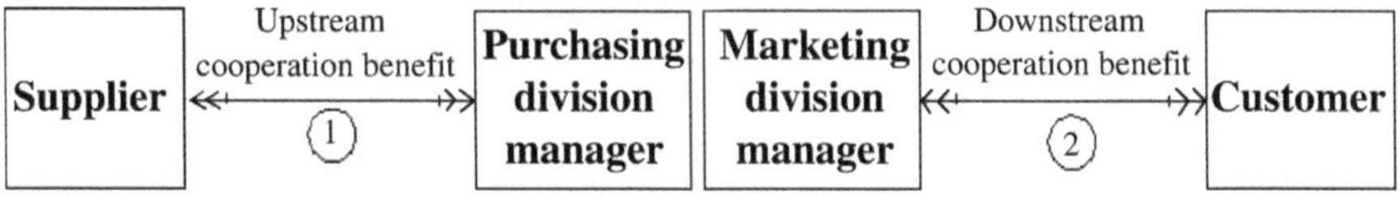

Figure 7.4 The structure of cooperation

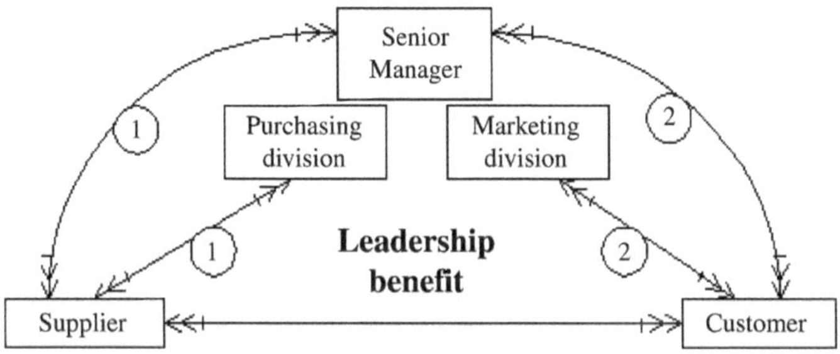

Figure 7.5 The structure of cooperation and leadership

benefit from the downstream cooperation (we use circled 2 to represent it); and the leadership benefit from the collaboration of three companies (*i.e.*, supplier, leader and customer) under the coordination of the leader company.

The supply chain considered in this chapter includes three companies but four or five players. We explain these two cases as follows.

Figure 7.4 shows the supply chain with four players (leader company includes two players: purchasing division manager and marketing division manager). In this case, benefits only occur from cooperation because no player holds the leadership to coordinate the whole supply chain.

On the other hand, Figure 7.5 shows the supply chain with five players (leader company includes three players: purchasing division manager, marketing division manager, and senior manager). Here, senior manager could be the vice-president of the leader company who is responsible for the supply chain. Unlike the two middle managers (*i.e.*, managers of purchasing and marketing divisions), the vice-president holds the leadership to coordinate the whole supply chain to obtain the leadership benefit. Additionally, she or he is able to replace purchasing/marketing manager to cooperate with supplier/customer in order to obtain upstream/downstream cooperation benefit. Therefore, from the standpoint of the leader company, the upstream cooperation benefit can be obtained by either purchasing manager or vice-president; similarly, the downstream cooperation benefit can be obtained by either marketing manger or vice-president; however, the leadership benefit can only be obtained by the vice-president. In this way, note that either purchasing manager or vice-president is employed to obtain upstream cooperation benefit, but not both. This is the same for downstream cooperation benefit. Finally, in the example of Toyota's new product development team discussed in Section 7.1, the senior product manager can act as the senior manager in Figure 7.5.

We assume that the cost of purchasing and marketing managers is the same, C_o (*e.g.*, salary of either manager). This is reasonable because both of them are middle-managers in the same organizational level of the leader company. As introduced in Figure 7.3 of Section 7.1, we assume that if both supplier and customer cooperative relations (*i.e.*, circled 1 and circled 2) exist, the leader company can obtain the leadership benefit from the coordination of the senior manager. Let the leadership benefit be T, and the cost of the senior manager be C_s. We assume $T > C_s > C_o$ (*i.e.*, senior manager is more expensive than the manager from purchasing or marketing division). We also assume that the occurrence probabilities of cooperation, *i.e.*, circled 1 and 2 are p_1 and p_2, respectively. Finally, to simplify the problem, we assume that for a given supply chain, the benefit from either cooperation, *i.e.*, circled 1 or 2, is the same, as E.

We have eight cooperation/leadership policies for the leader company. For conciseness, we incorporate two policies into one model. The eight policies and four models are as follows.

- Policy 1: Doing nothing. No cooperation and no leadership.
- Policy 2: Using senior manager only to obtain all benefits from cooperation and leadership.

$$R_1 = \max\{p_1 E + p_2 E + p_1 p_2 T - C_s, 0\}\,. \tag{7.1}$$

The benefit from policy 1 is 0. The net benefits of policy 2 are the expected benefits from two cooperations, $p_1 E + p_2 E$, plus the benefit from the leadership, $p_1 p_2 T$ (if both cooperations are present), minus the cost of the senior manager, C_s.

- Policy 3: Using purchasing manager only to obtain the benefit from upstream cooperation with supplier.
- Policy 4: Using purchasing manager firstly to obtain the benefit from upstream cooperation with supplier. Then, using senior manager to obtain downstream and leadership benefits.

$$\begin{aligned} R_2 = \max\{&p_1 E - C_o, p_1 E - C_o + p_1 \max\{p_2 E + p_2 T - C_s, 0\} \\ &+ (1 - p_1)\max\{p_2 E - C_s, 0\}\}\,. \end{aligned} \tag{7.2}$$

Policy 3 means that the supplier is the company that the leader company wants to cooperate with (*e.g.*, construct a long-term relation). The net benefit from policy 3 is $p_1 E - C_o$.

Policy 4 means that first the purchasing manager attempts to cooperate with the supplier, $p_1 E - C_o$ (same as policy 3). Then, the senior manager is used to obtain other benefits. There are two possibilities after the upstream cooperation attempted by the purchasing manager. First, if the upstream cooperation occurs (this happens with probability p_1, then the benefit from downstream cooperation is $p_2 E$ and the leadership benefit from the coordination of the senior manager is $p_2 T$. The cost of the senior manager is C_s. Therefore, we get net benefits $p_1 \max\{p_2 E + p_2 T - C_s, 0\}$. Second, if the upstream cooperation is not present (this happens with probability $1 - p_1$, the benefit from downstream cooperation remains $p_2 E$, but there is no leadership benefit and the cost remains C_s. Therefore, we get net benefits.

Taken together, the net benefits of policy 4 are $p_1E - C_o + p_1 \max\{p_2E + p_2T - C_s, 0\} + (1 - p_1)\max\{p_2E - C_s, 0\}$. It is easy to see that the benefit from policy 4 contains the benefit from policy 3. Thus, comparing with policy 4, policy 3 is often a suboptimal policy.

- Policy 5: Using marketing manager only to obtain the benefit from downstream cooperation with customer.
- Policy 6: Using marketing manager firstly to obtain the benefit from downstream cooperation with customer. Then, using senior manager to obtain upstream and leadership benefits.

$$R_3 = \max\left\{p_2E - C_o, p_2E - C_o + p_2 \max\{p_1E + p_1T - C_s, 0\} + (1 - p_2)\max\{p_1E - C_s, 0\}\right\}. \tag{7.3}$$

Policies 5 and 6 are similar and symmetric to policies 3 and 4. Thus, we obtain the net benefits as R_3.

- Policy 7: Using both purchasing and marketing managers to obtain benefits from both upstream and downstream cooperation.
- Policy 8: Using both purchasing and marketing managers to obtain benefits from both upstream and downstream cooperation firstly. Then, using senior manager to obtain the leadership benefit.

$$R_4 = \max\left\{p_1E + p_2E - 2C_o, p_1E + p_2E - 2C_o + p_1p_2(T - C_s)\right\}. \tag{7.4}$$

Policy 7 means that both purchasing and marketing managers are available for cooperation (see Figure 7.4), whereas, policy 8 means that all senior manager, purchasing and marketing managers are available (see Figure 7.5).

The net benefits from policy 7 are $p_1E + p_2E - 2C_o$. The net benefits from policy 8 are $p_1E + p_2E - 2C_o + p_1p_2(T - C_s)$. Since we assume $T > C_s$, comparing with policy 8, policy 7 is only a suboptimal policy.

7.4 Comparative Statics

Using a senior manager will result in a cost C_s but will allow the leader company to obtain a leadership benefit T. The cost and revenue trade-off makes policy selection sensitive to the cost of the purchasing or marketing division, *i.e.*, C_o. In this section, we seek insight into these trade-offs by analytically checking the optimal policy under different costs, C_o.

Without loss of generality, we assume $p_1 > p_2$. We also assume that the benefit from leadership T is large enough, $T \gg E$, to make $p_2(1 - p_1)T > p_1E$ possible. Thus, we get $p_2E + p_2T > p_1E + p_2E + p_1p_2T$.

According to the above assumptions, the cost of senior manager C_s should be a value in one of the following six intervals. We check the optimal decision of the leader company with different senior manager costs C_s.

$$0 \le C_s < p_2 E ;$$
$$p_2 E \le C_s < p_1 E ;$$
$$p_1 E \le C_s < p_1 E + p_2 E + p_1 p_2 T ;$$
$$p_1 E + p_2 E + p_1 p_2 T \le C_s < p_2 E + p_2 T ;$$
$$p_2 E + p_2 T \le C_s < p_1 E + p_1 T ;$$
$$p_1 E + p_1 T \le C_s < \infty .$$

Proposition 7.1. *Assume* $0 \le C_s < p_2 E$.
If $2C_o < \varsigma_1$, *select policy 8, where* $\varsigma_1 = C_s(1 - p_1 p_2)$*;*
otherwise, select policy 2.

If the cost of the senior manager C_s is very low, *e.g.*, $C_s < p_2 E$, then using a senior manager to get leadership benefits is always optimal. Among all eight policies, only policies 2 and 8 use a senior manager. Thus, the optimal policies are 2 and 8.

Comparing with policy 2, the benefit of policy 8 is that the senior manager is used only when both cooperations are present. If any or both cooperations are absent (with probability $1 - p_1 p_2$), the senior manager will not be used. Thus, policy 8 saves $C_s(1 - p_1 p_2)$ and needs costs of purchasing and marketing divisions $2C_o$. Conclusively, if $2C_o < \varsigma_1$, select policy 8; otherwise, policy 2.

Proposition 7.2. *Assume* $p_2 E \le C_s < p_1 E$.
If $2C_o < \min\{\varsigma_1, \varsigma_2\}$, *select policy 8, where*

$$\varsigma_2 = 2p_2 E - 2p_1 p_2 C_s - 2p_1 p_2 E + 2p_1 C_s ;$$

if $2C_o > \max\{\varsigma_1, \varsigma_3\}$, *select policy 2, where*

$$\varsigma_3 = 2p_1 p_2 E - 2p_2 E + 2C_s(1 - p_1) ;$$

otherwise, select policy 4.

Comparing values of R_1, R_2, R_3 and R_4 in Proposition 7.2 with those in Proposition 7.1, the only change is the value of R_2. Therefore, the optimal policies are policy 8, policy 2 from Proposition 7.1, and policy 4 from R_2. Note that policy 3 is only a suboptimal policy as discussed in Section 7.3.

For policy 8, firstly we compare it with policy 4. The benefit of policy 8 is that it saves the cost of the senior manager when the first cooperation is present (with probability p_1) but the other one is not, *i.e.*, $p_1(1 - p_2)C_s$. Policy 8 also gets benefits when the first cooperation (with probability p_1) is not present but the other one is, *i.e.*, $(1 - p_1)p_2 E$. However, policy 8 costs one C_o more than policy 4. Therefore, if $C_o < p_1(1-p_2)C_s + (1-p_1)p_2 E$, or $2C_o < 2p_2 E - 2p_1 p_2 C_s - 2p_1 p_2 E + 2p_1 C_s$, policy 8 is superior to policy 4. As mentioned in Proposition 7.1, if $2C_o < \varsigma_1$, policy 8 is superior to policy 2. Thus, if $2C_o < \min\{\varsigma_1, \varsigma_2\}$, policy 8 is the optimal policy.

For policy 2, first we compare it with policy 4. The benefit of policy 2 is that it saves the purchasing division cost C_o. By using policy 4, we use senior manager only if the first cooperation is present (with probability p_1). Thus, policy 4 saves $(1-p_1)C_s$, but loses $(1-p_1)p_2E$. That is, if $C_o > (1-p_1)(C_s-p_2E)$, or $2C_o > \varsigma_3$, policy 2 is superior to policy 4. As mentioned in Proposition 7.1, if $2C_o > \varsigma_1$, policy 2 is superior to policy 8. Thus, if $2C_o > \max\{\varsigma_1, \varsigma_3\}$, policy 2 is the optimal policy.

Proposition 7.3. *Assume* $p_1E \le C_s < p_1E + p_2E + p_1p_2T$.
If $2C_o < \min\{\varsigma_1, \varsigma_4, \varsigma_2\}$, *select policy 8, where*

$$\varsigma_4 = 2p_1E - 2p_1p_2C_s - 2p_1p_2E + 2p_2C_s\,;$$

if $2C_o > \max\{\varsigma_1, \varsigma_5, \varsigma_3\}$, *select policy 2, where*

$$\varsigma_5 = 2p_1p_2E - 2p_1E + 2C_s(1-p_2)\,;$$

if $\varsigma_2 < 2C_o < \varsigma_3$ *and* $C_s < E$, *select policy 4;*
otherwise, select policy 6.

Similarly, comparing with Proposition 7.2, the only change is the value of R_3. Thus, the optimal policy is either policy 8, policy 2, policy 4, or policy 6. Note that policy 5 is only a suboptimal policy.

For policies 8 and 2, the discussions are similar with Proposition 7.2. The only difference is that we also need to compare with the net value from policy 6, R_3. Since policy 6 is similar and symmetric to policy 4. Thus, we omit the detail here.

For policy 4, it is easy to show from Proposition 7.2 that if $2C_o > \varsigma_2$, then policy 4 is superior to policy 8; and if $2C_o < \varsigma_3$, then policy 4 is superior to policy 2. Thus, the only thing left is the comparison between policy 4 and policy 6. Policies 4 and 6 are the same in structure, the only difference is the probability. Comparing with policy 6, policy 4 has a high probability, p_1, to obtain benefits, E; but it also has a high probability, p_1, to waste the cost of the senior manager, C_s. Therefore, if $C_s < E$, policy 4 is superior to policy 6. Conclusively, If $\varsigma_2 < 2C_o < \varsigma_3$ and $C_s < E$, policy 4 is the optimal decision.

Proposition 7.4. *Assume* $p_1E + p_2E + p_1p_2T \le C_s < p_2E + p_2T$.
If $2C_o < \min\{\varsigma_2, \varsigma_4\}$, *select policy 8;*
if $2C_o > \varsigma_2$ *and* $C_s < E$, *select policy 4;*
otherwise, select policy 6.

Comparing with Proposition 7.3, the only change is the value of R_1; its value is zero. Policy 2 is no longer an optimal policy. Thus, the optimal policy is either policy 8, policy 4, or policy 6.

For policy 8, it is easy to show from Proposition 7.2 that if $2C_o < \varsigma_2$, then policy 8 is superior to policy 4; and from Proposition 7.3, if $2C_o < \varsigma_4$, then policy 8 is superior to policy 6. Therefore, if $2C_o < \min\{\varsigma_2, \varsigma_4\}$, policy 8 is the optimal decision.

The analysis of policy 4 is the same as in Proposition 7.3.

Proposition 7.5. *Assume* $p_2E + p_2T \le C_s < p_1E + p_1T$.
If $2C_o < \min\{\varsigma_6, \varsigma_4\}$, *select policy 8, where*

$$\varsigma_6 = 2p_2E + 2p_1p_2(T - C_s)\ ;$$

if $2C_o > \varsigma_4$ *and* $C_s < \varsigma_7$, *select policy 6, where*

$$\varsigma_7 = (p_2E - p_1E + p_2(p_1E + p_1T))/p_2\ ;$$

otherwise, select policy 3.

Comparing with Proposition 7.4, the only change is the value of R_2. Now the cost of the senior manager is high enough that makes policy 4 no longer an optimal policy. Thus, the optimal policy is either policy 8, policy 3, or policy 6.

For policy 8, the comparison with policy 6 is the same as before. Comparing with policy 3, the benefits of policy 8 is that it gets the benefits from another cooperation, p_2E. If both cooperations are present, this happens with probability p_1p_2, then the senior manager is used to get the leadership benefit T. The cost is the marketing division with probability p_2. Therefore, if $C_o < p_2E + p_1p_2(T - C_s)$, policy 8 is superior to policy 3. Summarily, if $2C_o < \min\{\varsigma_6, \varsigma_4\}$, policy 8 is the optimal decision.

For policy 6, the comparison with policy 8 is the same as before. Comparing with policy 6, the benefits of policy 3 are the benefits from the first cooperation, p_1E. On the other hand, the benefits of policy 6 relative to policy 3 are the benefits of the second cooperation, p_2E. If the second cooperation is present, this happens with probability p_2, then a senior manager is used to get the benefit from the first cooperation and also the benefit from the leadership T, minus the cost of the senior manager. Therefore, if $(p_2E + p_2(p_1E + p_1T - C_s)) - p_1E > 0$, that is, if $C_s < \varsigma_7$, policy 6 is superior to policy 3. Summarily, if $2C_o > \varsigma_4$ and $C_s < \varsigma_7$, policy 6 is the optimal decision.

Proposition 7.6. *Assume* $p_1E + p_1T \le C_s < \infty$.
If $2C_o < \varsigma_6$, *select policy 8;*
otherwise, select policy 3.

Comparing with Proposition 7.5, the only change is the value of R_3. Now the cost of the senior manager is high enough that makes policy 6 no longer an optimal policy. Since policy 5 has a low probability, it is an absolute suboptimal policy to policy 3. Thus, the optimal policy is either policy 8 or policy 3. The comparison between policies 8 and 3 is same as before.

So far our analysis has focused on the comparison of policies under different senior manager costs. In the following part of this section, we consider some of the empirical implications obtained from the prior comparative analysis.

First, there is no single policy which is optimal in all cases. It is intuitive to ask if there is a policy that is suboptimal in all cases? The finding from the analytical models is summarized in Corollary 7.1.

Corollary 7.1. *There is no single policy that is optimal in all cases, whereas there are policies (i.e., policies 1, 5, and 7) that are absolutely suboptimal.*

Corollary 7.1 asserts that, except obtaining the cooperative benefit, which holds a high probability, strategies without using a senior manager will never be optimal.

Second, since there are benefits from cooperation and leadership, an intuitive corollary is that the policy, *i.e.*, obtaining all three benefits, should be a candidate for the optimal policy.

Corollary 7.2. *Policy 8 is the only all-the-time candidate for Propositions 7.1–7.6.*

Corollary 7.2 tells us an important truth: policy 8 is a policy independent of the costs of managers (C_o, C_s). No matter how high the costs of the three managers, policy 8 uses both purchasing and marketing managers to obtain benefits from cooperation with both supplier and customer. Then, using a senior manager to obtain a leadership benefit is always a choice for the optimal strategic decision. The reason is the benefits brought by the cooperation and leadership.

Finally, the senior manager plays an important role in obtaining a leadership benefit. Thus, an intuitive corollary related to the senior manager is given as follows.

Corollary 7.3. *If the benefit from the leadership (T) is sufficiently large, then using a senior manager is always an optimal selection.*

Since T is very large, we can get $p_1 E + p_2 E + p_1 p_2 T > C_s$. From Propositions 7.1–7.3, optimal policies are 2, 4, 6 and 8. All these policies involve using a senior manager. This corollary shows the power of the leadership in a supply chain.

7.5 Conclusion

We consider cooperation and leadership relations of a supply chain in this chapter. A good example is the *keiretsu* of Japanese automobile industries. Supply chains with three companies (*i.e.*, supplier, customer and leader company) but four or five players are considered in this chapter. Four analytical models are developed to study the benefits from cooperation and leadership in a supply chain. We investigate conditions under which cooperation and leadership policies should be taken by the leader of the supply chain.

References

Ahmadjian CL, Lincoln JR (2001) Keiretsu, governance, and learning: case studies in change from the Japanese automotive industry. Organiz Sci 12:683–701

Anderson ET (2002) Sharing the wealth: when should firms treat customers as partners? Manage Sci 48:955–971

Brandenburger AM, Nalebuff BJ (1996) Co-opetition. Doubleday, New York
Chen IJ, Paulraj A (2004) Understanding supply chain management: critical research and a theoretical framework. Int J Prod Res 42:131–163
Clark KB, Fujimoto T (1991) Product development performance: strategy, organization, and management in the world auto industry. Harvard Business School Press, Boston
Dawande M, Geismar HN, Hall NG, Sriskandarajah C (2006) Supply chain scheduling: distribution systems. Prod Oper Manage 15:243–261
Karaesmen F, Buzacott JA, Dallery Y (2002) Integrating advance order information in make-to-stock production systems. IIE Trans 34:649–662
Liker JK (2004) The Toyota Way: 14 Management Principles from the World's Greatest Manufacturing. McGraw-Hill, New York
Majumder P, Srinivasan A (2006) Leader location, cooperation, and coordination in serial supply chains. Prod Oper Manage 15:22–39
McCarter MW, Northcraft GB (2007) Happy together? Insights and implications of viewing managed supply chains as a social dilemma. J Oper Manage 25:498–511
Nishiguchi T (1994) Strategic industrial sourcing: the Japanese advantage. Oxford University Press, New York
Sarmah SP, Acharya D, Goyal SK (2006) Buyer vendor coordination models in supply chain management. Eur J Oper Res 175:1–15
Schonberger RJ (2007) Japanese production management: an evolution – with mixed success. J Oper Manage 25:403–419
Smitka MJ (1991) Competitive ties: subcontracting in the Japanese automotive industry. Columbia University Press, New York

Chapter 8
Product Architecture and Product Development Process for Global Performance

In this chapter, we characterize the impact of product global performance on the choice of product architecture during the product development process. We classify product architectures into three categories: modular, hybrid, and integral. Existing research shows that the choice of product architecture during the new product development is a crucially strategic decision for a manufacturing firm. However, no single architecture is optimal in all cases; thus, analytic models are required to identify and discuss specific trade-offs associated with the choice of the optimal architecture under different circumstance. This chapter develops analytic models whose objectives are obtaining global performance of product through a modular/hybrid/integral architecture. Trade-offs between costs and expected benefits from different product architectures are analyzed and compared. Multifunction products and small size are used as examples to formalize the models and show the impact of the global performance characteristics. We also investigate how optimal architecture changes in response to the exogenous costs of system integrators. Some empirical implications obtained from this study show that if one considers global performance, modular architecture is an absolutely suboptimal decision and integral architecture is an all-the-time candidate for optimal architecture.

8.1 Introduction and Literature Review

Today, companies are competing in business environments in which customer demands are becoming increasingly heterogeneous and product life cycles are shortening. This asks companies to provide a wide variety of products at a short lead time with competitive prices (Hoch *et al.* 1999; Hopp and Xu 2005; Netessine and Taylor 2007). However, only increasing product variety does not guarantee stable long-term profits; it can in fact worsen a firm's competitiveness (Ramdas 2003; Gourville and Soman 2005) due to that product variety management is not an easy job.

As a result, to survive in such a business environment, a manufacturing firm requires abilities to cope with the problems brought by product variety. A lot of methods have been applied to respond to the challenge of how to efficiently provide a wide variety of customer-oriented products. These methods include operations flexibility such as Toyota production system (Ohno 1988; Fujimoto 1999; Liker 2004) and cellular manufacturing (Hyer and Brown 1999), resource flexibility such as workforce agility (Brusco and Johns 1998; Hopp and Van Oyen 2004) and flexible equipments (Benjaafar *et al.* 1998), research and development (R&D) such as new product development (Nonaka 1990; Clark and Fujimoto 1991; Lin *et al.* 2008), and others (Nonaka 1991; Nonaka and Takeuchi 1995).

In the new product development domain, there are many decisions on strategic and operational levels. Strategic decisions often have long-term horizons and evident impacts on the firm's strategic aim. Among various strategic decisions, an important one is the choice of product architecture. As will be introduced in the following sections of this chapter, product architecture has a direct influence on the product performance, which in turn has a direct influence on the revenue of manufacturing firms. Therefore, the choice of product architecture during the new product development is a crucially strategic decision for a manufacturing firm.

Excellent review papers that discuss product architecture can be found in the literature. Kirshnan and Ulrish (2001) give a comprehensive review of researches in product development. The review encompasses works in the academic fields of marketing, operations management, organizations, and engineering design. It focuses on product development projects within a single firm and investigates previous studies from a decision perspective, which is developed by the authors. Decisions are divided into four categories: concept development, supply chain design, product design, and production ramp-up and launch. Product architecture is one of five basic decisions of concept development. Kirshnan and Ulrish also indicate that perhaps the earliest discussions of the architecture of engineered systems are by Alexander (1964) and Simon (1969).

Product architecture can be defined as the way in which the functional elements of a product are allocated to physical components and the way in which these components interact (Ulrich and Eppinger 2004). Ulrich (1995) defines product architecture more explicitly as follows: (1) the arrangement of functional elements; (2) the mapping from functional elements to physical components; (3) the specification of the interfaces among interacting physical components.

From the marketing perspective, a product can be regarded as a bundle of customer attributes that reflect the requirements of customers (Green and Srinivasan 1990; Griffin and Hauser 1996). These attributes are represented by functional elements during the product design process and implemented by different components. The relationships between functional elements and physical components are determined by the selected product architecture.

There are two types of product architectures, *i.e.*, modularity and integrality. The literature (Ulrich 1995; Fine *et al.* 2005; Fixson 2005; Ramachandran and Krishnan 2007) has shown that the choice of the architecture for a product is important in managerial decision making and can be a key driver of the performance of manufacturing firms.

In a modular architecture, the mapping from functional elements to physical components is one-to-one and the interfaces among interacting physical components are loosely coupled. Most components of a modular product are interchangeable and the interfaces are standardized. Good examples of modular products include desktop computers, and bicycles. Mikkola (2006) proposes a way to measure the degree of modularization embedded in product architectures. Great benefits of modular architecture include products variety (Sanchez 1999; Ramdas 2003; Jiao *et al.* 2007), components commonality (Kim and Chhajed 2000; Baldwin and Clark 2000; Desai *et al.* 2001), upgradability (Ramachandran and Krishnan 2007), and others.

In spite of the mentioned advantages, modular architecture can only achieve local performance optimization. Global performance can only be optimized through an integral architecture (Ulrich 1995). Similarly, in his best-selling book, Fujimoto (2004) provides a large amount of evidence to show that many high-value added products adopt integral architecture. For example, many economical products (*e.g.*, economical motorcycle) adopt a modular architecture, whereas many analogous high-performance products (*e.g.*, luxurious motorcycle) adopt an integral architecture.

In an integral architecture, the mapping from functional elements to physical components is not one-to-one and the interfaces among interacting physical components are often tightly coupled. For an integral product, a change in some functional element or component will lead a change to other components in order for the overall product to work correctly. Good examples of integral products include luxurious motorcycles, game softwares, and automobiles. Chinese car-makers have a tendency now to convert automobiles into modular architecture (see Fujimoto 2004).

As discussed by previous studies (Ulrich 1995; Fine *et al.* 2005; Fixson 2005; Ramachandran and Krishnan 2007), a firm possesses substantial latitude in choosing a product architecture, and the choice is linked to the overall performance of the firm, which finally determines the revenues of the firm. Therefore, the choice of product architecture is extremely important.

Since no single architecture is optimal in all cases, analytic models are required to identify and discuss specific trade-offs associated with the choice of the optimal architecture under different circumstances. However, as indicated by Ulrich (1995), a single model of most of the trade-offs is unlikely (refer to many crucial questions indicated by Ulrich 1995), and he suggests that focused problems can probably be usefully isolated, analyzed and modeled. Recently, Ramachandran and Krishnan (2007) study a focused problem relative to the impact of product architecture and introduction timing on the launch of rapidly improving products by applying a modular upgrade policy. In this chapter, we develop a model to discuss a focused problem relative to the choice of product architecture. The objective of the model is to obtain global product performance through a modular/hybrid/integral product architecture. Trade-offs between costs and possible benefits from different product architectures are analyzed and compared. The researched problem in this chapter is a special case of the question "which global performance characteristics are of great value to customers and can therefore be optimized through an integral architecture?" (Ulrich 1995). We will give a detailed introduction of the problem in the next section.

This chapter is organized as follows. The research problem is introduced in Section 8.2. Analytical models are constructed in Section 8.3. Section 8.4 presents the result from comparative statics. Section 8.5 summarizes the models, and finally, a conclusion is given in Section 8.6.

8.2 The Research Problem

Ulrich (1995) defines product performance as how well the product implements its functional elements. Some typical product performance characteristics include mass, size, speed, life and others. There are two types of product performance characteristics, *i.e.*, local performance characteristics and global performance characteristics. Local performance characteristics relate only with partial components of a product and can be optimized through a modular architecture. For example, the speed of a computer is mainly determined by the ability of one component, *i.e.*, CPU. A good model that optimizes local performance can be found in Ramachandran and Krishnan (2007). In contrast, global performance characteristics relate with most components of a product and can only be optimized through an integral architecture. For example, mass and/or size of a product are determined by almost every component within the product. Function sharing and geometric nesting are design strategies that are frequently employed to minimize mass and/or size (Ulrich 1995). This chapter studies a special case of the geometric nesting strategy.

Recent publications (Ethiraj and Levinthal 2004; Ramachandran and Krishnan 2007; Rivkin and Siggelkow 2007) have only tended to offer modularity as a solution to local performance. Little attention has been paid to the problem of identifying what constitutes an appropriate product architecture for global performance. Ethiraj and Levinthal (2004) analyzed local performance by using three factors: innovation, strong and weak modular interfaces. Ramachandran and Krishnan (2007) discussed local performance by focusing launch timing of upgrade modular product models. Rivkin and Siggelkow (2007) studied local performance by investigating a shift in the pattern of the modular interface.

The factors considered by this chapter are similar with the work by Ethiraj and Levinthal (2004), and Rivkin and Siggelkow (2007). But the problem is more complicated. First, this chapter studies global performance, but not local performance. Second, more factors have been incorporated into this chapter. We not only consider innovation ability (probabilities of realization of local and global performances), strong and weak interfaces (with or without geometric nesting), a shift in the pattern of modular/integral interface (one, two and three function products), but also costs of designers and exogenous system integrators, and expected benefits from different product architecture.

One important factor that needs careful consideration is the hierarchy level of functions (Fixson 2005). Every function of a product can be decomposed into subfunctions, which in turn can be decomposed further into lower level subfunctions (Pahl and Beitz 1996). This hierarchy level determines the studied components. For example, consider a printer. Its main (highest level) function can be defined as "produce a hard copy of documents on physical print media such as papers or transparencies." In this hierarchy level, the printer itself should be the investigated object and all main components within the printer need to be considered. We also cannot identify whether the product is a laser type or an ink-jet type in this level. However, if one defines a detailed (lower level) function as "toner particles are melted by the heat of the fuser, causing them to bind to the print medium," then the mainly investigated components should be the toner, fuser and other related components. We can also identify that this is a laser printer in this level. Furthermore, if one defines a more detailed function as "generate heat," then, the hot roller (a part of a fuser) should be the investigated component. Therefore, when consider a product architecture, one is required to choose a specific hierarchy level that is meaningful. In this chapter, we consider the highest hierarchy level which directly relates to customers. In other words, those functional elements directly represent customer attributes that reflect the requirements of customers. Customers do not care about those technical solutions (*e.g.*, how do the toner particles melt in a laser-printer) on the lower hierarchy levels.

The problem considered in this chapter is a multifunction product design problem. We define multifunction product here as a device that performs a variety of functions that would otherwise be carried out by separate single-function devices. Following this definition, a multifunction product at least performs two functions. In fact, a multifunction product is developed as a combination of several single-function products. Good examples of such multifunction products include the DVD/VCR/HDD video recorder (see Figure 8.1), PC/CRT/Printer combo (see Figure 8.2), DVD/VCR/CRT-TV combo (see Figure 8.3), multifunction peripheral (see Figure 8.4), and others.

Figure 8.1 A DVD/VCR/HDD video recorder (Source: Panasonic)

Figure 8.2 A PC/CRT/Printer combo (Ulrich 1995)

Figure 8.3 A DVD/VCR/ CRT-TV combo (Source: Panasonic)

Figure 8.4 A multifunction peripheral (Source: Canon)

The main advantages of multifunction products include: (1) serves several functions, *e.g.*, a typical multifunction peripheral can usually act as a copier, a printer, a fax machine, and a scanner; (2) small space requirement,*e.g.*, comparing with three or four separate single-function devices, a multifunction peripheral has a small size and needs a lower space; (3) less expensive, *e.g.*, a multifunction peripheral is much cheaper than buying three or four separate single-function devices. In contrast, disadvantages of multifunction products are: (1) if the device breaks, all of its functions may lose at the same time; (2) a multifunction product usually can only implement one function at a time, *e.g.*, one cannot copy and print documents simultaneously on a multifunction peripheral; (3) single-function devices usually have higher performances than a multifunction peripheral.

Multifunction products now are acting as important alternatives in the markets. For example, multifunction peripherals are particularly popular for SOHO (small office/home office) users. We study the multifunction product from a relatively high hierarchy level. Since a multifunction product is developed as a combination of several single-function products, we assume that each single-function product serves as a component in the researched multifunction product. The characteristics of the architecture of a typical multifunction product are as follows.

- Mapping from functional elements to physical components.
  Each component only implements a functional element. Thus, the mapping from functional elements to physical components is one-to-one.

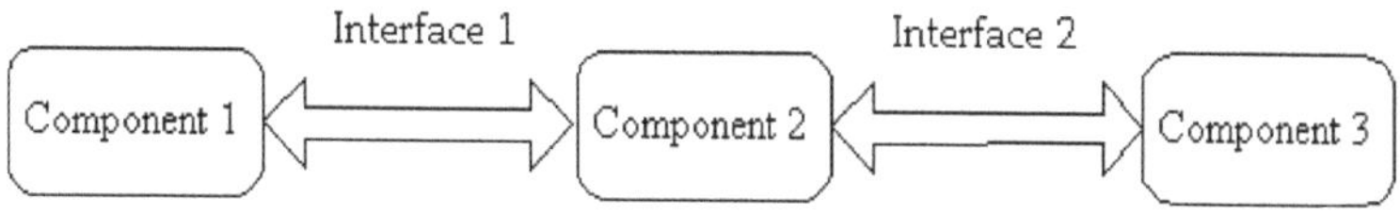

Figure 8.5 Three components with minimum number of interfaces

- Interfaces among physical components.
  Every interface is either decoupled or coupled. Moreover, if the number of the physical components in a multifunction product is M, then the number of interfaces within the product is at least $M - 1$ (when components are connected as a string by the interfaces, *e.g.*, see Figure 8.5) and at most $M(M - 1)/2$ (when every pair of components holds a interface between them, *e.g.*, see Figure 8.6).

As mentioned above, size is an important performance characteristic for a lot of multifunction products. For example, small size is always a sales point for multifunction peripherals. Makers like Canon and Brother often emphasize the small size of their multifunction peripherals in the sales promotion (*e.g.*, the product in Figure 8.4 is the world's smallest multifunction peripheral in Segment 1 class, *i.e.*, 11–20 ppm, investigated by Canon on December 4th, 2006). As discussed in the beginning of this section, size is typically a global performance characteristic and can only be optimized through an integral architecture. A frequently employed design strategy for minimizing the size of a product is geometric nesting.

As introduced by Ulrich (1995), "geometric nesting is a design strategy for efficient use of space and material and involves the interleaving and arrangement of components such that they occupy the minimum volume possible, or, in some cases, such that they occupy a volume with a particular desired shape." However, geometric nesting inevitably incurs the coupling of the interfaces between components, which often increases the costs of the product, particularly in the processes of product design and production. Therefore, optimizing global performance can get additional benefits but will also increase product costs.

Geometric nesting of components can also determine the architecture of the product. For example, if two physical components are geometrically nested, then the interface between them is tightly coupled, which is a hallmark of integral architecture. Therefore, different geometric nesting strategies result in different product

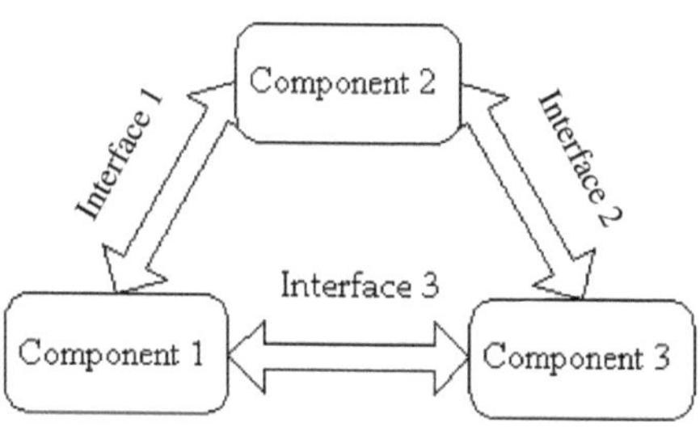

Figure 8.6 Three components with maximum number of interfaces

architectures. In this chapter, we formalize and analyze each of these proprietary geometric nesting strategies, and identify optimal choice for product architectures under different costs conditions. The detailed models are given in the next section.

8.3 The Models

In this section, we develop models to depict multifunction products that implement two, and three functions, respectively. Before describing the models, we firstly give the assumption of this chapter as follows.

Assumption 8.1. For every interface in a multifunction product, if the two physical components that are connected by this interface are geometrically nested, then the interface is coupled; otherwise, the interface is decoupled.

8.3.1 Two-function Products

There are many two-function products. A good example is the DVD/TV combo, a combination of a DVD player and a television.

A two-function product is developed from two single-function products. Each single-product serves as a component and implements a function within the two-function product. The two components can be geometrically nested or not. For example, consider two components in Figure 8.7; a two-function product can be developed from these two components without geometric nesting (see Figure 8.8). In Figure 8.8, because the mapping relation is one-to-one and because the interface is decoupled, the product architecture is modular. A good example of this type is a VHS/CRT-TV combo (see Figure 8.9).

On the other hand, minimizing the size of the two-function product will obtain additional benefits (*i.e.*, global performance). Thus, component 1 is redesigned into a specific shape to fit the second component. In this way, the two components occupy the minimum space (see Figure 8.10). In Figure 8.10, because the interface

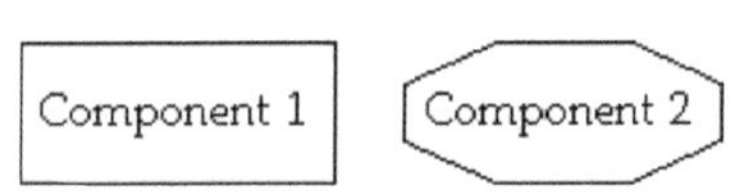

Figure 8.7 Two components of a two-function product

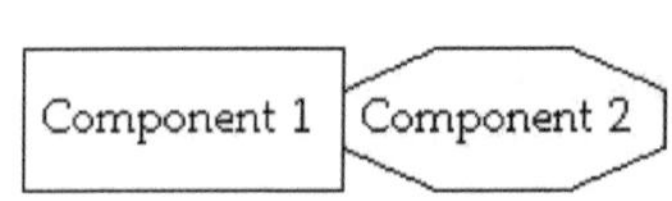

Figure 8.8 A two-function product without geometric nesting

Figure 8.9 A VHS/CRT-TV combo (Source: Panasonic)

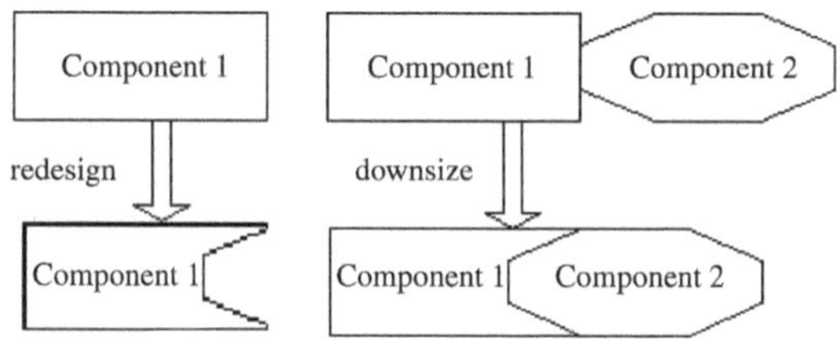

Figure 8.10 A two-function product with geometric nesting

between the two components is tightly coupled, the product architecture is integral. A good example of this type is a DVD/LCD-TV combo (see Figure 8.11). The combo in Figure 8.11 adopts a "slot-in" design strategy, which nests the DVD player completely into the LCD-TV to achieve minimum size.

The model for developing a new two-function product is simple. Assume that by using geometric nesting, *i.e.*, integral architecture, the costs of the geometric nesting is C, and the probability of realization of the geometric nesting is p (p can be obtained by using exponential or Weibull distribution), also suppose that the benefits brought by geometric nesting is E. The costs, C, can be considered as the costs mainly from the employment of system integrator (SI) (*e.g.*, a senior project manager, see Ulrich 1995). Unlike conventional product division's project managers who only hold knowledge of the product they are dedicated to, the SI holds knowledge of both components within the two-function product; thus, she or he is more costly than a conventional project manager.

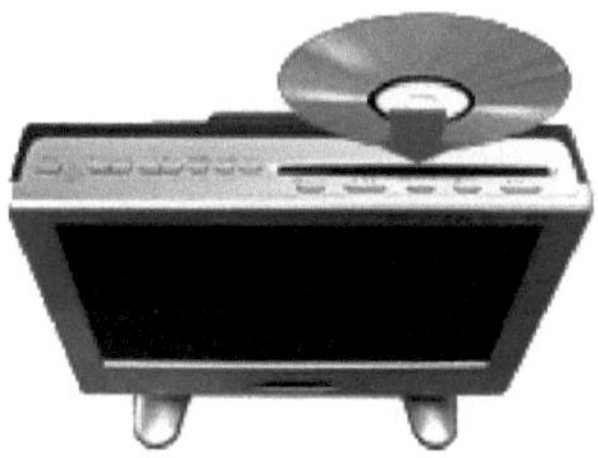

Figure 8.11 A DVD/LCD-TV combo (Source: Sanyo)

On the other hand, if the new product adopts a modular architecture without geometric nesting, then both the benefit and cost are 0. Thus, the revenue from this product is as follows:

$$R_1 = \max\{pE - C, 0\} . \tag{8.1}$$

8.3.2 Three-function Products

The three-function product is more complicated than the two-function product. A three-function product holds at least two (see Figure 8.5) and at most three interfaces (see Figure 8.6). We discuss both of them in the following subsections. For every interface within a product, there exist and only exist two design strategies: geometric nesting or not. Thus, if there are n interfaces within a product, the number of design strategies is 2^n.

8.3.2.1 Products with Two Interfaces

There are four (2^2) design strategies for products with two interfaces: modular (neither interface is geometrically nested), hybrid (only one interface is geometrically nested), and integral (both interfaces are geometrically nested). Examples are given in Figure 8.12.

We denote the interface between components i and j by I_{ij}. We also denote the design strategy for I_{ij} by I_{ij} (cou), and suppose $I_{ij}(\text{cou}) = 1$ if I_{ij} is coupled (*i.e.*, components i and j are geometrically nested) and I_{ij} (cou) $= 0$ if I_{ij} is decoupled.

For the product with two interfaces, there are usually two SIs, one for each interface. They handle the projects of geometric nesting. Suppose that the probabilities of realizations of the two geometric nestings are p_1, and p_2, respectively. For tractability, we assume that the costs of both SIs are identical, as C, respectively, and also

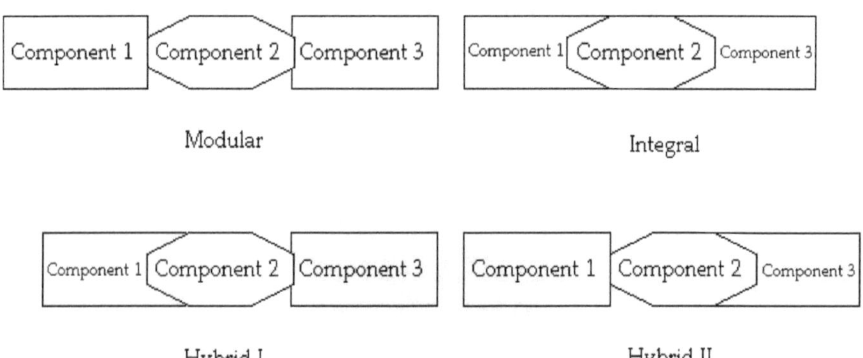

Figure 8.12 Design strategies for a three-function product with two interfaces

the expected benefits from both geometric nestings are identical, as E, respectively. Therefore, we construct the model as follows:

$$R_2 = \max\{(p_1 + p_2)E - 2C, p_1E - C, p_2E - C, 0\} . \tag{8.2}$$

In the above model, the first term is the net benefits from the integral architecture; the second and third terms are the net benefits from two hybrid architectures, respectively; and the final term is the net benefits of modular architecture.

8.3.2.2 Products with Three Interfaces

There are eight (2^3) design strategies for products with three interfaces: one modular (neither interface is geometrically nested), one integral (all interfaces are geometrically nested), and six hybrids (some, but not all interface are geometrically nested). The details are as follows:

$$\begin{aligned}\Omega = \{&(I_{12}(\text{cou}), I_{23}(\text{cou}), I_{13}(\text{cou}))|(0,0,0),(0,0,1),\\ &(0,1,1),(0,1,0),(1,0,0),(1,1,0),(1,0,1),(1,1,1)\} .\end{aligned}$$

The eight design strategies can be modeled theoretically like the model in Equation 8.2 in Section 8.3.2.1. However, rather than constructing theoretical models, we consider a more practical product development process. Consider developing a new three-function product. Among the three components, one usually serves as the core component which holds strong relations with the other two components (*i.e.*, strongly related interfaces). Conversely, the relation between the other two components is weak (*i.e.*, weakly related interface). For example, in a printer/copier/scanner multifunction peripheral, the copier is the core component, which holds strong relations with both printer and scanner, but the relation between printer and scanner is relatively weak. Actually, both printer and scanner can be nested completely or partially into the copier.

When developing such a three-function product, interfaces that hold strong relations are firstly considered to be geometrically nested. After both strongly related interfaces are geometrically nested, the weakly related, *i.e.*, the third, interface is considered to be nested to obtain global performance. For example, in the development process of the printer/copier/scanner multifunction peripheral, two strongly related interfaces (*i.e.*, printer-copier; scanner-copier) are firstly geometrically nested, then the weakly related interface (*i.e.*, printer-scanner) is geometrically nested to obtain the benefits of global performance, *i.e.*, minimum size, a sales point for multifunction peripherals.

As introduced in the beginning of Section 8.2, global performance characteristics such as minimum size relate with most of components of a product and can only be optimized through an integral architecture. Moreover, the benefits from the weakly related interface are usually limited and its realization is more difficult (*e.g.*, nesting printer and scanner can only reduce less space and is more difficult to realize than nesting printer and copier, or scanner and copier). Therefore, in this chapter, we assume the following.

Assumption 8.2. The global performance (*i.e.*, minimum size) can only be obtained if at least the two strongly related interfaces are geometrically nested (*i.e.*, tightly coupled).

We denote the benefits of the global performance by T.

The nestings of the first and second interfaces can be generated by two SIs who hold correct knowledge (*i.e.*, one holds knowledge of copier and printer for interface 1, and the other holds knowledge of copier and scanner for interface 2). However, using a dedicated SI to generate the third, *i.e.*, the weak, interface is not economical because its benefits are limited and it's difficult to generate.

Generation of the third interface is a job of the final process of the product development. In fact, the final process of the development involves more tuning activities than a mere generation of the third interface. To optimize the global performance, most jobs of the final process include tuning, adjusting, coordinating and integrating the three interfaces and the whole product system. Sometimes, there are conflicts between already designed interfaces 1 and 2, thus one job of the final process may be to resolve these conflicts. For example, consider developing a three-function product (*e.g.*, printer/copier/scanner multifunction peripheral) from two already marketed two-function products (*e.g.*, printer/copier, and copier/scanner multifunction peripherals), although the technologies of the old products (*i.e.*, interface between printer and copier, and interface between copier and scanner) can be transferred into the new three-function product, the development process is not merely a generation of interface between printer and scanner, but involves other more important jobs such as the required changes of the two old interfaces, the coordination of the three interfaces, and the tuning and integration of the whole product system.

All these jobs related to the generation of the third interface and integration of the whole product system are usually performed by a so-called heavyweight system integrator (HSI), the leader of the development team (see Wheelwright and Clark 1992; Ulrich 1995). The HSI not only holds knowledge of all three components (*i.e.*, she or he is able to generate all three interfaces), but also has abilities to integrate the whole product system. In other words, the HSI can replace the two SIs to generate the benefits, E, from interfaces 1 and 2, respectively, and only she or he can generate the benefits, W, from interface 3. Moreover, only the HSI can reap the benefits, T, of global performance by integration of the whole product system.

Suppose that the probability of realizing geometric nesting of interface 3 is p_3, and suppose that the cost of the HSI is C_s. We assume that $T > C_s > C$ (benefits from the HSI is bigger than her or his costs, and she or he is more expensive than a SI), $E > W$ (since benefits from interface 3 are limited), and both p_1, $p_2 \gg p_3$ (since the realization of interface 3 is more difficult).

For the first and second interfaces, because both of them can be generated by the HSI or SIs who hold correct knowledge, the product design problem can be considered as how to decide which integrator should be used to generate which interface. We develop models for each case in which different integrators generate interfaces 1 and 2.

- HSI only

$$R_3 = \max\{(p_1 + p_2)E + p_3 W + p_1 p_2 T - C_s, 0\} . \tag{8.3}$$

In this case, all jobs are implemented by the HSI. The first term of the model in Equation 8.3 is the expected net benefits from the product, which exhibits an integral architecture. $(p_1 + p_2)E + p_3 W$ are expected benefits from the three interfaces, $p_1 p_2 T$ are expected benefits from the global performance, which can only be reaped if both strongly related interfaces 1 and 2 are geometrically nested (see Assumption 8.2). C_s is the costs of the HSI.
On the other hand, the second term, *i.e.*, 0, is the revenue of the product which exhibits a modular architecture.

- HSI and a SI only

$$\begin{aligned} R_4 = \max\big\{&p_1 E - C, p_1 E - C + p_1 \max\{p_2 E + p_3 W + p_2 T - C_s, 0\} \\ &+ (1 - p_1)\max\{p_2 E + p_3 W - C_s, 0\}\big\} . \end{aligned} \tag{8.4}$$

In this case, there are two integrators, a SI and the HSI. The SI in the model in Equation 8.4 is dedicated to interface 1. Since only one SI is available, the second interface will be held by the HSI.
The SI generates the first interface and the expected benefits are $p_1 E - C$. If the first interface is geometrically nested (this occurs with probability p_1), then the HSI holds the second and third interfaces and integrates the whole product system; the net benefits from the HSI are $p_2 E + p_3 W + p_2 T - C_s$. On the other hand, if the first interface is decoupled (this occurs with probability $1 - p_1$), then the HSI holds the second and third interfaces and cannot achieve the global performance, the net benefits from the HSI are only $p_2 E + p_3 W - C_s$.

$$\begin{aligned} R_5 = \max\big\{&p_2 E - C, p_2 E - C + p_2 \max\{p_1 E + p_3 W + p_1 T - C_s, 0\} \\ &+ (1 - p_2)\max\{p_1 E + p_3 W - C_s, 0\}\big\} . \end{aligned} \tag{8.5}$$

The model in Equation 8.5 is similar and symmetric to model in Equation 8.4. The only change is that the available integrators are the HSI and a SI who dedicates to interface 2.

- HSI and both SIs

$$\begin{aligned} R_6 = \max\big\{&p_1 E + p_2 E - 2C, p_1 E + p_2 E - 2C + p_1 p_2 (p_3 W + T - C_s) \\ &+ (1 - p_1 p_2)\max(p_3 W - C_s, 0)\big\} . \end{aligned} \tag{8.6}$$

In this case, all three integrators are available. Firstly, the two strongly related interfaces are generated by the two SIs. If both interfaces are geometrically nested (this occurs with probability $p_1 p_2$), then HSI integrates the whole product system to optimize the global performance and reap benefits. However, if either strongly related interfaces cannot be geometrically nested (this occurs with probability $1 - p_1 p_2$), HSI can only reap the benefit from the third interface.

In the following of this chapter, we analyze the developed models of this section and investigate how optimal product architecture changes in response to the exogenous costs of integrators.

8.4 Comparisons and Implications

In this section, we chose to analyze trade-off relations between costs and benefits to facilitate the decision of the optimal product architecture that leads to maximum benefits. Since the model of two-function products, *i.e.*, the model in Equation 8.1, is very simple, we only discuss the models of three-function products.

Without loss generality, assume that $p_1 > p_2$, which can be interpreted that interface 1 holds stronger relation than interface 2, so that it is easier to be geometrically nested.

8.4.1 Three-function Products with Two Interfaces

The cost of a SI, *i.e.*, C, must be a value within one of the following three intervals: $0 \leq C < p_2E$; $p_2E \leq C < p_1E$; $p_1E \leq C < \infty$. We will check the optimal decision for the three different costs C in the following proposition.

Proposition 8.1.
If $0 \leq C < p_2E$, *select integral architecture;*
if $p_2E \leq C < p_1E$, *select hybrid architecture, i.e., Hybrid-I in Figure 8.12;*
if $p_1E \leq C < \infty$, *select modular architecture.*

Proposition 8.1 is simple and intuitive. If the cost of a SI is very low (*e.g.*, $0 \leq C < p_2E$), then using two SI to nest the two interfaces is the optimal decision, which results in $p_1E + p_2E - 2C$ of expected net benefits.

On the other hand, if the cost of a SI is within $p_2E \leq C < p_1E$, then geometric nesting of interface 2 is not profitable since its return is p_2E. Thus, a SI is used to nest interface 1, which results in $p_1E - C$ of expected net benefits.

Finally, if the cost of a SI is very expensive (*e.g.*, $p_1E \leq C < \infty$), then not only interface 2, but interface 1 also is not profitable since its return is p_1E. Thus, in this case modular architecture is the optimal decision.

8.4.2 Three-function Products with Three Interfaces

Using the HSI results in a cost C_s but will allow a benefit T from global performance. The cost and revenue trade-off makes the decision of product architecture sensitive to the costs of the SIs, *i.e.*, C. This section seeks insights into these trade-offs by analytically checking the optimal product architecture under different costs, C. Assume that the global performance benefit T is large enough, *i.e.*, $T \gg E$, to make that $p_2(1-p_1)T > p_1 E$ possible. Thus, $p_2 E + p_2 T > p_1 E + p_2 E + p_1 p_2 T$.

According to the above assumption, the cost of HSI, *i.e.*, C_s, must be a value within one of the following six intervals:

$$
\begin{aligned}
&0 \le C_s < p_2 E \; ; \\
&p_2 E \le C_s < p_1 E \; ; \\
&p_1 E \le C_s < p_1 E + p_2 E + p_1 p_2 T \; ; \\
&p_1 E + p_2 E + p_1 p_2 T \le C_s < p_2 E + p_2 T \; ; \\
&p_2 E + p_2 T \le C_s < p_1 E + p_1 T \; ; \\
&p_1 E + p_1 T \le C_s < \infty \; .
\end{aligned}
$$

We will check the optimal decision for the six different costs C_s in the following problems.

To simplify the analysis, we ignore the term of $p_3 W$, because of three reasons. First, the weakly related interface (*i.e.*, the third interface) is hard to realize (so p_3 is low). Second, the benefit from the third interface is limited (so W is small). Finally, according to Assumption 8.2, the global performance can be obtained without the realization of the third interface. Therefore, the third interface has a very little influence to the whole product architecture and performance.

In the following discussions, we assume that if both interfaces 1 and 2 are geometrically nested, then the product architecture is integral (this happens with probability $p_1 p_2$). On the other hand, if neither of them is geometrically nested, then the product architecture is modular (this happens with probability $(1-p_1)(1-p_2)$). Otherwise, the product architecture is hybrid (this happens with probability $p_1(1-p_2) + p_2(1-p_1)$). Thus, according to the realizations of interfaces 1 and 2, the product architectures in each of the following propositions can be modular, hybrid, or integral.

- For $0 \le C_s < p_2 E$, we get the following problem.

Problem 8.1.

$$
\begin{aligned}
&\max R = \{R_3, R_4, R_5, R_6\} \; ; \\
&s.t.\; 0 \le C_s < p_2 E \; .
\end{aligned}
$$

If the cost of HSI C_s is very low, *e.g.*, $C_s < p_2 E$, then using the HSI to get global performance benefit is always optimal (*i.e.*, integral architecture). Moreover, if the cost of SI (C) is not expensive, then both SIs should be used to support the HSI. Therefore, optimal models should be those found in Equations 8.3 and 8.6.

Comparing with the model in Equation 8.3, the benefit of Equation 8.6 is that the HSI is used only when both strongly related interfaces are geometrically nested. If any or both are not nested (this happens with probability $1-p_1 p_2$), the HSI will not be used (since we ignore $p_3 W$). Thus, the model in Equation 8.6 saves $C_s(1-p_1 p_2)$ and needs the costs of two SIs $2C$, if $2C < C_s(1-p_1 p_2)$, using all three integrators; otherwise, using only HSI.

We summarize this conclusion in the following proposition.

Proposition 8.2. *Assume* $0 \leq C_s < p_2 E$.

Select integral architecture, and
if $2C_o < Z_1$, using all three integrators to realize it (*i.e.*, R_6),
where $Z_1 = C_s(1 - p_1 p_2)$;
Otherwise, using HSI only to realize it (*i.e.*, R_3).

- For $p_2 E \leq C_s < p_1 E$, we get the following problem.

Problem 8.2.

$$\max R = \{R_3, R_4, R_5, R_6\} ;$$
$$s.t.\ p_2 E \leq C_s < p_1 E .$$

Comparing values of R_3, R_4, R_5, R_6 in Problem 8.2 with those in Problem 8.1, the only change is the value of R_4. Therefore, the optimal models are those in Equations 8.6 and 8.3 from Proposition 8.2, and Equation 8.4 from R_4.

For the model in Equation 8.6, we first compare it with Equation 8.4. The benefit of Equation 8.6 is that it saves the cost of the HSI when the first interface is geometrically nested (this happens with probability p_1) but the second interface is not, *i.e.*, $p_1(1 - p_2)C_s$. The model in Equation 8.6 also gets benefits when the first interface is not nested (this happens with probability $1 - p_1$) but the second one is, *i.e.*, $(1 - p_1)p_2 E$. However, the model in Equation 8.6 costs one C more than in Equation 8.4. Therefore, if $C < p_1(1 - p_2)C_s + (1 - p_1)p_2 E$, or $2C < 2p_2 E - 2p_1 p_2 C_s - 2p_1 p_2 E + 2p_1 C_s$, the model in Equation 8.6 is superior to Equation 8.4. As mentioned in Proposition 8.2, if $2C < Z_1$, Equation 8.6 is superior to Equation 8.3. Thus, if $2C < \min\{Z_1, Z_2\}$, the model in Equation 8.6 is the optimal model, where $Z_2 = 2p_2 E - 2p_1 p_2 C_s - 2p_1 p_2 E + 2p_1 C_s$.

For the model represented by Equation 8.3, we first compare it with the model in Equation 8.4. The benefit of Equation 8.3 is that it saves a SI cost C. By using the model in Equation 8.4, the HSI is only used if the first interface is geometrically nested (this happens with probability p_1). Thus, Equation 8.4 saves $(1 - p_1)C_s$, but loses $(1 - p_1)p_2 E$. That is, if $C > (1 - p_1)(C_s - p_2 E)$, or $2C > Z_3$, where $Z_3 = 2p_1 p_2 E - 2p_2 E + 2C_s(1 - p_1)$, Equation 8.3 is superior to Equation 8.4. As mentioned in Proposition 8.2, if $2C > Z_1$, the model in Equation 8.3 is superior to the model in Equation 8.6. Thus, if $2C > \max\{Z_1, Z_3\}$, the model in Equation 8.3 is the optimal decision.

We summarize this conclusion in the following proposition

Proposition 8.3. *Assume* $p_2 E \leq C_s < p_1 E$.

If $2C < \min\{Z_1, Z_2\}$, *use all three integrators to realize integral architecture* R_6.

If $2C > \max\{Z_1, Z_3\}$, *use HSI only to realize integral architecture* R_3.

Otherwise, use a SI to realize the first interface firstly, then use HSI to realize integral architecture R_4.

- For $p_1 E \leq C_s < p_1 E + p_2 E + p_1 p_2 T$, we get the following problem.

Problem 8.3.

$$\max R = \{R_3, R_4, R_5, R_6\} \; ;$$
$$s.t.\; p_1 E \leq C_s < p_1 E + p_2 E + p_1 p_2 T \; .$$

Similarly, comparing with Proposition 8.3, the only change is the value of R_5. Thus, the optimal model is either Equation 8.6, 8.3, 8.4, or 8.5.

For the model in Equations 8.6 and 8.3, the discussions are similar with Proposition 8.3. The only difference is that we also need to compare with the net benefit from the model in Equation 8.5, R_5. Since the model in Equation 8.5 is similar and symmetric to the model in Equation 8.4, we thus omit the detail here.

For the model in Equation 8.4, it is easy to show from Proposition 8.3 that if $2C > Z_2$, then Equation 8.4 is superior to Equation 8.6; and if $2C < Z_3$, then the model in Equation 8.4 is superior to Equation 8.3. Thus, the only thing left is the comparison between the models in Equations 8.4 and 8.5. The models in Equations 8.4 and 8.5 are the same in structure; the only difference is the probability. Comparing with the model in Equation 8.5, Equation 8.4 has a high probability p_1, to get benefits E, but it also has a high probability, p_1, to waste the cost of the HSI, C_s. Therefore, if $C_s < E$, the model in Equation 8.4 is superior to Equation 8.6. Conclusively, if $Z_2 < 2C < Z_3$ and $C_s < E$, the model in Equation 8.4 is the optimal decision.

We summarize this conclusion in the following proposition.

Proposition 8.4. *Assume* $p_1 E \leq C_s < p_1 E + p_2 E + p_1 p_2 T$.

If $2C < \min\{Z_1, Z_4, Z_2\}$, *use all three integrators to realize integral architecture* R_6, *where* $Z_4 = 2p_1 E - 2p_1 p_2 C_s - 2p_1 p_2 E + 2p_2 C_s$.

If $2C > \max\{Z_1, Z_5, Z_3\}$, *use HSI only to realize integral architecture* R_3, *where* $Z_5 = 2p_1 p_2 E - 2p_1 E + 2C_s(1 - p_2)$.

If $Z_2 < 2C < Z_3$ *and* $C_s < E$, *use a SI to realize the first interface firstly, then use HSI to realize integral architecture* R_4.

Otherwise, use a SI to realize the second interface firstly, then use HSI to realize integral architecture R_5.

- For $p_1 E + p_2 E + p_1 p_2 T \leq C_s < p_2 E + p_2 T$, we get the following problem.

Problem 8.4.

$$\max R = \{R_3, R_4, R_5, R_6\} \; ;$$
$$s.t.\; p_1 E + p_2 E + p_1 p_2 T \leq C_s < p_2 E + p_2 T \; .$$

Comparing with Proposition 8.4, the only change is the value of R_3, where its value is zero. The model in Equation 8.3 is no longer an optimal model. Thus, the optimal model is either Equation 8.6, 8.4, or 8.5.

For the model in Equation 8.6, it is easy to show from Proposition 8.3 that if $2C < Z_2$, then Equation 8.6 is superior to the model in Equation 8.4; and from Proposition 8.4, if $2C < Z_4$, then Equation 8.6 is superior to the model in Equation 8.5. Therefore, if $2C < \min\{Z_2, Z_4\}$, Equation 8.6 is the optimal decision.

The analysis of Equation 8.4 is the same as in Proposition 8.4. We summarize this conclusion in the following proposition.

Proposition 8.5. *Assume* $p_1E + p_2E + p_1p_2T \leq C_s < p_2E + p_2T$.

If $2C < \min\{Z_2, Z_4\}$, *use all three integrators to realize integral architecture* R_6.

If $2C > Z_2$ *and* $C_s < E$, *use a SI to realize the first interface firstly, then use HSI to realize integral architecture* R_4.

Otherwise, use a SI to realize the second interface firstly, then use HSI to realize integral architecture R_5.

- For $p_2E + p_2T \leq C_s < p_1E + p_1T$, we get the following problem.

Problem 8.5.

$$\max R = \{R_3, R_4, R_5, R_6\} ;$$
$$s.t.\ p_2E + p_2T \leq C_s < p_1E + p_1T .$$

Comparing with Proposition 8.5, the only change is the value of R_4. Now the cost of the HSI is high enough that makes using the HSI in the model in Equation 8.4 no longer an optimal decision. Thus, the optimal model is either the model in Equation 8.6, the model in Equation 8.4 but without the HSI, or the model in Equation 8.5.

For the model in Equation 8.6, the comparison with the model in Equation 8.5 is the same as before. Comparing with the model in Equation 8.4, the benefit of the model in Equation 8.6 is that it gets benefits from the second interface, p_2E. If both interfaces are geometrically nested, this happens with probability p_1p_2, then the HSI is used to get the global performance benefits T. The cost is a SI with probability p_2. Therefore, if $C < p_2E + p_1p_2(T - C_s)$, the model in Equation 8.6 is superior to the model in Equation 8.4. Summarily, if $2C < \min\{Z_6, Z_4\}$, the model in Equation 8.6 is the optimal decision.

For the model in Equation 8.5, the comparison with the model in Equation 8.6 is the same as before. Comparing with the model in Equation 8.5, the benefits of the model in Equation 8.4 are the benefits from the first interface, p_1E. On the other hand, the benefits of the model in Equation 8.5 relative to the model in Equation 8.4 are the benefits of the second interface, p_2E, and if the second interface is geometrically nested, this happens with probability p_2, then the HSI is used to get the benefit from the first interface as well as the the global performance benefit T. Minus the cost of the HSI. Therefore, if $(p_2E + p_2(p_1E + p_1T - C_s)) - p_1E > 0$, that is, if $C_s < Z_7$, the model in Equation 8.5 is superior to the model in Equation 8.4. Summarily, if $2C > Z_4$ and $C_s < Z_7$, the model in Equation 8.5 is the optimal decision.

We summarize this conclusion in the following proposition.

Proposition 8.6. *Assume* $p_2E + p_2T \le C_s < p_1E + p_1T$.

If $2C < \min\{Z_6, Z_4\}$, *use all three integrators to realize integral architecture* R_6, *where* $Z_6 = 2p_2E + 2p_1p_2(T - C_s)$.

If $2C > Z_4$ *and* $C_s < Z_7$, *use a SI to realize the second interface firstly, then use HSI to realize integral architecture* R_5, *where*

$$Z_7 = (p_2E - p_1E + p_2(p_1E + p_1T))/p_2 .$$

Otherwise, select a hybrid architecture: use a SI only to realize the first interface R_4.

- For $p_1E + p_1T \le C_s < \infty$, we get the following problem.

Problem 8.6.

$$\max R = \{R_3, R_4, R_5, R_6\} ;$$
$$s.t.\ p_1E + p_1T \le C_s < \infty .$$

Comparing with Proposition 8.6, the only change is the value of R_5. Now the cost of the HSI is high enough that makes using a HSI in the model in Equation 8.5 no longer an optimal decision. In this case, since the model in Equation 8.5 has a low probability, it is an absolute suboptimal decision to the model in Equation 8.4. Thus, the optimal decision is either the model in Equation 8.6 or the model in Equation 8.4. The comparison between the models in Equations 8.6 and 8.4 is the same as before.

We summarize this conclusion in the following proposition.

Proposition 8.7. *Assume* $p_1E + p_1T \le C_s < \infty$.

If $2C < Z_6$, *use all three integrators to realize integral architecture* R_6.

Otherwise, select a hybrid architecture: use a SI only to realize the first interface R_4.

8.4.3 Implications

So far the analysis has focused on the comparison of product architecture decisions under different costs. In this section, we consider some of the empirical implications obtained from the comparative analysis in the prior sections.

First, Ulrich (1995) asserts that "no single architecture is optimal in all cases." It is intuitive to ask if there is architecture that is suboptimal in all cases? The finding from the analytical models is summarized in Corollary 8.1.

Corollary 8.1. *There is no single architecture that is optimal in all cases, whereas there are architectures that are always suboptimal.*

Corollary 8.1 can be shown directly from Proposition 8.1, in which no architecture is always optimal in all three cases and Hybrid-II (see Figure 8.12) is suboptimal in all cases.

Second, previous studies have summarized many benefits of modular architecture, which include products variety, components commonality, upgradability, and others. However, as shown in the above analytical models and indicated by Ulrich (1995), modular architecture cannot achieve global performance, thus we summarize this as the following corollary.

Corollary 8.2. *Under consideration of the global performance, modular architecture is an absolutely suboptimal decision.*

Corollary 8.2 can simply be verified by checking Propositions 8.2–8.7; among them, modular architecture is never a candidate for the optimal architecture.

Third, opposite to Corollary 8.2, global performance can only be optimized through an integral architecture. Similarly, Fujimoto (2004) gives a large number of evidences to show that many high-value added products adopt integral architecture. Therefore, we give an intuitive corollary as follows.

Corollary 8.3. *To obtain global performance, achieving integral architecture is an all-the-time candidate for optimal architecture.*

Corollary 8.3 can also be verified by checking Propositions 8.2–8.7; among them, using all three integrators to realize an integral architecture is the only all-the-time candidate for optimal architecture. Corollary 8.3 indicates an important truth: the model in Equation 8.6 is independent of the cost C_s. No matter how high the cost of the HSI may be, the model in Equation 8.6, *i.e.*, using all three integrators, is always a choice for optimal strategic decision. The reason is the global performance benefit brought by the integral architecture.

Finally, we consider the product development process. The HSI plays an important role in obtaining global performance. Thus, a corollary related to HSI is given as follows.

Corollary 8.4. *If the global performance benefit T is sufficiently large, then using a HSI to reap it is always an optimal decision.*

Since T is very large, we can get $p_1E + p_2E + p_2p_2T > C_s$. From Propositions 8.2–8.4, we can find all optimal decisions involve using a HSI.

8.5 A Summary of the Model

In this section, we summarize the model analyzed in the last section into a decision tree. Decision tree is a powerful tool which is frequently used in the areas of data

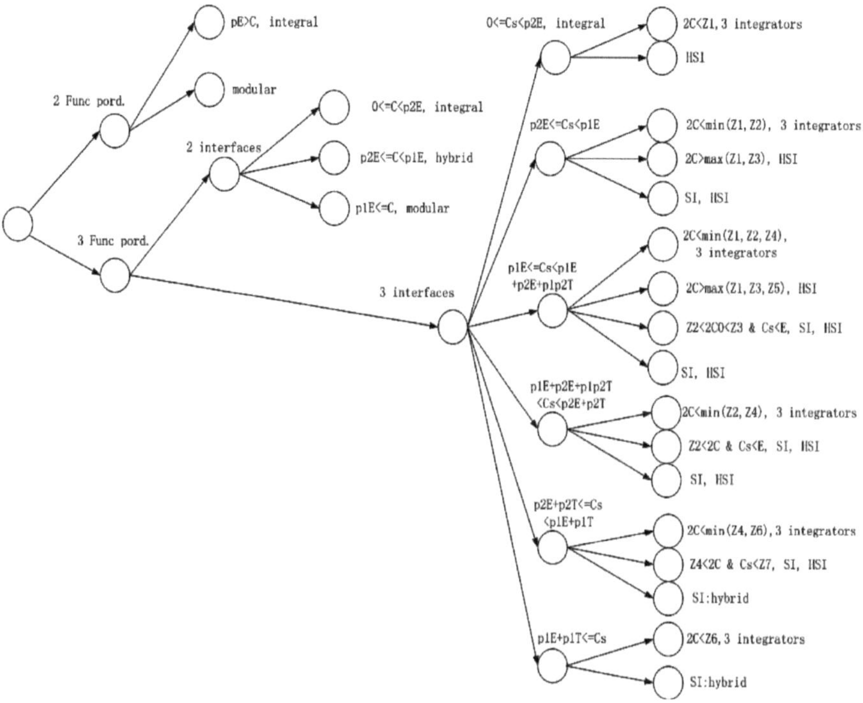

Figure 8.13 A decision tree model

mining and probability theory. One of the advantages of decision tree is that it can describe complex models into a simple tree structure. The decision tree is depicted in the following Figure 8.13.

We use a simple example to illustrate the decision tree model. The example is a three-function three-interface product. Because most data related to this research are secrets of companies, we use simulated data to show the model. The data is given as follows.

The benefit from the global performance T is 100. The benefit from the geometric nesting E is 35. The costs of the HSI C_s and system integrator C are 40 and 20, respectively. The probabilities of realization of the two geometric nestings p_1, and p_2 are 0.8 and 0.6, respectively.

Because $p_1E = 0.8\times 35 = 28$ and $p_1E + p_2E + p_2p_2T = 0.8\times 35 + 0.6\times 35 + 0.8\times 0.6\times 100 = 97$, thus $p_1E \le C_s < p_1E + p_2E + p_2p_2T$. Proposition 8.4 should be used to decide the product architecture. Because $Z_1 = C_s(1 - p_1p_2) = 20.8$, $Z_3 = 2p_1p_2E - 2p_2E + 2C_s(1-p_1) = 7.6$, $Z_5 = 2p_1p_2E - 2p_1E + 2C_s(1-p_2) = 9.6$, and $2C = 40$; thus, $2C > \max\{Z_1, Z_5, Z_3\}$. In this way, the HSI only is used to realize the integral product architecture. The profit from this strategy is $R_3 = (p_1 + p_2)E + p_1p_2T - C_s = 57$.

8.6 Conclusion

By using examples of multifunction products and small size, this chapter has shown the impact of global performance on the choice of product architecture during the product development process. We have focused on analytic models whose objectives are obtaining global performance of the product through a modular/hybrid/integral architecture. Trade-offs between costs and expected benefits from different product architectures have been analyzed and compared. We also have given some empirical implications obtained from comparative study. However, the models developed in this chapter are only dedicated to products which have a simple architecture structure, for example, multifunction products. Therefore, we suggest future research on models that can analyze products that have a complex architecture structure, for example, automobiles. Also, many other factors such as customer demand, risk assessment and so on need to be considered in the future study.

References

Alexander C (1964) Notes on the synthesis of form. Harvard University Press, Cambridge, MA

Baldwin CY, Clark KB (2000) Design rules: the power of modularity, vol 1. MIT Press, Cambridge, MA

Benjaafar S, Sheikhzadeh M, Gupta D (1998) Machine sharing in manufacturing systems: total flexibility versus chaining. Int J Flexible Manuf Syst 10:351–378

Brusco MJ, Johns TR (1998) Staffing a multiskilled workforce with varying levels of productivity: an analysis of cross-training policies. Decis Sci 29:499–515

Clark KB, Fujimoto T (1991) Product development performance. Harvard Business School Press, Boston

Desai P, Kekre S, Radhakrishnan S, Srinivasan K (2001) Product differentiation and commonality in design: balancing revenue and cost drivers. Manage Sci 47:37–51

Ethiraj SK, Levinthal D (2004) Modularity and innovation in complex systems. Manage Sci 50:159–173

Fine CH, Golany B, Naseraldin H (2005) Modeling tradeoffs in three-dimensional concurrent engineering: a goal programming approach. J Oper Manage 23:389–403

Fixson SK (2005) Product architecture assessment: a tool to link product, process, and supply chain design decisions. J Oper Manage 23:345–369

Fujimoto T (1999) The evolution of a manufacturing system at Toyota. Oxford University Press, New York

Fujimoto T (2004) Manufacturing philosophy. Nihon Keizai Shinbunsha, Tokyo

Gourville JT, Soman D (2005) Overchoice and assortment type: when and why variety backfires. Market Sci 24:382–395

Green PE, Srinivasan V (1990) Conjoint analysis in marketing: new developments with implications for research and practice. J Market 54:3–19

Griffin A, Hauser JR (1996) Integrating R&D and marketing: a review and analysis of the literature. J Prod Innov Manage 13:191–215

Hoch SJ, Bradlow ET, Wansink B (1999) The variety of an assortment. Market Science, 18:527–546

Hopp WJ, Van Oyen MP (2004) Agile workforce evaluation: a framework for cross-training and coordination. IIE Trans 36:919–940

Hopp WJ, Xu X (2005) Product line selection and pricing with modularity in design. Manuf Service Oper Manage 7:172–187

Hyer NL, Brown KA (1999) The discipline of real cells. J Oper Manage 17:557–574

Jiao J, Simpson TW, Siddique Z (2007) Product family design and platform-based product development: a state-of-the art review. J Int Manuf 18(1):5–29

Kim K, Chhajed D (2000) Commonality in product design: cost saving, valuation change and cannibalization. Eur J Oper Res 125:602–621

Liker JK (2004) The Toyota Way: 14 Management Principles from the World's Greatest Manufacturing. McGraw-Hill, New York

Lin J, Chai KH, Wong YS, Brombacher AC (2008) A dynamic model for managing overlapped iterative product development. Eur J Oper Res 185:378–392

Mikkola JH (2006) Capturing the degree of modularity embedded in product architectures. J Prod Innov Manage 23:128–146

Netessine S, Taylor TA (2007) Product line design and production technology. Market Sci 26:101–117

Nonaka I (1990) Redundant, overlapping organizations: a Japanese approach to managing the innovation process. California Manage Rev 32:27–38

Nonaka I (1991) The knowledge-creating company. Harvard Business Rev November-December: 96–104

Nonaka I, Takeuchi H (1995) The knowledge-creating company. Oxford University Press, New York

Ohno T (1988) Toyota production system: beyond large-scale production. Productivity, New York

Pahl G, Beitz W (1996) Engineering design: a systematic approach. Springer, London

Ramachandran K, Krishnan V (2008) Design architecture and introduction timing for rapidly improving industrial products. Manuf Service Oper Manage 10:149–171

Ramdas K (2003) Managing product variety: an integrative review and research directions. Product Oper Manage 12:79–101

Rivkin JW, Siggelkow N (2007) Patterned interactions in complex systems: implications for exploration. Manage Sci 53:1068–1085

Sanchez R (1999) Modular architectures in the marketing process. J Market 63:92–111

Simon HA (1969) The sciences of the artificial. MIT Press, Cambridge, MA

Ulrich KT (1995) The role of product architecture in the manufacturing firms. Res Policy 24:419–440

Ulrich KT, Eppinger SD (2004) Product design and development, 3rd edn. McGraw-Hill, New York

Wheelwright SC, Clark KB (1992) Revolutionizing product development quantum leaps in speed, efficiency and quality. Free Press, New York

Chapter 9
Application of Cluster Analysis to Cellular Manufacturing

The primary motivation for adopting cellular manufacturing is the globalization and intense competition in the current marketplace. The initial step in the design of a cellular manufacturing system is the identification of part families and machine groups and forming manufacturing cells so as to process each part family within a machine group with minimum intercellular movements of parts. One methodology to manufacturing cells is the use of similarity coefficients in conjunction with clustering procedures. In this chapter, we give a comprehensive overview and discussion for similarity coefficients developed to date for use in solving the cell formation problem. Despite previous studies indicated that the similarity coefficients-based method (SCM) is more flexible than other cell formation methods, none of the studies has explained the reason why SCM is more flexible. This chapter tries to explain the reason explicitly. To summarize various similarity coefficients, we develop a taxonomy to clarify the definition and usage of various similarity coefficients in designing cellular manufacturing systems. Existing similarity (dissimilarity) coefficients developed so far are mapped onto the taxonomy. Additionally, production information based similarity coefficients are discussed and a historical evolution of these similarity coefficients is outlined. Although many similarity coefficients have been proposed, very fewer comparative studies have been done to evaluate the performance of various similarity coefficients. In this chapter, we compare the performance of twenty well-known similarity coefficients. More than two hundred numerical cell formation problems, which are selected from the literature or generated deliberately, are used for the comparative study. Nine performance measures are used for evaluating the goodness of cell formation solutions. Two characteristics, discriminability and stability of the similarity coefficients are tested under different data conditions. From the results, three similarity coefficients are found to be more discriminable; Jaccard is found to be the most stable similarity coefficient. Four similarity coefficients are not recommendable due to their poor performances.

9.1 Introduction

Group technology (GT) is a manufacturing philosophy that has attracted a lot of attention because of its positive impacts in the batch-type production. Cellular manufacturing (CM) is one of the applications of GT principles to manufacturing. In the design of a CM system, similar parts are groups into families and associated machines into groups so that one or more part families can be processed within a single machine group. The process of determining part families and machine groups is referred to as the cell formation (CF) problem.

CM has been considered as an alternative to conventional batch-type manufacturing where different products are produced intermittently in small lot sizes. For batch manufacturing, the volume of any particular part may not be enough to require a dedicated production line for that part. Alternatively, the total volume for a family of similar parts may be enough to efficiently utilize a machine-cell (Miltenburg and Zhang 1991).

It has been reported (Seifoddini 1989a) that employing CM may help overcome major problems of batch-type manufacturing including frequent setups, excessive in-process inventories, long through-put times, complex planning and control functions, and provides the basis for implementation of manufacturing techniques such as just-in-time (JIT) and flexible manufacturing systems (FMS).

A large number of studies related to GT/CM have been performed both in academia and industry. Reisman *et al.* (1997) gave a statistical review of 235 articles dealing with GT and CM over the years 1965 through 1995. They reported that the early (1966–1975) literature dealing with GT/CM appeared predominantly in book form. The first written material on GT was Mitrofanov (1966) and the first journal paper that clearly belonged to CM appeared in 1969 (Optiz *et al.* 1969). Reisman *et al.* (1997) also reviewed and classified these 235 articles on a five-point scale, ranging from pure theory to bona fide applications. In addition, they analyzed seven types of research processes used by authors.

There are many researchable topics related to cellular manufacturing. Wemmerlöv and Hyer (1987) presented four important decision areas for group technology adoption: applicability, justification, system design, and implementation. A list of some critical questions was given for each area.

Applicability, in a narrow sense, can be understood as feasibility (Wemmerlöv and Hyer 1987). Shafer *et al.* (1995) developed a taxonomy to categorize manufacturing cells. They suggested three general cell types: process cells, product cells, and other types of cells. They also defined four shop layout types: product cell layouts, process cell layouts, hybrid layouts, and mixture layouts. Despite the growing attraction of cellular manufacturing, most manufacturing systems are hybrid systems (Wemmerlöv and Hyer 1987; Shambu and Suresh 2000). A hybrid CM system is a combination of both a functional layout and a cellular layout. Some hybrid CM systems are unavoidable, since some processes such as painting or heat treatment are frequently more efficient and economic to keep the manufacturing facilities in a functional layout.

Implementation of a CM system contains various aspects such as human, education, environment, technology, organization, management, evaluation and even culture. Unfortunately, only a few papers have been published related to these areas. Research reported on the human aspect can be found in Fazakerley (1976), Burbidge *et al.* (1991), Beatty (1992), and Sevier (1992). Some recent studies on implementation of CM systems are Silveira (1999), and Wemmerlöv and Johnson (1997, 2000).

The problem involved in justification of cellular manufacturing systems has received a lot of attention. Much of the research was focused on the performance comparison between cellular layout and functional layout. A number of researchers support the relative performance supremacy of cellular layout over functional layout, while others doubt this supremacy. Agrawal and Sarkis (1998) gave a review and analysis of comparative performance studies on functional and CM layouts. Shambu and Suresh (2000) studied the performance of hybrid CM systems through a computer simulation investigation.

System design is the most researched area related to CM. Research topics in this area include cell formation (CF), cell layout (Kusiak and Heragu 1987; Balakrishnan and Cheng 1998; Liggett 2000), production planning (Mosier and Taube 1985a; Singh 1996), and others (Lashkari *et al.* 2004; Solimanpur *et al.* 2004). CF is the first, most researched topic in designing a CM system. Many approaches and methods have been proposed to solve the CF problem. Among these methods, production flow analysis (PFA) is the first one, which was used by Burbidge (1971) to rearrange a machine-part incidence matrix on trial and error until an acceptable solution is found. Several review papers have been published to classify and evaluate various approaches for CF; some of them will be discussed in this chapter. Among various cell formation models, those based on the similarity coefficient method (SCM) are more flexible in incorporating manufacturing data into the machine-cells formation process (Seifoddini 1989a). In this chapter, an attempt has been made to develop a taxonomy for a comprehensive review of almost all similarity coefficients used for solving the cell formation problem.

Although numerous CF methods have been proposed, fewer comparative studies have been done to evaluate the robustness of various methods. Part of the reason is that different CF methods include different production factors, such as machine requirement, setup times, utilization, workload, setup cost, capacity, part alternative routings, and operation sequences. Selim *et al.* (1998) emphasized the necessity to evaluate and compare different CF methods based on the applicability, availability, and practicability. Previous comparative studies include Mosier (1989), Chu and Tsai (1990), Shafer and Meredith (1990), Miltenburg and Zhang (1991), Shafer and Rogers (1993), Seifoddini and Hsu (1994), and Vakharia and Wemmerlöv (1995).

Among the above seven comparative studies, Chu and Tsai (1990) examined three array-based clustering algorithms: rank order clustering (ROC) (King 1980), direct clustering analysis (DCA) (Chan and Milner 1982), and bond energy analysis (BEA) (McCormick *et al.* 1972). Shafer and Meredith (1990) investigated six cell formation procedures: ROC, DCA, cluster identification algorithm (CIA) (Kusiak and Chow 1987), single linkage clustering (SLC), average linkage clustering (ALC), and an operation sequences-based similarity coefficient (Vakharia and Wemmerlöv 1990). Miltenburg and Zhang (1991) compared nine cell formation procedures. Some of the compared procedures are combinations of two different algorithms A1/A2. A1/A2 denotes using A1 (algorithm 1) to group machines and using A2 (algorithm 2) to group parts. The nine procedures include: ROC,

SLC/ROC, SLC/SLC, ALC/ROC, ALC/ALC, modified ROC (MODROC) (Chandrasekharan and Rajagopalan 1986b), ideal seed non-hierarchical clustering (ISNC) (Chandrasekharan and Rajagopalan 1986a), SLC/ISNC, and BEA.

The other four comparative studies evaluated several similarity coefficients. We will discuss them in the later sections.

9.2 Background

This section gives a general background of machine part CF models and detailed algorithmic procedures of the SCM.

9.2.1 Machine-part Cell Formation

The CF problem can be defined as: "if the number, types, and capacities of production machines, the number and types of parts to be manufactured, and the routing plans and machine standards for each part are known, which machines and their associated parts should be grouped together to form cell?" (Wei and Gaither 1990). Numerous algorithms, heuristic or non-heuristic, have emerged to solve the cell formation problem. A number of researchers have published review studies for existing CF literature (refer to King and Nakornchai 1982; Kumar and Vannelli 1983; Mosier and Taube 1985a; Wemmerlöv and Hyer 1986; Chu and Pan 1988; Chu 1989; Lashkari and Gunasingh 1990; Kamrani *et al.* 1993; Singh 1993; Offodile *et al.* 1994; Reisman *et al.* 1997; Selim *et al.* 1998; Mansouri *et al.* 2000). Some timely reviews are summarized as follows.

Singh (1993) categorized numerous CF methods into the following subgroups: part coding and classifications, machine-component group analysis, similarity coefficients, knowledge-based, mathematical programming, fuzzy clustering, neural networks, and heuristics.

Offodile *et al.* (1994) employed a taxonomy to review the machine-part CF models in CM. The taxonomy is based on the Mehrez *et al.* (1988) five-level conceptual scheme for knowledge representation. Three classes of machine-part grouping techniques have been identified: visual inspection, part coding and classification, and analysis of the production flow. They used the production flow analysis segment to discuss various proposed CF models.

Reisman *et al.* (1997) gave a most comprehensive survey. A total of 235 CM papers were classified based on seven alternative, but not mutually exclusive, strategies used in Reisman and Kirshnick (1995).

Selim *et al.* (1998) developed a mathematical formulation and a methodology-based classification to review the literature on the CF problem. The objective function of the mathematical model is to minimize the sum of costs for purchasing machines, variable cost of using machines, tooling cost, material handling cost, and amortized worker training cost per period. The model is combinatorially complex and will not be solvable for any real problem. The classification used in this chapter is based on the type of general solution methodology. More than 150 works have been reviewed and listed in the reference.

9.2.2 Similarity Coefficient Methods (SCM)

A large number of similarity coefficients have been proposed in the literature. Some of them have been utilized in connection with CM. SCM rely on similarity measures in conjunction with clustering algorithms. It usually follows a prescribed set of steps (Romesburg 1984), the main one is as follows.

1. Form the initial machine-part incidence matrix, whose rows are machines and columns stand for parts. The entries in the matrix are 0s or 1s, which indicate a part need or need not a machine for a production. An entry a_{ik} is defined as follows.

$$a_{ik} = \begin{cases} 1 & \text{if part } k \text{ visits machine } i \ , \\ 0 & \text{otherwise} \ . \end{cases} \quad (9.1)$$

   The following are definitions:
   i machine index ($i = 1, \ldots, M$)
   k part index ($k = 1, \ldots, P$)
   M number of machines
   P number of parts.

2. Select a similarity coefficient and compute similarity values between machine (part) pairs and construct a similarity matrix. An element in the matrix represents the sameness between two machines (parts).
3. Use a clustering algorithm to process the values in the similarity matrix, which results in a diagram called a tree, or dendrogram, that shows the hierarchy of similarities among all pairs of machines (parts). Find the machines groups (part families) from the tree or dendrogram, check all predefined constraints such as the number of cells, cell size, *etc.*

9.3 Why Present a Taxonomy on Similarity Coefficients?

Before answering the question, "why present a taxonomy on similarity coefficients," we need to answer the following question first: "why are similarity coefficient methods more flexible than other cell formation methods?"

In this section, we present past review studies on similarity coefficients, discuss their weaknesses and confirm the need of a new review study from the viewpoint of the flexibility of SCM.

9.3.1 Past Review Studies on SCM

Although a large number of similarity coefficients exist in the literature, very few review studies have been performed on similarity coefficients. Three review papers on similarity coefficients (Shafer and Rogers 1993a; Sarker 1996; Mosier *et al.* 1997) are available in the literature.

Shafer and Rogers (1993a) provided an overview of similarity and dissimilarity measures applicable to cellular manufacturing. They introduced general measures of association firstly. Then, similarity and distance measures for determining part families or clustering machine types are discussed. Finally, they concluded the paper with a discussion of the evolution of similarity measures applicable to cellular manufacturing.

Sarker (1996) reviewed a number of commonly used similarity and dissimilarity coefficients. In order to assess the quality of solutions to the cell formation problem, several different performance measures are enumerated, and some experimental results provided by earlier researchers are used to evaluate the performance of reviewed similarity coefficients.

Mosier *et al.* (1997) presented an impressive survey of similarity coefficients in terms of structural form, and in terms of the form and levels of the information required for computation. They particularly emphasized the structural forms of various similarity coefficients and made an effort for developing a uniform notation to convert the originally published mathematical expression of reviewed similarity coefficients into a standard form.

9.3.2 Objective of this Study

The three previous review studies provide important insights from different viewpoints. However, we still need an updated and more comprehensive review to achieve the following objectives.

- Develop an explicit taxonomy. To the best of our knowledge, none of the previous articles have developed or employed an explicit taxonomy to categorize various similarity coefficients. We discuss in detail the important role of taxonomy in the Section 9.3.3.
  Neither Shafer and Rogers (1993a) nor Sarker (1996) provided a taxonomic review framework. Sarker (1996) enumerated a number of commonly used similarity and dissimilarity coefficients; Shafer and Rogers (1993a) classified similarity coefficients into two groups based on measuring the resemblance between (1) part pairs, or (2) machine pairs.
- Give a more comprehensive review. Only a few similarity coefficient-related studies have been reviewed by previous articles.
  Shafer and Rogers (1993a) summarized 20 or more similarity coefficients related research; most of the similarity coefficients reviewed in the Sarker (1996) paper

need prior experimental data; Mosier *et al.* (1997) made some efforts to abstract the intrinsic nature inherent in different similarity coefficients. Only a few similarity coefficient-related studies have been cited in their paper.

Owing to the accelerated growth of the amount of research reported on similarity coefficients subsequently, and owing to the discussed objectives above, there is a need for a more comprehensive review of research to categorize and summarize various similarity coefficients that have been developed in the past years.

9.3.3 Why SCM Are More Flexible

The cell formation problem can be extraordinarily complex, because of various different production factors, including alternative process routings, operational sequences, production volumes, machine capacities, tooling times and others. Numerous cell formation approaches have been developed, these approaches can be classified into the following three groups:

- mathematical programming (MP) models;
- (meta-)heuristic algorithms (HA); and
- similarity coefficient methods (SCM).

Among these approaches, SCM is the application of cluster analysis to cell formation procedures. Since the basic idea of GT depends on the estimation of the similarities between part pairs and cluster analysis is the most basic method for estimating similarities, it is concluded that the SCM-based method is one of the most basic methods for solving CF problems.

Despite previous studies (Seifoddini 1989a), which indicated that SCM-based approaches are more flexible in incorporating manufacturing data into the machine-cells formation process, none of the previous articles has explained the reason why SCM-based methods are more flexible than other approaches such as MP and HA. We try to explain the reason as follows.

For any concrete cell formation problem, there is generally no "correct" approach. The choice of the approach is usually based on the tool availability, analytical tractability, or simply personal preference. There are, however, two effective principles that are considered reasonable and generally accepted for large and complex problems. They are as follows.

Principle 9.1. Decompose the complex problem into several small conquerable problems. Solve small problems, and then reconstitute the solutions.

All three groups of cell formation approaches (MP, HA, SCM) mentioned above can use Principle 9.1 for solving complex cell formation problems. However, the difficulty for this principle is that a systematic mean must be found for dividing one complex problem into many small conquerable problems, and then reconstituting the solutions. It is usually not easy to find such systematic means.

Principle 9.2. It usually needs a complicated solution procedure to solve a complex cell formation problem. The second principle is to decompose the complicated solution procedure into several small tractable stages.

Comparing with MP and HA-based methods, the SCM-based method is more suitable for Principle 9.2. We use a concrete cell formation model to explain this conclusion. Assume there is a cell formation problem that incorporates two production factors: production volume and operation time of parts.

1. MP, HA. By using MP, HA-based methods, the general way is to construct a mathematical or non-mathematical model that takes into account production volume and operation time, and then the model is analyzed, optimal or heuristic solution procedure is developed to solve the problem. The advantage of this way is that the developed model and solution procedure are usually unique for the original problem. So, even if they are not the "best" solutions, they are usually "very good" solutions for the original problem. However, there are two disadvantages inherent in the MP, HA-based methods.
   Firstly, extension of an existing model is usually difficult work. For example, if we want to extend the above problem to incorporate other production factors such as alternative process routings and operational sequences of parts, what we need to do is to extend the old model to incorporate additional production factors or construct a new model to incorporate all required production factors: production volumes, operation times, alternative process routings and operational sequences. Without further information, we do not know which one is better; in some cases extending the old one is more efficient and economical, in other cases constructing a new one is more efficient and economical. However, in most cases both extension and construction are difficult and costly works.
   Secondly, no common or standard ways exist for MP, HA to decompose a complicated solution procedure into several small tractable stages. To solve a complex problem, some researchers decompose the solution procedure into several small stages. However, the decomposition is usually based on the experience, ability and preference of the researchers. There are, however, no common or standard ways that exist for decomposition.
2. SCM. SCM is more flexible than MP, HA-based methods, because it overcomes the two mentioned disadvantages of MP, HA. We have introduced in Section 2.2 that the solution procedure of SCM usually follows a prescribed set of steps:

   Step 1 Get input data.
   Step 2 Select a similarity coefficient.
   Step 3 Select a clustering algorithm to get machine cells.

   Thus, the solution procedure is composed of three steps, this overcomes the second disadvantage of MP, HA. We show how to use SCM to overcome the first disadvantage of MP, HA as follows.
   An important characteristic of SCM is that the three steps are independent from each other. That means the choice of the similarity coefficient in step 2 does not influence the choice of the clustering algorithm in step 3. For example, if

we want to solve the production volumes and operation times considered in the cell formation problem mentioned before, after getting the input data we select a similarity coefficient that incorporates production volumes and operation times of parts; finally we select a clustering algorithm (for example ALC algorithm) to get machine cells. Now we want to extend the problem to incorporate additional production factors: alternative process routings and operational sequences. We re-select a similarity coefficient that incorporates all required four production factors to process the input data, and since step 2 is independent from step 3, we can easily use the ALC algorithm selected before to get new machine cells. Thus, comparing with MP, HA-based methods, SCM is very easy to extend a cell formation model.

Therefore, according to the above analysis, SCM-based methods are more flexible than MP, HA-based methods for dealing with various cell formation problems. To take full advantage of the flexibility of SCM and to facilitate the selection of similarity coefficients in step 2, we need an explicit taxonomy to clarify and classify the definition and usage of various similarity coefficients. Unfortunately, none such taxonomies have been developed in the literature, so in the next section we will develop a taxonomy to summarize various similarity coefficients.

9.4 Taxonomy for Similarity Coefficients Employed in Cellular Manufacturing

Different similarity coefficients have been proposed by researchers in different fields. A similarity coefficient indicates the degree of similarity between object pairs. A tutorial of various similarity coefficients and related clustering algorithms are available in the literature (Anderberg 1973; Bijnen 1973; Sneath and Sokal 1973; Arthanari and Dodge 1981; Romesburg 1984; Gordon 1999). In order to classify similarity coefficients applied in CM, a taxonomy is developed and shown in Figure 9.1. The objective of the taxonomy is to clarify the definition and usage of various similarity or dissimilarity coefficients in designing CM systems. The taxonomy is a five-level framework numbered from level 0 to 4. Level 0 represents the root of the taxonomy. The detail of each level is described as follows.

Level 1 *l* 1 categorizes existing similarity coefficients into two distinct groups: problem-oriented similarity coefficients (*l* 1.1) and general-purpose similarity coefficients (*l* 1.2). Most of the similarity coefficients introduced in the field of numerical taxonomy are classified in *l* 1.2 (general-purpose), which are widely used in a number of disciplines, such as psychology, psychiatry, biology, sociology, the medical sciences, economics, archeology and engineering. The characteristic of this type of similarity coefficient is that they always maximize the similarity value when two objects are perfectly similar.

On the other hand, problem-oriented (*l* 1.1) similarity coefficients aim at evaluating the predefined specific "appropriateness" between object pairs. This type of

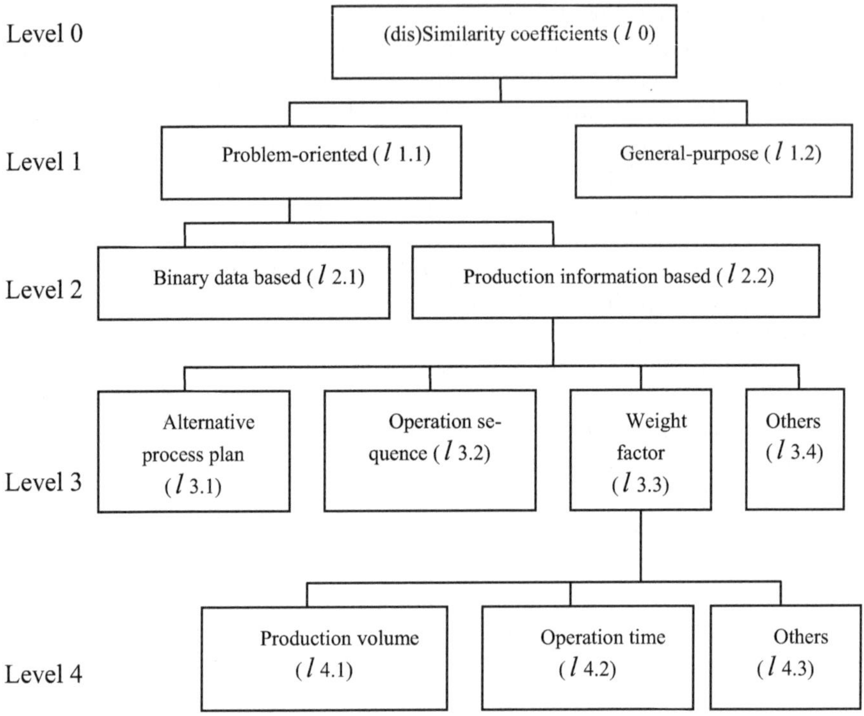

Figure 9.1 A taxonomy for similarity coefficients

similarity coefficient is designed specially to solve specific problems, such as CF. They usually include additional information and do not need to produce maximum similarity value even if the two objects are perfectly similar. Two less similar objects can produce a higher similarity value due to their "appropriateness" and more similar objects may produce a lower similarity value due to their "inappropriateness."

We use three similarity coefficients to illustrate the difference between the problem-oriented and general-purpose similarity coefficients. Jaccard is the most commonly used general-purpose similarity coefficient in the literature. The Jaccard similarity coefficient between machine i and machine j is defined as follows.

$$s_{ij} = \frac{a}{a+b+c}, \quad 0 \leq s_{ij} \leq 1 \tag{9.2}$$

where

a is the number of parts visited both machines,
b is the number of parts visited machine i but not j, and
c is the number of parts visited machine j but not i.

Two problem-oriented similarity coefficients, MaxSC (Shafer and Rogers 1993b) and commonality score (CS) (Wei and Kern 1989), are used to illustrate this comparison. MaxSC between machine i and machine j is defined as follows:

$$ms_{ij} = \max\left[\frac{a}{a+b}, \frac{a}{a+c}\right], 0 \le ms_{ij} \le 1 \tag{9.3}$$

and CS between machine i and machine j is calculated as follows:

$$c_{ij} = \sum_{k=1}^{P} \delta(a_{ik}, a_{jk}) \tag{9.4}$$

where

$$\delta(a_{ik}, a_{jk}) = \begin{cases} (P-1), & \text{if } a_{ik} = a_{jk} = 1 \\ 1, & \text{if } a_{ik} = a_{jk} = 0 \\ 0, & \text{if } a_{ik} \ne a_{jk}. \end{cases} \tag{9.5}$$

$$a_{ik} = \begin{cases} 1, & \text{if machine } i \text{ uses part } k, \\ 0, & \text{otherwise}. \end{cases} \tag{9.6}$$

The k part index ($k = 1, \ldots, P$) is the kth part in the machine-part matrix.

We use Tables 9.1 and 9.2 to illustrate the "appropriateness" of problem-oriented similarity coefficients. Table 9.1 is a machine-part incidence matrix whose rows represent machines and columns represent parts. The Jaccard coefficient s_{ij}, MaxSC coefficient ms_{ij} and commonality score c_{ij} of machine pairs in Table 9.1 are calculated and given in Table 9.2.

The characteristic of general-purpose similarity coefficients is that they always maximize similarity value when two objects are perfectly similar. Among the four machines in Table 9.1, we find that machine 2 is a perfect copy of machine 1, and they should have the highest value of similarity. We also find that the degree of

Table 9.1 Illustrative machine-part matrix for the "appropriateness"

	Part	$p1$	$p2$	$p3$	$p4$	$p5$	$p6$	$p7$	$p8$	$p9$	$p10$	$p11$	$p12$	$p13$	$p14$
Machine	$m1$	1	1	1											
	$m2$	1	1	1											
	$m3$	1	1	1	1										
	$m4$	1	1	1	1	1	1	1							

Table 9.2 Similarity values of Jaccard, MaxSC and CS of Table 9.1

		Similarity values, s_{ij}, ms_{ij} and c_{ij}			
		$i = 1, j = 2$	$i = 3, j = 4$	$i = 1$ or 2, $j = 3$	$i = 1$ or 2, $j = 4$
Jaccard	s_{ij}	1	4/7	3/4	3/7
MaxSC	ms_{ij}	1	1	1	1
CS	c_{ij}	50	59	49	46

similarity between machines 3 and 4 is lower than that of machines 1 and 2. The results of Jaccard in Table 9.2 reflect our finds clearly. That is, $\max(s_{ij}) = s_{12} = 1$, and $s_{12} > s_{34}$.

Problem-oriented similarity coefficients are designed specially to solve CF problems. CF problems are multi-objective decision problems. We define the "appropriateness" of two objects as the degree of possibility to achieve the objectives of CF models by grouping the objects into the same cell. Two objects will obtain a higher degree of "appropriateness" if they facilitate achieving the predefined objectives, and *vice versa*. As a result, two less similar objects can produce a higher similarity value due to their "appropriateness" and more similar objects may produce a lower similarity value due to their "inappropriateness." Since different CF models aim at different objectives, the criteria of "appropriateness" are also varied. In short, for problem-oriented similarity coefficients, rather than evaluating the similarity between two objects, they evaluate the "appropriateness" between them.

MaxSC is a problem-oriented similarity coefficient (Shafer and Rogers 1993b). The highest value of MaxSC is given to two machines if the machines process exactly the same set of parts or if one machine processes a subset of the parts processed by the other machine. In Table 9.2, all machine pairs obtain the highest MaxSC value even if not all of them are perfectly similar. Thus, in the procedure of cell formation, no difference can be identified from the four machines by MaxSC.

CS is another problem-oriented similarity coefficient (Wei and Kern 1989). The objective of CS is to recognize not only the parts that need both machines, but also the parts on which the machines both do not process. Some characteristics of CS have been discussed by Yasuda and Yin (2001). In Table 9.2, the highest CS is produced between machine 3 and machine 4, even if the degree of similarity between them is lower and even if machines 1 and 2 are perfectly similar. The result $s_{34} > s_{12}$ illustrates that two less similar machines can obtain a higher similarity value due to the higher "appropriateness" between them.

Therefore, it is concluded that the definition of "appropriateness" is very important for every problem-oriented similarity coefficient, it determines the quality of CF solutions by using these similarity coefficients.

Level 2 In Figure 9.1, problem-oriented similarity coefficients can be further classified into binary data based (*l* 2.1) and production information-based (*l* 2.2) similarity coefficients. Similarity coefficients in *l*2.1 only consider assignment information, that is, a part need or need not have a machine to perform an operation. The assignment information is usually given in a machine-part incidence matrix, such as Table 9.1. An entry of "1" in the matrix indicates that the part needs a operation by the corresponding machine. The characteristic of *l* 2.1 is similar to *l* 1.2, which also uses binary input data. However, as we mentioned above, they are essentially different in the definition for assessing the similarity between object pairs.

Level 3 In the design of CM systems, many manufacturing factors should be involved when the cells are created, *e.g.*, machine requirement, machine setup times, utilization, workload, alternative routings, machine capacities, operation sequences, setup cost and cell layout (Wu and Salvendy 1993). Choobineh and Nare (1999) described a sensitivity analysis for examining the impact of ignored manufacturing factors on a CMS design. Due to the complexity of CF problems, it is impossible to take into consideration all of the real-life production factors by a single approach. A number of similarity coefficients have been developed in the literature to incorporate different production factors. In this chapter, we use the three most researched manufacturing factors (alternative process routing *l*3.1, operation sequence *l* 3.2 and weighted factors *l* 3.3) as the base to perform the taxonomic review study.

Level 4 Weighted similarity coefficient is a logical extension or expansion of the binary data-based similarity coefficient. Merits of the weighted factor-based similarity coefficients have been reported by previous studies (Mosier and Taube 1985b; Mosier 1989; Seifoddini and Djassemi 1995). This kind of similarity coefficient attempts to adjust the strength of matches or misses between object pairs to reflect the resemblance value more realistically and accurately by incorporating object attributes.

The taxonomy can be used as an aid to identify and clarify the definition of various similarity coefficients. In the next section, we will review and map similarity coefficients related researches based on this taxonomy.

9.5 Mapping SCM Studies onto the Taxonomy

In this section, we map existing similarity coefficients onto the developed taxonomy and review academic studies through 5 tables. Tables 9.3 and 9.4 are general-purpose (*l* 1.2) similarity/dissimilarity coefficients, respectively. Table 9.5 gives expressions of some binary data-based (*l*2.1) similarity coefficients, while Table 9.6 summarizes problem-oriented (*l*1.1) similarity coefficients. Finally, SCM related academic researches are illustrated in Table 9.7.

Among the similarity coefficients in Table 9.3, eleven of them have been selected by Sarker and Islam (1999) to address the issues relating to the performance of them along with their important characteristics, appropriateness and applications to manufacturing and other related fields. They also presented numerical results to demonstrate the closeness of the eleven similarity and eight dissimilarity coefficients that are presented in Table 9.4. Romesburg (1984) and Sarker (1996) provided detailed definitions and characteristics of these eleven similarity coefficients, namely Jaccard (Romesburg 1984), Hamann (Holley and Guilford 1964), Yule (Bishop *et al.* 1975), simple matching (Sokal and Michener 1958), Sorenson (Romesburg 1984), Rogers and Tanimoto (1960), Sokal and Sneath (Romesburg 1984), Rusell and Rao (Romesburg 1984), Baroni-Urbani and Buser (1976), Phi (Romesburg 1984), and Ochiai (Romesburg 1984). In addition to these eleven similarity coefficients, Table 9.3 also introduces several other similarity coefficients, namely PSC (Waghodekar and Sahu 1984), dot-product, Kulczynski, Sokal and Sneath 2, Sokal and Sneath 4, and relative matching (Islam and Sarker 2000). Relative matching coefficient was developed

Table 9.3 Definitions and ranges of some selected general-purpose similarity coefficients (l 1.2)

Similarity coefficient	Definition S_{ij}	Range
1. Jaccard	$a/(a+b+c)$	0–1
2. Hamann	$[(a+d)-(b+c)]/[(a+d)+(b+c)]$	−1–1
3. Yule	$(ad-bc)/(ad+bc)$	−1–1
4. Simple matching	$(a+d)/(a+b+c+d)$	0–1
5. Sorenson	$2a/(2a+b+c)$	0–1
6. Rogers and Tanimoto	$(a+d)/[a+2(b+c)+d]$	0–1
7. Sokal and Sneath	$2(a+d)/[2(a+d)+b+c]$	0–1
8. Rusell and Rao	$a/(a+b+c+d)$	0–1
9. Baroni-Urbani and Buser	$[a+(ad)^{1/2}]/[a+b+c+(ad)^{1/2}]$	0–1
10. Phi	$(ad-bc)/$ $[(a+b)(a+c)(b+d)(c+d)]^{1/2}$	−1–1
11. Ochiai	$a/[(a+b)(a+c)]^{1/2}$	0–1
12. PSC	$a^2/[(b+a)*(c+a)]$	0–1
13. Dot-product	$a/(b+c+2a)$	0–1
14. Kulczynski	$1/2[a/(a+b)+a/(a+c)]$	0–1
15. Sokal and Sneath 2	$a/[a+2(b+c)]$	0–1
16. Sokal and Sneath 4	$1/4[a/(a+b)+a/(a+c)+$ $d/(b+d)+d/(c+d)]$	0–1
17. Relative matching	$[a+(ad)^{1/2}]/[a+b+c+d+(ad)^{1/2}]$	0–1

a number of parts that visit both machines
b number of parts that visit machine i but not j
c number of parts that visit machine j but not i
d number of parts that visit neither machine

recently, and considers a set of similarity properties such as no mismatch, minimum match, no match, complete match and maximum match.

Table 9.4 shows eight most commonly used general-purpose (l1.2) dissimilarity coefficients. The dissimilarity coefficient does reverse to those similarity coefficients in Table 9.1. In Table 9.4, d_{ij} is the original definition of these coefficients, in order to show the comparison more explicitly, we modify these dissimilarity coefficients and use binary data to express them. The binary data based definition is represented by d'_{ij}.

Table 9.5 presents some selected similarity coefficients in group l2.1. The expressions in Table 9.5 are similar to that of Table 9.3. However, rather than judging the similarity between two objects, problem-oriented similarity coefficients evaluate a predetermined "appropriateness" between two objects. Two objects that have the highest "appropriateness" maximize similarity value even if they are less similar than some other object pairs.

Table 9.6 is a summary of problem-oriented (l 1.1) similarity coefficients developed so far for dealing with CF problems. This table is the tabulated expression of the proposed taxonomy. Previously developed similarity coefficients are mapped into the table, additional information such as solution procedures, novel characteristics are also listed in the "Notes/Keywords" column.

Table 9.4 Definitions of some selected general-purpose dissimilarity coefficients (l 1.2)

Dissimilarity coefficient	Definition d_{ij}	Definition d'_{ij}
1. Minkowski	$\left(\sum_{k=1}^{M} \lvert a_{ki} - a_{kj} \rvert^r\right)^{1/r}$	$(b+c)^{1/r}$
2. Euclidean	$\left(\sum_{k=1}^{M} \lvert a_{ki} - a_{kj} \rvert^2\right)^{1/2}$	$(b+c)^{1/2}$
3. Manhattan (City Block)	$\sum_{k=1}^{M} \lvert a_{ki} - a_{kj} \rvert$	$b+c$
4. Average Euclidean	$\left(\sum_{k=1}^{M} \lvert a_{ki} - a_{kj} \rvert^2 / M\right)^{1/2}$	$\left(\frac{b+c}{a+b+c+d}\right)^{1/2}$
5. Weighted Minkowski	$\left(\sum_{k=1}^{M} w_k \lvert a_{ki} - a_{kj} \rvert^r\right)^{1/r}$	$[w_k(b+c)]^{1/r}$
6. Bray–Curtis	$\sum_{k=1}^{M} \lvert a_{ki} - a_{kj} \rvert / \sum_{k=1}^{M} \lvert a_{ki} + a_{kj} \rvert$	$\frac{b+c}{2a+b+c}$
7. Canberra Metric	$\frac{1}{M} \sum_{k=1}^{M} \left(\frac{\lvert a_{ki} - a_{kj} \rvert}{a_{ki} + a_{kj}}\right)$	$\frac{b+c}{a+b+c+d}$
8. Hamming	$\sum_{k=1}^{M} \delta(a_{kl}, a_{kj})$	$b+c$

$$\delta(a_{kl}, a_{kj}) = \begin{cases} 1, & \text{if } a_{kl} \neq a_{kj}; \\ 0, & \text{otherwise}. \end{cases}$$

r positive integer
d_{ij} dissimilarity between i and j
d'_{ij} dissimilarity by using binary data
k attribute index ($k = 1, \ldots, M$)

Table 9.5 Definitions and ranges of some selected problem-oriented binary data based similarity coefficients (l 2.1)

Coefficient/Resource	Definition S_{ij}	Range
1. Chandrasekharan and Rajagopalan (1986b)	$a/\text{Min}[(a+b),(a+c)]$	0–1
2. Kusiak *et al.* (1986)	a	Integer
3. Kusiak (1987)	$a+d$	Integer
4. Kaparthi *et al.* (1993)	$a'/(a+b)'$	0–1
5. MaxSC/Shafer and Rogers (1993b)	$\max[a/(a+b), a/(a+c)]$	0–1
6. Baker and Maropoulos (1997)	$a/\text{Max}[(a+b),(a+c)]$	0–1

a': number of matching ones between the matching exemplar and the input vector
$(a+b)'$: number of ones in the input vector

Table 9.6 Summary of developed problem-oriented (dis)similarity coefficients (SC) for cell formation (*l* 1.1)

No	Resource/coefficient		Production information (*l* 2.2)							Notes/keywords
						Weights (*l* 3.3)				
	Author(s)/(SC)	Year	Binary data based (*l* 2.1)	Alternative proc. (*l* 3.1)	Operation seq. (*l* 3.2)	Prod. vol. (*l* 4.1)	Oper. time (*l* 4.2)	Others (*l* 4.3)	Others (*l* 3.4)	
1	De Witte	1980				Y			MM	3 SC created; graph theory
2	Waghodekar and Sahu (PSC and SCTF)	1984	Y							*l* 1.2; 2 SC created
3	Mosier and Taube	1985b				Y				2 SC created
4	Selvam and Balasubramanian	1985			Y	Y				Heuristic
5	Chandrasekharan and Rajagopalan	1986b	Y							*l* 2.1; hierarchical algorithm
6	Dutta *et al.*	1986							CS; NC	5 D developed
7	Faber and Carter (MaxSC)	1986	Y							*l* 2.1; graph
8	Kusiak *et al.*	1986	Y							*l* 2.1; 3 distinct integer models
9	Kusiak	1987	Y							*l* 2.1; APR by p-median
10	Seifoddini	87/88			Y	Y				
11	Steudel and Ballakur	1987					Y			Dynamic programming
12	Choobineh	1988			Y					Mathematical model
13	Gunasingh and Lashkari	1989						T		Math; compatibility index
14	Wei and Kern	1989	Y							*l* 2.1; heuristic
15	Gupta and Seifoddini	1990			Y	Y	Y			Heuristic
16	Tam	1990			Y					*k* nearest neighbor
17	Vakharia and Wemmerlöv	1987; 1990			Y					Heuristic
18	Offodile	1991						Y		Parts coding and classification
19	Kusiak and Cho	1992	Y							*l* 2.1; 2 SC proposed
20	Zhang and Wang	1992							Y	Combine SC with fuzziness
21	Balasubramanian and Panneerselvam	1993			Y	Y			MHC	D; covering technique
22	Ho *et al.*	1993			Y					Compliant index
23	Gupta	1993		Y	Y	Y	Y			Heuristic
24	Kaparthi *et al.*	1993	Y							*l* 2.1; improved neural network

Table 9.6 continued

No	Resource/coefficient		Production information (l 2.2)							Notes/keywords
				Weights (l 3.3)						
	Author(s)/(SC)	Year	Binary data based (l 2.1)	Alternative proc. (l 3.1)	Operation seq. (l 3.2)	Prod. vol. (l 4.1)	Oper. time (l 4.2)	Others (l 4.3)	Others (l 3.4)	
25	Luong	1993							CS	Heuristic
26	Ribeiro and Pradin	1993	Y							D, l 1.2; knapsack
27	Seifoddini and Hsu	1994						Y		Comparative study
28	Akturk and Balkose	1996			Y					D; multi-objective model
29	Ho and Moodie (POSC)	1996							FPR	Heuristic; mathematical
30	Ho and Moodie (GOSC)	1996				Y				SC between two part groups
31	Suer and Cedeno	1996						C		
32	Viswanathan	1996	Y							l 2.1; modify p-median
33	Baker and Maropoulos	1997	Y							l 2.1; black box algorithm
34	Lee *et al.*	1997			Y	Y				APR by genetic algorithm
35	Won and Kim	1997		Y						Heuristic
36	Askin and Zhou	1998			Y					Shortest path
37	Nair and Narendran	1998			Y					Non-hierarchical
38	Jeon *et al.*	1998b		Y						Mathematical
39	Kitaoka *et al.* (double centering)	1999	Y							l 2.1; quantification model
40	Nair and Narendran	1999							WL	Mathematical; non-hierarchical
41	Nair and Narendran	1999			Y	Y			WL	Mathematical; non-hierarchical
42	Seifoddini and Tjahjana	1999							BS	
43	Sarker and Xu	2000			Y					Three-phases algorithm
44	Won	2000a		Y						Modify p-median
45	Yasuda and Yin	2001							CS	D; heuristic

APR alternative process routings
C cost of unit part
D dissimilarity coefficient
MHC material handling cost
NC number of cell
T tooling requirements of parts
BS batch size
CS cell size
FPR flexible processing routing
MM multiple machines available for a machine type
SC similarity coefficient
WL workload

Table 9.7 Literature of cell formation research in conjunction with similarity coefficients (SC)

Articles Author(s)	Year	SC used	Description/keywords
McAuley	1972	Jaccard	First study of SC on cell formation
Carrie	1973	Jaccard	Apply SC on forming part families
Rajagopalan and Batra	1975	Jaccard	Graph theory
Waghodekar and Sahu	1984	Jaccard; PSC; SCTF	Propose MCSE method
Kusiak	1985	Minkowski (D)	p-median; heuristics
Chandrasekharan and Rajagopalan	1986a	Minkowski (D)	Non-hierarchical algorithm
Han and Ham	1986	Manhattan (D)	Classification and coding system
Seifoddini and Wolfe	1986	Jaccard	Bit-level data storage technique
Chandrasekharan and Rajagopalan	1987	Manhattan (D)	Develop ZODIAC algorithm
Marcotorchino	1987	Jaccard; Sorenson	Create a block seriation model
Seifoddini and Wolfe	1987	Jaccard	Select threshold on material handling cost
Chandrasekharan and Rajagopalan	1989	Jaccard; simple matching; Manhattan (D)	An analysis of the properties of datasets
Mosier	1989	7 Similarity coefficients	Comparative study
Seifoddini	1989a	Jaccard	SLC *vs.* ALC
Seifoddini	1989b	Jaccard	Improper machine assignment
Srinivasan *et al.*	1990	Kusiak (1987)	An assignment model
Askin *et al.*	1991	Jaccard	Hamiltonian path; TSP
Chow	1991	CS	Unjustified claims of LCC
Gongaware and Ham	1991	—*	Classification and coding; multi-objective model
Gupta	1991	Gupta and Seifoddini (1990)	Comparative study on chaining effect
Logendran	1991	Jaccard; Kusiak (1987)	Identification of key machine
Srinivasan and Narendran	1991	Kusiak (1987)	A non-hierarchical clustering algorithm
Wei and Kern	1991	CS	Reply to Chow (1991)
Chow and Hawaleshka	1992	CS	Define machine unit concept
Shiko	1992	Jaccard	Constrained hierarchical
Chow and Hawaleshka	1993a	CS	Define machine unit concept
Chow and Hawaleshka	1993b	CS	A knowledge-based approach
Kang and Wemmerlöv	1993	Vakharia and Wemmerlöv (1987, 1990)	Heuristic; Alternative operations of parts
Kusiak *et al.*	1993	Hamming (D)	Branch-Bound and A* approaches

Table 9.7 continued

Articles Author(s)	Year	SC used	Description/keywords
Offodile	1993	Offodile (1991)	Survey of robotics and GT; robot selection model
Shafer and Rogers	1993a	Many	Review of similarity coefficients
Shafer and Rogers	1993b	16 Similarity coefficients	Comparative study
Vakharia and Kaku	1993	Kulczynski	Long-term demand change
Ben-Arieh and Chang	1994	Manhattan (D)	Modify p-median algorithm
Srinivasan	1994	Manhattan (D)	Minimum spanning trees
Balakrishnan and Jog	1995	Jaccard	TSP algorithm
Cheng *et al.*	1995	Hamming (D)	Quadratic model; A* algorithm
Kulkarni and Kiang	1995	Euclidean (D)	Self-organizing neural network
Murthy and Srinivasan	1995	Manhattan (D)	Heuristic; Consider fractional cell formation
Seifoddini and Djassemi	1995	Jaccard	Merits of production volume consideration
Vakharia and Wemmerlöv	1995	8 Dissimilarity coefficients	Comparative study
Wang and Roze	1995	Jaccard; Kusiak (1987), CS	An experimental study
Balakrishnan	1996	Jaccard	CRAFT
Cheng *et al.*	1996	Hamming (D)	Truncated tree search algorithm
Hwang and Ree	1996	Jaccard	Define compatibility coefficient
Lee and Garcia-Diaz	1996	Hamming (D)	Use a three-phase network-flow approach
Leem and Chen	1996	Jaccard	Fuzzy set theory
Lin *et al.*	1996	Bray-Curtis (D)	Heuristic; workload balance within cells
Sarker	1996	Many	Review of similarity coefficient
Al-Sultan and Fedjki	1997	Hamming (D)	Genetic algorithm
Askin *et al.*	1997	MaxSC	Consider flexibility of routing and demand
Baker and Maropoulos	1997	Jaccard *et al.* (1997)	Black box clustering algorithm
Cedeno and Suer	1997	—	Approach to "remainder clusters"
Masnata and Settineri	1997	Euclidean (D)	Fuzzy clustering theory
Mosier *et al.*	1997	Many	Review of similarity coefficients
Offodile and Grznar	1997	Offodile (1991)	Parts coding and classification analysis
Wang and Roze	1997	Jaccard and Kusiak (1987), CS	Modify p-median model

Table 9.7 continued

Articles Author(s)	Year	SC used	Description/keywords
Cheng *et al.*	1998	Manhattan (D)	TSP by genetic algorithm
Jeon *et al.*	1998a	Jeon *et al.* (1998b)	p-median
Onwubolu and Mlilo	1998	Jaccard	A new algorithm (SCDM)
Srinivasan and Zimmers	1998	Manhattan (D)	Fractional cell formation problem
Wang	1998	—	A linear assignment model
Ben-Arieh and Sreenivasan	1999	Euclidean (D)	A distributed dynamic algorithm
Lozano *et al.*	1999	Jaccard	Tabu search
Sarker and Islam	1999	Many	Performance study
Baykasoglu and Gindy	2000	Jaccard	Tabu search
Chang and Lee	2000	Kusiak (1987)	Multisolution heuristic
Josien and Liao	2000	Euclidean (D)	Fuzzy set theory
Lee-post	2000	Offodile (1991)	Use a simple genetic algorithm
Won	2000a	Won and Kim (1997)	Alternative process plan with p-median model
Won	2000b	Jaccard, Kusiak (1987)	Two-phase p-median model
Dimopoulos and Mort	2001	Jaccard	Genetic algorithm
Samatova *et al.*	2001	5 Dissimilarity coefficients	Vector perturbation approach

* No specific SC mentioned

Finally, Table 9.7 is a brief description of the published CF studies in conjunction with similarity coefficients. Most studies listed in this table do not develop new similarity coefficients. However, all of them use similarity coefficients as a powerful tool for coping with cell formation problems under various manufacturing situations. This table also shows the broad range of applications of similarity coefficient-based methods.

9.6 General Discussion

We give a general discussion of production information based similarity coefficients (*l* 2.2) and an evolutionary timeline in this section.

9.6.1 Production Information-based Similarity Coefficients

- Alternative process routings
  In most cell formation methods, parts are assumed to have a unique part process plan. However, it is well known that alternatives may exist in any level of a pro-

cess plan. In some cases, there may be many alternative process plans for making a specific part, especially when the part is complex (Qiao *et al.* 1994). Explicit consideration of alternative process plans invoke changes in the composition of all manufacturing cells so that lower capital investment in machines, more independent manufacturing cells and higher machine utilization can be achieved (Hwang and Ree 1996).

Gupta (1993) was the first person that incorporated alternative process routings into similarity coefficients. His similarity coefficient also includes other production information such as operation sequences, production volumes and operation times. The similarity coefficient assigns pair-wise similarity among machines with usage factors of all alternative process routings. The usage factors are determined by satisfying production and capacity constraints. The production volumes and operation times are assumed to be known with certainty.

An alternative process routings considered similarity coefficient was developed by Won and Kim (1997) and slightly modified by Won (2000a). In the definition of the similarity coefficient, if machine i is used by some process routing of part j, then the number of parts processed by machine i is counted as one for that part even if the remaining process routings of part j also use machine i. The basic idea is that in the final solution only one process routing is selected for each part. The p-median approach was used by Won (2000a) to associate the modified similarity coefficient.

A similarity coefficient that considers the number of alternative process routings when available during machine failure was proposed by Jeon *et al.* (1998b). The main characteristic of the proposed similarity coefficient is that it draws on the number of alternative process routings during machine failure when alternative process routings are available instead of drawing on operations, sequence, machine capabilities, production volumes, processing requirements and operational times. Based on the proposed similarity coefficient, the p-median approach was used to form part families.

- Operation sequences
  The operation sequence is defined as an ordering of the machines on which the part is sequentially processed (Vakharia and Wemmerlöv 1990). A lot of similarity coefficients have been developed to consider operation sequence.

Selvam and Balasubramanian (1985) are the first who incorporated alternative process routings into similarity coefficients. Their similarity coefficient is very simple and intuitive. The value of the similarity coefficient is determined directly by the production volume of parts moves between machines.

Seifoddini (1987, 1988) modified the Jaccard similarity coefficient to take into account the production volume of parts moves between machine pairs. A simple heuristic algorithm was used by the author to form machine cells.

Choobineh (1988) gave a similarity coefficient between parts j and k which is based on the common sequences of length 1 through L between the two parts. To select the value L, one has to balance the need to uncover the natural strength of the relationships among the parts and the computational efforts necessary to calculate the sequences of length 1 through L. In general, the higher the value of L, the more discriminating power similarity coefficient will have.

Gupta and Seifoddini (1990) proposed a similarity coefficient incorporating operation sequence, production volume and operation time simultaneously. From the definition, each part that is processed by at least one machine from a pair of machines contributes towards their similarity coefficient value. A part that is processed by both machines increases the coefficient value for the two machines, whereas a part that is processed on one machine tends to reduce it.

The similarity coefficient developed by Tam (1990) is based on Levenshtein's distance measure of two sentences. The distance between two sentences is defined as the minimum number of transformations required to derive one sentence from the other. Three transformations are defined. The similarity coefficient between two operation sequences x and y is defined as the smallest number of transformations required to derive y from x.

Vakharia and Wemmerlöv (1990) proposed a similarity coefficient based on operation sequences to integrate the intracellular flow with the cell formation problem by using clustering methodology. The similarity coefficient measures the proportion of machine types used by two part families in the same order.

Balasubramanian and Panneerselvam (1993) developed a similarity coefficient which needs following input data: (1) operation sequences of parts; (2) additional cell arrangements; (3) production volume per day and the bulk factor; (4) guidelines for computing excess moves; (5) actual cost per move.

The Ho *et al.* (1993) similarity coefficient calculates a compliant index first. The compliant index of the sequence of a part compared with a flow path is determined by the number of operations in the sequence of the part that have either an "in-sequence" or "by-passing" relationship with the sequence of the flow path. There are two kinds of compliant indices: forward compliant index and backward index. These two compliant indexes can be calculated by comparing the operation sequence of the part with the sequence of the flow path forwards and backwards.

As mentioned in Section 9.6.1, Gupta (1993) proposed a similarity coefficient, which incorporates several production factors such as operation sequences, production volumes, and alternative process routings.

Akturk and Balkose (1996) revised the Levenshtein distance measure to penalize the backtracking parts neither does award the commonality. If two parts have no common operations, then a dissimilarity value is found by using the penalizing factor.

The Lee *et al.* (1997) similarity coefficient takes the direct and indirect relations between the machines into consideration. The direct relation indicates that two machines are connected directly by parts, whereas the indirect relation indicates that two machines are connected indirectly by other machines.

Askin and Zhou (1998) proposed a similarity coefficient, which is based on the longest common operation subsequence between part types and used to group parts into independent, flow-line families.

Nair and Narendran (1998) gave a similarity coefficient as the ratio of the sum of the moves common to a pair of machines and the sum of the total number of moves to and from the two machines. Laterally, they extended the coefficient to incorporate the production volume of each part (Nair and Narendran 1999).

Sarker and Xu (2000) developed an operation sequence-based similarity coefficient. The similarity coefficient was applied in a p-median model to group the parts to form part families with similar operation sequences.

- Weight factors
  Weighted similarity coefficient is a logical extension or expansion of the binary data-based similarity coefficient. The two most researched weight factors are production volume and operation time.

De Witte (1980) is the first person who incorporated production volume into similarity coefficient. In order to analyze the relations between machine types, the author has used three different similarity coefficients. Absolute relations, mutual interdependence relations and relative single interdependence relations between machine pairs are defined by similarity coefficients SA, SM and SS, respectively.

The Mosier and Taube (1985b) similarity coefficient is a simple weighted adaptation of McAuley's Jaccard similarity coefficient with an additional term whose purpose is to trap the coefficient between -1.0 and $+1.0$. Production volumes of parts have been incorporated into the proposed similarity coefficient.

Ho and Moodie (1996) developed a similarity coefficient, namely the group-operation similarity coefficient (GOSC), to measure the degree of similarity between two part groups. The calculation of GOSC considers the demand quantities of parts. A part with a larger amount of demand will have a heavier weight. This is reasonable since if a part comprises the majority of a part group, then it should contribute more in the characterization of the part group it belongs to.

The operation time is considered firstly by Steudel and Ballakur (1987). Their similarity coefficient is based on the Jaccard similarity coefficient and calculates the operation time by multiplying each part's operation time by the production requirements for the part over a given period of time. Operation set-up time is ignored in the calculation since set-up times can usually be reduced after the cells are implemented. Hence set-up time should not be a factor in defining the cells initially.

Other production volume/operation time-considered studies include Selvam and Balasubramanian (1985), Seifoddini (1987/1988), Gupta and Seifoddini (1990), Balasubramanian and Panneerselvam (1993), Gupta (1993), Lee *et al.* (1997) and Nair and Narendran (1999). Their characteristics have been discussed in Sections 9.6.1 and 9.6.2.

9.6.2 Historical Evolution of Similarity Coefficients

Shafer and Rogers (1993a) delineated the evolution of similarity coefficients until the early 1990s. Based on their work and Table 9.6, we depict the historical evolution of similarity coefficients over the last three decades.

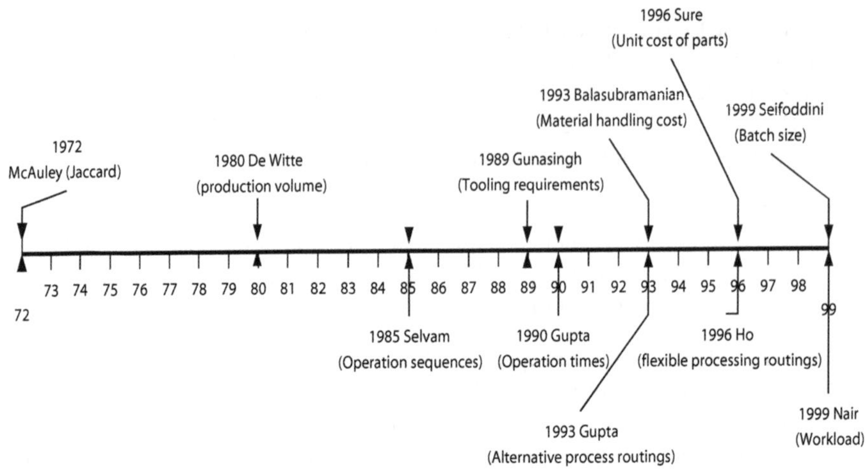

Figure 9.2 Evolutionary timeline of the similarity coefficient

McAuley (1972) was the first person who used the Jaccard similarity coefficient to form machine cells. The first weighted factor that was considered by researchers is the production volume of parts (De Witte 1980; Mosier and Taube 1985b). Operation sequences, one of the most important manufacturing factors, was incorporated in 1985 (Selvam and Balasubramanian). In the late 1980s and early 1990s, other weighted manufacturing factors such as tooling requirements (Gunasingh and Lashkari 1989) and operation times (Gupta and Seifoddini 1990) were taken into consideration.

Alternative process routings of parts is another important manufacturing factor in the design of a CF system. Although it was firstly studied by Kusiak (1987), it was not combined into the similarity coefficient definition until Gupta (1993). Material handling cost was also considered in the early 1990s (Balasubramanian and Panneerselvam 1993). In the middle of 1990s, flexible processing routings (Ho and Moodie 1996) and unit cost of parts (Sure and Cedeno 1996) were incorporated. Finally, some impressive progresses that have been achieved in the late 1990s were workload (Nair and Narendran 1999) and batch size (Seifoddini and Tjahjana 1999) consideration in the definition of similarity coefficients.

The similarity coefficient's evolutionary timeline is given in Figure 9.2.

9.7 Comparative Study of Similarity Coefficients

9.7.1 Objective

Although a large number of similarity coefficients exist in the literature, only a handful have been used for solving CF problems. Among various similarity coefficients,

Jaccard similarity coefficient (Jaccard 1908) is the most used similarity coefficient in the literature (Table 9.7). However, contradictory viewpoints among researchers have been found in previous studies: some researchers advocated the dominant power of Jaccard similarity coefficient, whereas others emphasized the drawbacks of Jaccard similarity coefficient and recommended other similarity coefficients. Moreover, several researchers believed that there is no difference between Jaccard and other similarity coefficients; they considered that none of the similarity coefficients seems to perform always well under various cell formation situations.

Therefore, a comparative research is crucially necessary to evaluate various similarity coefficients. Based on the comparative study, even if we cannot find a dominant similarity coefficient for all cell formation situations, at least we need to know which similarity coefficient is more efficient and more appropriate for some specific cell formation situations.

In this chapter, we investigate the performance of 20 well-known similarity coefficients. A large number of numerical datasets, which are taken from the open literature or generated specifically, are tested on nine performance measures.

9.7.2 Previous Comparative Studies

Four studies that have focused on comparing various similarity coefficients and related cell formation procedures have been published in the literature.

Mosier (1989) applied a mixture model experimental approach to compare seven similarity coefficients and four clustering algorithms. Four performance measures were used to judge the goodness of solutions: simple matching measure, generalized matching measure, product moment measure and intercellular transfer measure. As pointed out by Shafer and Rogers (1993), the limitation of this study is that three of the four performance measures are for measuring how closely the solution generated by the cell formation procedures matched the original machine-part matrix. However, the original machine-part matrix is not necessarily the best or even a good configuration. Only the last performance measure, intercellular transfer measure is for considering specific objectives associated with the CF problem.

Shafer and Rogers (1993) compared sixteen similarity coefficients and four clustering algorithms. Four performance measures were used to evaluate the solutions. Eleven small, binary machine-part group technology datasets mostly from the literature were used for the purpose of comparison. However, small and/or "well-structured" datasets may not have sufficient discriminatory power to separate "good" from "inferior" techniques. Further, results based on a small number of datasets may have little general reliability due to clustering results' strong dependency on the input data (Vakharia and Wemmerlöv 1995; Milligan and Cooper 1987; Anderberg 1973).

Seifoddini and Hsu (1994) introduced a new performance measure: grouping capability index (GCI). The measure is based on exceptional elements and has been widely used in subsequent research. However, only three similarity coefficients have been tested in their study.

Vakharia and Wemmerlöv (1995) studied the impact of dissimilarity measures and clustering techniques on the quality of solutions in the context of cell formation. Twenty-four binary datasets were solved to evaluate eight dissimilarity measures and seven clustering algorithms. Some important insights have been provided by this study, such as dataset characteristics, stopping parameters for clustering, performance measures, and the interaction between dissimilarity coefficients and clustering procedures. Unfortunately, similarity coefficients have not been discussed in this research.

9.8 Experimental Design

9.8.1 Tested Similarity Coefficients

Twenty well-known similarity coefficients (Table 9.8) are compared in this chapter. Among these similarity coefficients, several of them have never been studied by previous comparative researches.

Table 9.8 Definitions and ranges of selected similarity coefficients

Coefficient	Definition S_{ij}	Range
1. Jaccard	$a/(a+b+c)$	0–1
2. Hamann	$[(a+d)-(b+c)]/[(a+d)+(b+c)]$	−1–1
3. Yule	$(ad-bc)/(ad+bc)$	−1–1
4. Simple matching	$(a+d)/(a+b+c+d)$	0–1
5. Sorenson	$2a/(2a+b+c)$	0–1
6. Rogers and Tanimoto	$(a+d)/[a+2(b+c)+d]$	0–1
7. Sokal and Sneath	$2(a+d)/[2(a+d)+b+c]$	0–1
8. Rusell and Rao	$a/(a+b+c+d)$	0–1
9. Baroni-Urbani and Buser	$[a+(ad)^{1/2}]/[a+b+c+(ad)^{1/2}]$	0–1
10. Phi	$(ad-bc)/[(a+b)(a+c)(b+d)(c+d)]^{1/2}$	−1–1
11. Ochiai	$a/[(a+b)(a+c)]^{1/2}$	0–1
12. PSC	$a^2/[(b+a)*(c+a)]$	0–1
13. Dot-product	$a/(b+c+2a)$	0–1
14. Kulczynski	$1/2[a/(a+b)+a/(a+c)]$	0–1
15. Sokal and Sneath 2	$a/[a+2(b+c)]$	0–1
16. Sokal and Sneath 4	$1/4[a/(a+b)+a/(a+c)+d/(b+d)+d/(c+d)]$	0–1
17. Relative matching	$[a+(ad)^{1/2}]/[a+b+c+d+(ad)^{1/2}]$	0–1
18. Chandrasekharan and Rajagopalan (1986b)	$a/\text{Min}[(a+b),(a+c)]$	0–1
19. MaxSC	$\text{Max}[a/(a+b),a/(a+c)]$	0–1
20. Baker and Maropoulos	$a/\text{Max}[(a+b),(a+c)]$	0–1

a number of parts that visit both machines
b number of parts that visit machine i but not j
c number of parts that visit machine j but not i
d number of parts that visit neither machine

9.8.2 Datasets

It is desirable to judge the effectiveness of various similarity coefficients under varying datasets conditions. The tested datasets are classified into two distinct groups: selected from the literature and generated deliberately. Previous comparative studies used either of them to evaluate the performance of various similarity coefficients. Unlike those studies, this chapter uses both types of datasets to evaluate twenty similarity coefficients.

- Datasets selected from the literature
  In the previous comparative studies, Shafer and Rogers (1993), and Vakharia and Wemmerlöv (1995) took 11 and 24 binary datasets from the literature, respectively. The advantage of the datasets from the literature is that they stand for a variety of CF situations. In this chapter, 70 datasets are selected from the literature. Table 9.9 shows the details of the 70 datasets.
- Datasets generated deliberately
  From the computational experience with a wide variety of CF datasets, one finds that it may not always be possible to obtain a good GT solution, if the original CF problem is not amenable to well-structural dataset (Chandrasekharan and Rajagopalan 1989). Hence, it is important to evaluate the quality of solutions of various structural datasets. Using datasets that are generated deliberately is a shortcut to evaluate the GT solutions obtained by various similarity coefficients. The generation process of datasets is often controlled by using experimental factors. In this chapter, we use two experimental factors to generate datasets.

Ratio of non-zero element in cells (REC)

Density is one of the most used experimental factors (Miltenburg and Zhang 1991). However, in our opinion, density is an inappropriate factor for being used to control the generation process of cell formation datasets. We use following Figure 9.5 to illustrate this problem.

Cell formation data are usually presented in a machine-part incidence matrix such as Figure 9.3 (a). The matrix contains 0s and 1s elements that indicate the machine requirements of parts (to show the matrix clearly, 0s are usually not shown). Rows represent machines and columns represent parts. A "1" in the ith row and jth column represents that the jth part needs an operation on the ith machine; similarly, a "0" in the ith row and jth column represents the fact that the ith machine is not needed to process the jth part.

For Figure 9.3 (a), we assume that two machine-cells exist. The first cell is constructed by machines 2, 4, 1 and parts 1, 3, 7, 6, 10; the second cell is constructed by machines 3, 5 and parts 2, 4, 8, 9, 5, 11. Without loss of generality, we use Figure 9.3 (b) to represent Figure 9.3 (a). The two cells in Figure 9.3 (a) are now shown as capital letter "A", we call "A" as the inside cell region. Similarly, we call "B" as the outside cell region.

Table 9.9 Datasets from literature

Dataset	Size	Number of cells
1. Singh and Rajamani 1996	4 × 4	2
2. Singh and Rajamani 1996	4 × 5	2
3. Singh and Rajamani 1996	5 × 6	2
4. Waghodekar and Sahu 1984	5 × 7	2
5. Waghodekar and Sahu 1984	5 × 7	2
6. Chow and Hawaleshka 1992	5 × 11	2
7. Chow and Hawaleshka 1993a	5 × 13	2
8. Chow and Hawaleshka 1993b	5 × 13	2
9. Seifoddini 1989b	5 × 18	2
10. Seifoddini 1989b	5 × 18	2
11. Singh and Rajamani 1996	6 × 8	2
12. Chen *et al.* 1996	7 × 8	3
13. Boctor 1991	7 × 11	3
14. Islam and Sarker 2000	8 × 10	3
15. Seifoddini and Wolfe 1986	8 × 12	3
16. Chandrasekharan and Rajagopalan 1986a	8 × 20	2, 3
17. Chandrasekharan and Rajagopalan 1986b	8 × 20	2, 3
18. Faber and Carter 1986	9 × 9	2
19. Seifoddini and Wolfe 1986	9 × 12	3
20. Chen *et al.* 1996	9 × 12	3
21. Hon and Chi 1994	9 × 15	3
22. Selvam and Balasubramanian 1985	10 × 5	2
23. Mosier and Taube 1985a	10 × 10	3
24. Seifoddini and Wolfe 1986	10 × 12	3
25. McAuley 1972	12 × 10	3
26. Seifoddini 1989a	11 × 22	3
27. Hon and Chi 1994	11 × 22	3
28. De Witte 1980	12 × 19	2, 3
29. Irani and Khator 1986	14 × 24	4
30. Askin and Subramanian 1987	14 × 24	4
31. King 1980 (machine 6, 8 removed)	14 × 43	4, 5
32. Chan and Milner 1982	15 × 10	3
33. Faber and Carter 1986	16 × 16	2, 3
34. Sofianopoulou 1997	16 × 30	2, 3
35. Sofianopoulou 1997	16 × 30	2, 3
36. Sofianopoulou 1997	16 × 30	2, 3
37. Sofianopoulou 1997	16 × 30	2, 3
38. Sofianopoulou 1997	16 × 30	2, 3
39. Sofianopoulou 1997	16 × 30	2, 3
40. Sofianopoulou 1997	16 × 30	2, 3
41. Sofianopoulou 1997	16 × 30	2, 3
42. Sofianopoulou 1997	16 × 30	2, 3
43. Sofianopoulou 1997	16 × 30	2, 3
44. King 1980	16 × 43	4, 5
45. Boe and Cheng 1991 (mach. 1 removed)	19 × 35	4
46. Shafer and Rogers 1993	20 × 20	4
47. Shafer and Rogers 1993	20 × 20	4
48. Shafer and Rogers 1993	20 × 20	4
49. Mosier and Taube 1985b	20 × 20	3, 4
50. Boe and Cheng 1991	20 × 35	4

Table 9.9 continued

Dataset	Size	Number of cells
51. Ng 1993	20 × 35	4
52. Kumar and Kusiak 1986	23 × 20	2, 3
53. McCormick *et al.* 1972	24 × 16	6
54. Carrie 1973	24 × 18	3
55. Chandrasekharan and Rajagopalan 1989	24 × 4	7
56. Chandrasekharan and Rajagopalan 1989	24 × 40	7
57. Chandrasekharan and Rajagopalan 1989	24 × 40	7
58. Chandrasekharan and Rajagopalan 1989	24 × 40	7
59. Chandrasekharan and Rajagopalan 1989	24 × 40	7
60. Chandrasekharan and Rajagopalan 1989	24 × 40	7
61. Chandrasekharan and Rajagopalan 1989	24 × 40	7
62. McCormick *et al.* 1972	27 × 27	8
63. Carrie 1973	28 × 46	3, 4
64. Lee *et al.* 1997	30 × 40	6
65. Kumar and Vannelli 1987	30 × 41	2, 3, 9
66. Balasubramanian and Panneerselvam 1993	36 × 21	7
67. King and Nakornchai 1982	36 × 90	4, 5
68. McCormick *et al.* 1972	37 × 53	4,5,6
69. Chandrasekharan and Rajagopalan 1987	40 × 100	10
70. Seifoddini and Tjahjana 1999	50 × 22	14

There are three densities that are called problem density (PD), non-zero elements inside cells density (ID) and non-zero elements outside cells density (OD). The calculations of these densities are as follows:

$$\mathrm{PD} = \frac{\textit{total number of non-zero elements in regions } A + B}{\textit{total number of elements in regions } A + B} \tag{9.7}$$

$$\mathrm{ID} = \frac{\textit{total number of non-zero elements in regions } A}{\textit{total number of elements in regions } A} \tag{9.8}$$

$$\mathrm{OD} = \frac{\textit{total number of non-zero elements in regions } B}{\textit{total number of elements in regions } B} . \tag{9.9}$$

In the design of cellular manufacturing systems, what we are concerned about is to find out appropriate machine-part cells: the region A. In practice, region B is only a virtual region that does not exist in real job shops. For example, if Figure 9.3 (a) is applied to a real-life job shop, Figure 9.3 (c) is a possible layout. There is no region B that exists in a real-life job shop. Therefore, we conclude that region B-based densities are meaningless. Since PD and OD are based on B, this drawback weakens the quality of generated datasets in the previous comparative studies.

To overcome the above shortcoming, we introduce a ratio to replace the density used by previous researchers. The ratio is called the ratio of non-zero element in cells (REC) and is defined as follows:

$$REC = \frac{\textit{total number of non-zero elements}}{\textit{total number of elements in region } A} . \tag{9.10}$$

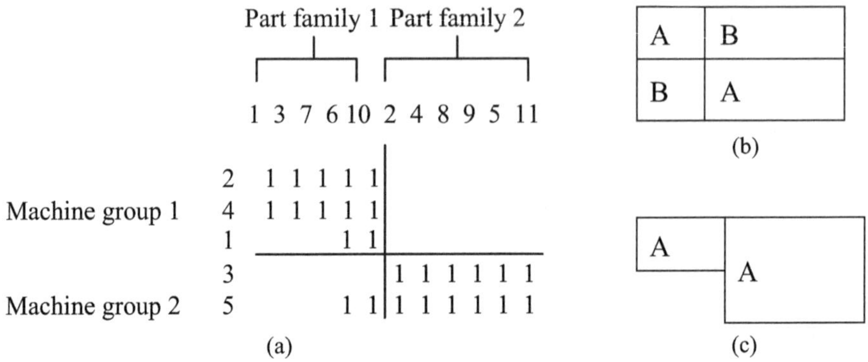

Figure 9.3 (a–c) Illustration of three densities used by previous studies

The definition is intuitive. REC can also be used to estimate the productive capacity of machines. If REC is bigger than 1, current machine capacity cannot respond to the productive requirements of parts. Thus, additional machines need to be considered. Therefore, REC can be used as a sensor to assess the capacity of machines.

Ratio of exceptions (RE)

The second experimental factor is the ratio of exceptions (RE). An exception is defined as a "1" in the region B (an operation outside the cell). We define RE as follows:

$$RE = \frac{\textit{total number of non-zero elements in region B}}{\textit{total number of non-zero elements}} . \tag{9.11}$$

RE is used to judge the "goodness" of machine-part cells and distinguish well-structured problems from ill-structured problems.

In this chapter, three levels of REC, from sparse cells (0.70) to dense cells (0.90), and eight levels of RE, from well-structured cells (0.05) to ill-structured cells (0.40), are examined. Certain 24 (3 × 8) combinations exist for all levels of the two experimental factors. For each combination, five 30 × 60-sized (30 machines by 60 parts) problems are generated. The generation process of the five problems is similar to using the random number. Therefore, a total of 120 test problems for all 24 combinations are generated, where each problem is made up of six equally sized cells. The levels of REC and RE are shown in Table 9.10.

Table 9.10 Test levels of REC and RE

Level	1	2	3	4	5	6	7	8
REC	0.70	0.80	0.90	–	–	–	–	–
RE	0.05	0.10	0.15	0.20	0.25	0.30	0.35	0.40

9.8.3 Clustering Procedure

The most well-known clustering procedures that have been applied to cell formation are single linkage clustering (SLC) algorithm, complete linkage clustering (CLC) algorithm and average linkage clustering (ALC) algorithm. These three procedures have been investigated by a large number of studies. A summary of the past comparative results is shown in Table 9.11.

Table 9.11 Comparative results of SLC, ALC and CLC

Procedure	Advantage	Drawback
SLC	Simplicity; minimal computational requirement; tends to minimize the degree of adjusted machine duplication (Vakharia and Wemmerlöv 1995).	Largest tendency to chain; leads to the lowest densities and the highest degree of single part cells (Seifoddini 1989a; Gupta 1991; Vakharia and Wemmerlöv 1995).
CLC	Simplicity; minimal computational requirement (does the reverse of SLC).	Performed as the worst procedure (Vakharia and Wemmerlöv 1995; Yasuda and Yin 2001).
ALC	Performed as the best procedure; produces the lowest degree of chaining; leads to the highest cell densities; indifferent to choice of similarity coefficients; few single part cells (Tarsuslugil and Bloor 1979; Seifoddini 1989a; Vakharia and Wemmerlöv 1995; Yasuda and Yin 2001).	Requires the highest degree of machine duplication; requires more computation (Vakharia and Wemmerlöv 1995).

Due to that ALC has the advantage of showing the greatest robustness regardless of similarity coefficients, in this section, we select ALC as the clustering algorithm to evaluate the twenty similarity coefficients (Table 9.8).

The ALC algorithm usually works as follows:

Step 1 Compute similarity coefficients for all machine pairs and store the values in a similarity matrix.

Step 2 Join the two most similar objects (two machines, a machine and a machine group or two machine groups) to form a new machine group.

Step 3 Evaluate the similarity coefficient between the new machine group and other remaining machine groups (machines) as follows:

$$S_{tv} = \frac{\sum\limits_{i \in t} \sum\limits_{j \in v} S_{ij}}{N_t N_v} \tag{9.12}$$

where i is the machine in the machine group t; j is the machine in the machine group v; and N_t is the number of machines in group t. N_v is the number of machines in group v.

Step 4 When all machines are grouped into a single machine group, or predetermined number of machine groups has been obtained, go to step 5; otherwise, go back to step 2.

Step 5 Assign each part to the cell, in which the total number of exceptions is minimum.

9.8.4 Performance Measures

A number of quantitative performance measures have been developed to evaluate the final cell formation solutions. Sarker and Mondal (1999) reviewed and compared various performance measures.

Nine performance measures are used in this study to judge final solutions. These measures provide different viewpoints by judging solutions from different aspects.

1. Number of exceptional elements (EE)
   Exceptional elements are the source of intercellular movements of parts. One objective of cell formation is to reduce the total cost of material handling. Therefore, EE is the most simple and intuitive measure for evaluating the cell formation solution.
2. Grouping efficiency
   Grouping efficiency is one of the first measures developed by Chandrasekharan and Rajagopalan (1986a,b). Grouping efficiency is defined as a weighted average of two efficiencies η_1 and η_2:

   $$\eta = w\eta_1 + (1 - w)\eta_2 \tag{9.13}$$

   where

   $$\eta_1 = \frac{o - e}{o - e + v}$$
   $$\eta_2 = \frac{MP - o - v}{MP - o - v + e} .$$

   M is defined as the number of machines, P the number of parts, o the number of operations (1s) in the machine-part matrix $\{a_{ik}\}$, e the number of exceptional elements in the solution, and v the number of voids in the solution.
   A value of 0.5 is recommended for w. η_1 is defined as the ratio of the number of 1s in the region A (Figure 9.3 (b)) to the total number of elements in the region A (both 0s and 1s). Similarly, η_2 is the ratio of the number of 0s in the region B to the total number of elements in the region B (both 0s and 1s). The weighting factor allows the designer to alter the emphasis between utilization and intercellular movement. The efficiency ranges from 0 to 1.
   Group efficiency has been reported to have a lower discriminating power (Chandrasekharan and Rajagopalan 1987). Even an extremely bad solution with large number of exceptional elements has an efficiency value as high as 0.77.

3. Group efficacy
   To overcome the problem of group efficiency, Kumar and Chandrasekharan (1990) introduced a new measure, group efficacy.

$$\tau = (1-\varphi)/(1+\phi) \tag{9.14}$$

   where φ is the ratio of the number of exceptional elements to the total number of elements; ϕ is the ratio of the number of 0s in the region A to the total number of elements.
4. Machine utilization index (grouping measure, GM)
   This was first proposed by Miltenburg and Zhang (1991), which is used to measure machine utilization in a cell. The index is defined as follows:

$$\eta_g = \eta_u - \eta_m \tag{9.15}$$

   where $\eta_u = d/(d+v)$ and $\eta_m = 1-(d/o)$. d is the number of 1s in the region A, η_u is the measure of utilization of machines in a cell, and η_m is the measure of intercellular movements of parts. η_g ranges from -1 to 1, and η_u and η_m range from 0 to 1. A bigger value of machine utilization index η_g is desired.
5. Clustering measure (CM)
   This measure tests how closely the 1s gather around the diagonal of the solution matrix, the definition of the measure is as follows (Singh and Rajamani 1996).

$$\eta_c = \frac{\left\{\sum_{i=1}^{M}\sum_{k=1}^{P}\left(\sqrt{\delta_h^2(a_{ik}) + \delta_v^2(a_{ik})}\right)\right\}}{\sum_{i=1}^{M}\sum_{k=1}^{P} a_{ik}} \tag{9.16}$$

   where $\delta_h(a_{ik})$ and $\delta_v(a_{ik})$ are horizontal and vertical distances between a non-zero entry a_{ik} and the diagonal, respectively.

$$\delta_h = i - \frac{k(M-1)}{(P-1)} - \frac{(P-M)}{(P-1)} \tag{9.17}$$

$$\delta_v = k - \frac{i(P-1)}{(M-1)} - \frac{(P-M)}{(M-1)}. \tag{9.18}$$

6. Grouping index (GI)
   Nair and Narendran (1996) indicated that a good performance measure should be defined with reference to the block diagonal space. And the definition should ensure equal weigh to voids (0s in the region A) and exceptional elements. They introduced a measure, incorporating the block diagonal space, weighting factor and correction factor.

$$\gamma = \frac{1 - \dfrac{qv + (1-q)(e-A)}{B}}{1 + \dfrac{qv + (1-q)(e-A)}{B}} \tag{9.19}$$

where B is the block diagonal space and q is a weighting factor ranges between 0 and 1. $A = 0$ for $e \leq B$ and $A = e - B$ for $e > B$. For convenience, Equation 9.19 could be written as follows:

$$\gamma = \frac{1-\alpha}{1+\alpha} \tag{9.20}$$

where

$$\alpha = \frac{qv + (1-q)(e-A)}{B} \tag{9.21}$$

and both α and γ range from 0 to 1.

7. Bond energy measure (BEM)
McCormick *et al.* (1972) used the BEM to convert a binary matrix into a block diagonal form. This measure is defined as follows:

$$\eta_{BE} = \frac{\sum_{i=1}^{M}\sum_{k=1}^{P-1} a_{ik}a_{i(k+1)} + \sum_{i=1}^{M-1}\sum_{k=1}^{P} a_{ik}a_{(i+1)k}}{\sum_{i=1}^{M}\sum_{k=1}^{P} a_{ik}} . \tag{9.22}$$

Bond energy is used to measure the relative "clumpiness" of a clustered matrix. Therefore, the closer the 1s are, the larger the bond energy measure will be.

8. Grouping capability index (GCI)
Hsu (1990) showed that neither group efficiency nor group efficacy is consistent in predicting the performance of a cellular manufacturing system based on the structure of the corresponding machine-part matrix (Seifoddini and Djassemi 1996). Hsu (1990) considered the GCI as follows:

$$\text{GCI} = 1 - \frac{e}{o} . \tag{9.23}$$

Unlike group efficiency and group efficacy, GCI excludes zero entries from the calculation of grouping efficacy.

9. Alternative routing grouping efficiency (ARG efficiency)
ARG was propose by Sarker and Li (1998). ARG evaluates the grouping effect in the presence of alternative routings of parts. The efficiency is defined as follows:

$$\eta_{\text{ARG}} = \frac{\left(1-\frac{e}{o'}\right)\left(1-\frac{v}{z'}\right)}{\left(1+\frac{e}{o'}\right)\left(1+\frac{v}{z'}\right)} = \left(\frac{o'-e}{o'+e}\right)\left(\frac{z'-v}{z'+v}\right) \tag{9.24}$$

where o' is the total number of 1s in the original machine-part incidence matrix with multiple process routings, and z' is the total number of 0s in the original machine-part incidence matrix with multiple process routings. ARG efficiency can also be used to evaluate CF problems that have no multiple process routings of parts. The efficiency ranges from 0 to 1 and is independent of the size of the problem.

9.9 Comparison and Results

Two key characteristics of similarity coefficients are tested in this study: discriminability and stability. In this study, we compare the similarity coefficients by using the following steps.

Comparative steps

1. Computation

1.1. At first, solve each problem in the datasets by using 20 similarity coefficients; compute performance values by nine performance measures. Thus, we obtain at least a total of $\delta \times 20 \times 9$ solutions. δ is the number of the problems (some datasets from literature are multiple problems due to the different number of cells, see the item NC of Table 9.6).

1.2. Average performance values matrix: create a matrix whose rows are problems and columns are nine performance measures. An element in row i and column j indicates, for problem i and performance measure j, the average performance value produced by 20 similarity coefficients.

2. Based on the results of step 1, construct two matrices whose rows are 20 similarity coefficients and columns are nine performance measures, an entry SM_{ij} in the matrixes indicates:

2.1. Discriminability matrix: the number of problems to which the similarity coefficient i gives the best performance value for measure j.

2.2. Stability matrix: the number of problems to which the similarity coefficient i gives the performance value of measure j with at least average value (better or equal than the value in the matrix of step 1.2).

3. For each performance measure, find the top 5 values in the above two matrices. The similarity coefficients corresponding to these values are considered to be the most discriminable/stable similarity coefficients for this performance measure.

4. Based on the results of step 3, for each similarity coefficient, find the number of times that it has been selected as the most discriminable/stable coefficient for the total nine performance measures.

We use small examples here to show the comparative steps.

Step 1.1. A total of 214 problems were solved. One hundred and twenty problems were deliberately generated; 94 problems were from the literature, see Table 9.4 (some datasets were multiple problems due to the different number of cells). A total of 38,520 (214 × 20 × 9) performance values were obtained by using 20 similarity coefficients and nine performance measures. For example, by using the Jaccard similarity coefficient, the nine performance values of the problem McCormick *et al.* (1972, item 62 in Table 9.9) are as follows in Table 9.12.

Step 1.2. The average performance values matrix contained 214 problems (rows) and nine performance measures (columns). An example of a row (problem in McCormick *et al.* 1972) is in Table 9.13.

We use the Jaccard similarity coefficient and the 94 problems from literature to explain the following steps 2, 3, and 4.

Table 9.12 The performance values of McCormick *et al.* (1972) by using Jaccard similarity coefficient

	EE	Grouping efficiency	Group efficacy	GM	CM	GI	BEM	GCI	ARG
Jaccard	87	0.74	0.45	0.25	7.85	0.44	1.07	0.6	0.32

Table 9.13 The average performance values of 20 similarity coefficients, for the problem McCormick *et al.* (1972)

	EE	Grouping efficiency	Group efficacy	GM	CM	GI	BEM	GCI	ARG
Average values	94.7	0.77	0.45	0.28	8.06	0.4	1.06	0.57	0.31

Step 2.1. (Discriminability matrix) Among the 94 problems and for each performance measure, the numbers of problems to which Jaccard gave the best values are shown in Table 9.14. For example, the 60 in the column EE means that comparing with other 19 similarity coefficients, Jaccard produced minimum exceptional elements to 60 problems.

Table 9.14 The number of problems to which Jaccard gave the best performance values

	EE	Grouping efficiency	Group efficacy	GM	CM	GI	BEM	GCI	ARG
Jaccard	60	51	55	62	33	65	41	60	57

Step 2.2. (Stability matrix) Among the 94 problems and for each performance measure, the numbers of problems to which Jaccard gave the value with at least average value (matrix of step 1.2) are shown in Table 9.15. For example, the meaning of 85 in the column EE is as follows: comparing with the average exceptional elements of 94 problems in the matrix of step 1.2, the number of problems to which Jaccard produced fewer exceptional elements is 85.

Table 9.15 The number of problems to which Jaccard gave the best performance values

	EE	Grouping efficiency	Group efficacy	GM	CM	GI	BEM	GCI	ARG
Jaccard	85	85	85	89	69	91	75	88	73

Step 3. For example, for the exceptional elements, the similarity coefficients that corresponded to the top 5 values in the discriminability matrix are Jaccard, Sorenson, Rusell and Rao, Dot-product, Sokal and Sneath 2, Relative matching, and

Baker and Maropoulos. These similarity coefficients are considered as the most discriminable coefficients for the performance measure – exceptional elements. The same procedures are conducted on the other performance measures and stability matrix.

Step 4. Using the results of step 3, Jaccard has been selected 5 out of 6 times as the most discriminable/stable similarity coefficient. That means, among nine performance measures, Jaccard is the most discriminable/stable similarity coefficient for 5 out of 6 performance measures. The result is shown in the "Literature" column in Table 9.16.

The results are shown in Table 9.16 and Figures 9.4, 9.5, and 9.6 (in the figures, the horizontal axes are 20 similarity coefficients and the vertical axes are nine performance measures). The tables and figures show the number of performance measures for which these 20 similarity coefficients have been regarded as the most discriminable/stable coefficients. The columns of the table represent different conditions of datasets. The column "Literature" includes all 94 problems from literature; the column ""“all random" includes all 120 deliberately generated problems. The deliberately generated problems are further investigated by using different levels of REC and RE.

Table 9.16 Comparative results under various conditions

No.	Similarity coefficient	Literature		All random		REC 0.7		0.8		0.9		RE 0.05–0.15		0.2–0.3		0.35–0.4	
		D	S	D	S	D	S	D	S	D	S	D	S	D	S	D	S
1	Jaccard	5	6	6	9	8	9	8	9	9	9	9	9	9	9	8	9
2	Hamann	0	0	2	1	1	1	2	3	7	7	9	9	1	0	2	2
3	Yule	4	4	2	6	3	7	5	7	7	8	9	9	2	6	6	7
4	Simple matching	0	0	2	0	1	0	3	5	6	8	9	9	0	0	2	2
5	Sorenson	6	4	9	8	7	9	8	9	9	9	9	9	9	9	7	7
6	Rogers and Tanimoto	0	0	2	1	2	2	4	4	6	7	9	9	1	2	2	2
7	Sokal and Sneath	0	0	0	0	2	1	5	6	6	8	9	9	1	1	2	2
8	Rusell and Rao	4	4	5	3	5	5	9	8	8	6	9	9	9	8	6	6
9	Baroni-Urban and Buser	5	6	1	3	3	7	9	7	7	8	9	9	4	7	2	6
10	Phi	5	5	6	6	9	7	8	8	7	8	9	9	9	8	7	7
11	Ochiai	1	4	8	7	9	7	8	8	9	9	9	9	9	9	7	7
12	PSC	2	2	9	8	9	9	9	8	9	9	9	9	9	9	8	9
13	Dot-product	3	5	9	8	7	9	8	9	9	9	9	9	9	9	7	7
14	Kulczynski	2	5	8	7	8	8	8	8	9	9	9	9	9	9	7	7
15	Sokal and Sneath 2	4	5	6	8	9	9	7	9	9	9	9	9	9	9	9	9
16	Sokal and Sneath 4	5	5	7	6	8	7	8	8	7	8	9	9	8	8	7	7
17	Relative matching	5	4	4	8	7	9	9	9	9	9	9	9	5	9	6	8
18	Chandrasekharan and Rajagopalan	2	5	8	6	9	8	8	8	7	7	9	9	9	9	6	7
19	MaxSC	1	4	8	6	9	8	8	8	7	7	9	9	9	9	6	7
20	Baker and Maropoulos	5	3	6	9	7	9	8	9	9	9	9	9	6	9	6	8

D discriminability
S stability

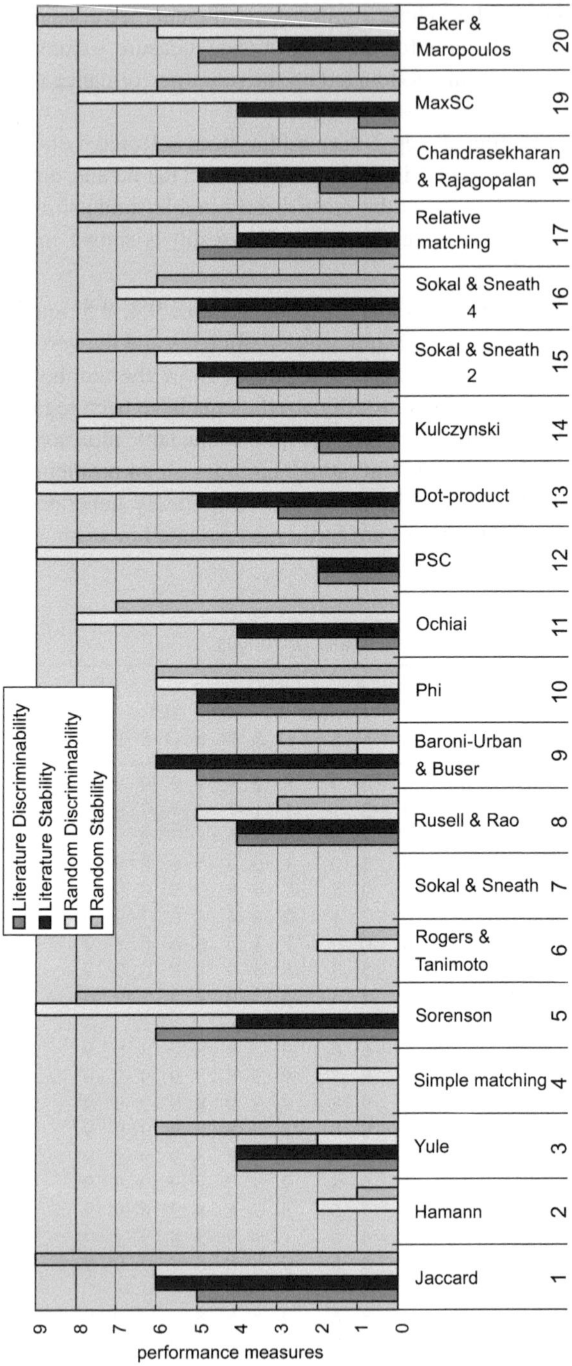

Figure 9.4 Performance for all tested problems

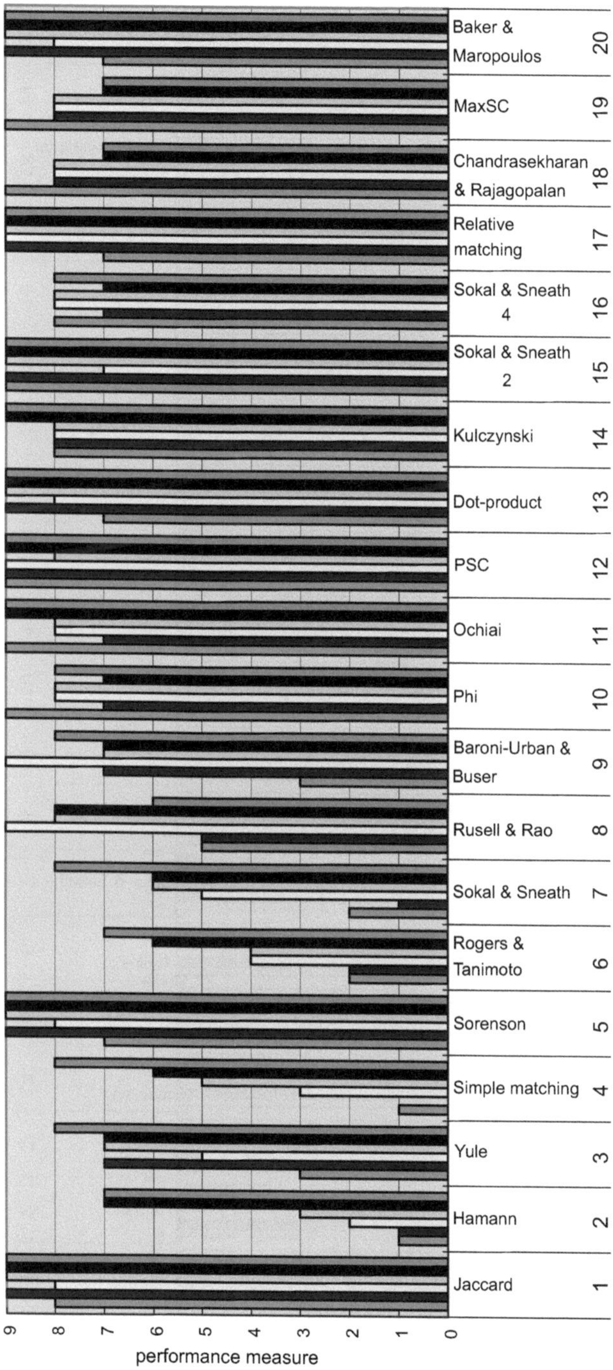

Figure 9.5 Performance under different REC

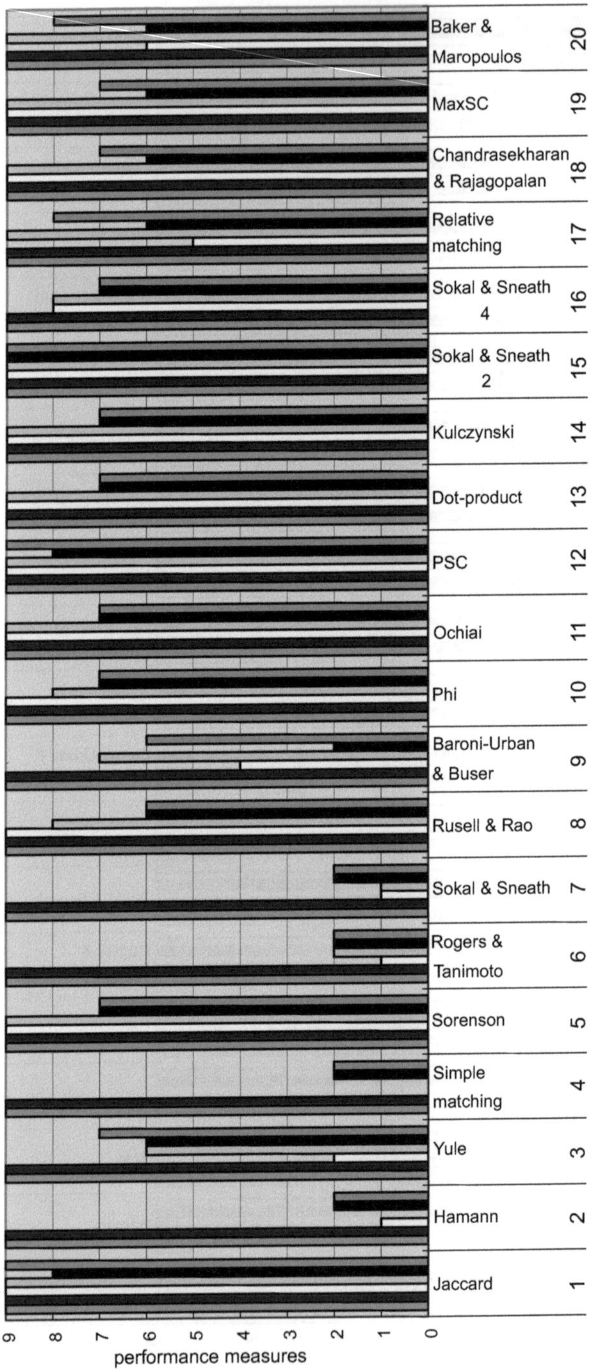

Figure 9.6 Performance under different RE

"Literature" and "All random" columns in Table 9.16 (also Figure 9.4) give the performance results of all 214 tested problems. We can find that Jaccard and Sorenson are the two best coefficients. On the other hand, four similarity coefficients, Hamann, Simple matching, Rogers and Tanimoto, and Sokal and Sneath are inefficient in both discriminability and stability.

The REC columns in Table 9.11 (also Table 9.2) show the performance results under the condition of different REC ratios. We can find that almost all similarity coefficients perform well under a high REC ratio. However, four similarity coefficients, Hamann, Simple matching, Rogers and Tanimoto, and Sokal and Sneath, again produce bad results under the low REC ratio.

The RE columns in Table 9.14 (also Figure 9.6) give the performance results under the condition of different RE ratios. All similarity coefficients perform best under a low RE ratio (datasets are well-structured). Only a few similarity coefficients perform well under a high RE ratio (datasets are ill-structured), Sokal and Sneath 2 is very good for all RE ratios. Again, the four similarity coefficients mentioned above perform badly under high RE ratios.

In summary, three similarity coefficients, Jaccard, Sorenson, and Sokal and Sneath 2 perform best among 20 tested similarity coefficients. Jaccard emerges from the 20 similarity coefficients for its stability. For all problems, from the literature or deliberately generated and for all levels of both REC and RE ratios, the Jaccard similarity coefficient is constantly the most stable coefficient among all 20 similarity coefficients. Another finding in this study is four similarity coefficients, Hamann, Simple matching, Rogers and Tanimoto, and Sokal and Sneath are inefficient under all conditions. Therefore, these similarity coefficients are not recommendable for use in cell formation applications.

9.10 Conclusions

In this chapter various similarity coefficients to the cell formation problem were investigated and reviewed. Previous review studies were discussed and the need for this review was identified. The reason why the similarity coefficient-based method (SCM) is more flexible than other cell formation methods was explained through a simple example. We also proposed a taxonomy which is combined by two distinct dimensions. The first dimension is the general-purpose similarity coefficients and the second is the problem-oriented similarity coefficients. The difference between two dimensions is discussed through three similarity coefficients. Based on the framework of the proposed taxonomy, existing similarity (dissimilarity) coefficients developed so far were reviewed and mapped onto the taxonomy. The details of each production information-based similarity coefficient were simply discussed and an evolutionary timeline was drawn based on reviewed similarity coefficients. Although a number of similarity coefficients have been proposed, very few comparative studies have been done to evaluate the performance of various similarity coefficients. This chapter evaluated the performance of 20 well-known similarity coefficients; 94 problems from literature and 120 problems generated deliberately were solved by using the 20 similarity coefficients. To control the generation process of datasets, experimental factors have been discussed. Two experimental factors were proposed and used for generating experimental problems. Nine performance measures were used to judge the solutions of the tested problems. The numerical results showed that three similarity coefficients are more efficient and four similarity coefficients are inefficient for solving the cell formation problems. Another finding is that the Jaccard similarity coefficient is the most stable similarity coefficient. For further studies, we suggest comparative studies in consideration of some production factors, such as production volumes, operation sequences, *etc.*, of parts.

References

Agrawal A, Sarkis J (1998) A review and analysis of comparative performance studies on functional and cellular manufacturing layouts. Comput Ind Eng 34:77–89

Akturk MS, Balkose HO (1996) Part-machine grouping using a multi-objective cluster analysis. Int J Prod Res 34:2299–2315

Al-Sultan KS, Fedjki CA (1997) A genetic algorithm for the part family formation problem. Prod Plan Control 8:788–796

Anderberg MR (1973) Cluster analysis for applications. Academic, New York

Arthanari TS, Dodge Y (1981) Mathematical programming in statistics. Wiley, New York

Askin RG, Cresswell SH, Goldberg JB, Vakharia AJ (1991) A Hamiltonian path approach to reordering the part-machine matrix for cellular manufacturing. Int J Prod Res 29:1081–1100

Askin RG, Selim HM, Vakharia AJ (1997) A methodology for designing flexible cellular manufacturing systems. IIE Trans 29:599–610

Askin RG, Subramanian SP (1987) A cost-based heuristic for group technology configuration. Int J Prod Res 25(1):101–113

Askin RG, Zhou M (1998) Formation of independent flow-line cells based on operation requirements and machine capabilities. IIE Trans 30:319–329

Baker RP, Maropoulos PG (1997) An automatic clustering algorithm suitable for use by a computer-based tool for the design, management and continuous improvement of cellular manufacturing systems. Comput Integr Manuf Syst 10:217–230

Balakrishnan J (1996) Manufacturing cell formation using similarity coefficients and pair-wise interchange: formation and comparison. Prod Plan Control 7:11–21

Balakrishnan J, Cheng CH (1998) Dynamic layout algorithms: a state-of-the-art survey. Omega 26:507–521

Balakrishnan J, Jog PD (1995) Manufacturing cell formation using similarity coefficients and a parallel genetic TSP algorithm: formulation and comparison. Math Comput Model 21:61–73

Balasubramanian KN, Panneerselvam R (1993) Covering technique-based algorithm for machine grouping to form manufacturing cells. Int J Prod Res 31:1479–1504

Baroni-Urbani C, Buser MW (1976) Similarity of binary data. Syst Zoo 25:251–259

Baykasoglu A, Gindy NNZ (2000) MOCACEF 1.0: multiple objective capability based approach to form part-machine groups for cellular manufacturing applications. Int J Prod Res 38:1133–1161

Beatty CA (1992) Implementing advanced manufacturing technologies: rules of the road. Sloan Manage Rev Summer:49–60

Ben-Arieh D, Chang PT (1994) An extension to the p-median group technology algorithm. Comput Oper Res 21:119–125

Ben-Arieh D, Sreenivasan R (1999) Information analysis in a distributed dynamic group technology method. Int J Prod Econ 60–61:427–432
Bijnen EJ (1973) Cluster analysis. Tilburg University Press, The Netherlands
Bishop YMM, Fienberg SE, Holland PW (1975) Discrete multivariate analysis: theory and practice. MIT Press, Cambridge MA
Boctor FF (1991) A linear formulation of the machine-part cell formation problem. Int J Prod Res 29(2):343–356
Boe WJ, Cheng CH (1991) A close neighbour algorithm for designing cellular manufacturing systems. Int J Prod Res 29(10):2097–2116
Burbidge JL (1971) Production flow analysis. Prod Eng 50:139–152
Burbidge JL, Falster P, Rhs JO (1991) Why is it difficult to sell group technology and just-in-time to industry? Prod Plan Control 2:160–166
Carrie AS (1973) Numerical taxonomy applied to group technology and plant layout. Int J Prod Res 11:399–416
Cedeno AA, Suer GA (1997) The use of a similarity coefficient-based method to perform clustering analysis to a large set of data with dissimilar parts. Comput Ind Eng 33:225–228
Chan HM, Milner DA (1982) Direct clustering algorithm for group formation in cellular manufacture. J Manuf Syst 1(1):65–75
Chandrasekharan MP, Rajagopalan R (1986a) An ideal seed non-hierarchical clustering algorithm for cellular manufacturing. Int J Prod Res 24:451–464
Chandrasekharan MP, Rajagopalan R (1986b) MODROC: an extension of rank order clustering for group technology. Int J Prod Res 24:1221–1233
Chandrasekharan MP, Rajagopalan R (1987) ZODIAC: an algorithm for concurrent formation of part families and machine cells. Int J Prod Res 25:451–464
Chandrasekharan MP, Rajagopalan R (1989) GROUPABILITY: an analysis of the properties of binary data matrices for group technology. Int J Prod Res 27:1035–1052
Chang PT, Lee ES (2000) A multisolution method for cell formation – exploring practical alternatives in group technology manufacturing. Comput Math Appl 40:1285–1296
Chen DS, Chen HC, Part JM (1996) An improved ART neural net for machine cell formation. J Mater Process Technol 61:1–6
Cheng CH, Goh CH, Lee A (1995) A two-stage procedure for designing a group technology system. Int J Oper Prod Manage 15:41–50
Cheng CH, Gupta YP, Lee WH, Wong KF (1998) A TSP-based heuristic for forming machine groups and part families. Int J Prod Res 36:1325–1337
Cheng CH, Madan MS, Motwani J (1996) Designing cellular manufacturing systems by a truncated tree search. Int J Prod Res 34:349–361
Choobineh F (1988) A framework for the design of cellular manufacturing systems. Int J Prod Res 26:1161–1172
Choobineh F, Nare A (1999) The impact of ignored attributes on a CMS design. Int J Prod Res 37:3231–3245
Chow WS (1991) Discussion: a note on a linear cell clustering algorithm. Int J Prod Res 29:215–216
Chow WS, Hawaleshka O (1992) An efficient algorithm for solving the machine chaining problem in cellular manufacturing. Comput Ind Eng 22:95–100
Chow WS, Hawaleshka O (1993a) Minimizing intercellular part movements in manufacturing cell formation. Int J Prod Res 31:2161–2170
Chow WS, Hawaleshka O (1993b) A novel machine grouping and knowledge-based approach for cellular manufacturing. Eur J Oper Res 69:357–372
Chu CH (1989) Cluster analysis in manufacturing cellular formation. Omega 17:289–295
Chu CH, Pan P (1988) The use of clustering techniques in manufacturing cellular formation. In: Proceedings of the International Industrial Engineering Conference, Orlando, Florida, pp 495–500

Chu CH, Tsai M (1990) A comparison of three array-based clustering techniques for manufacturing cell formation. Int J Prod Res 28(8):1417–1433

De Witte J (1980) The use of similarity coefficients in production flow analysis. Int J Prod Res 18:503–514

Dimopoulos C, Mort N (2001) A hierarchical clustering methodology based on genetic programming for the solution of simple cell-formation problems. Int J Prod Res 39:1–19

Dutta SP, Lashkari RS, Nadoli G, Ravi T (1986) A heuristic procedure for determining manufacturing families from design-based grouping for flexible manufacturing systems. Comput Ind Eng 10:193–201

Faber Z, Carter MW (1986) A new graph theory approach for forming machine cells in cellular production systems. In: A Kusiak (ed) Flexible manufacturing systems: methods and studies. North-Holland: Elsevier Science, Amsterdam, pp 301–315

Fazakerley GM (1976) A research report on the human aspects of group technology and cellular manufacture. Int J Prod Res 14:123–134

Gongaware TA, Ham I (1991) Cluster analysis applications for group technology manufacturing systems. In: Proceedings of the 9th North American Manufacturing Research Conference, pp 503–508

Gordon AD (1999) Classification, 2nd edn. Chapman and Hall, London

Gunasingh KR, Lashkari RS (1989) The cell formation problem in cellular manufacturing systems – a sequential modeling approach. Comput Ind Eng 16:469–476

Gupta T (1991) Clustering algorithms for the design of a cellular manufacturing system – an analysis of their performance. Comput Ind Eng 20:461–468

Gupta T (1993) Design of manufacturing cells for flexible environment considering alternative routeing. Int J Prod Res 31:1259–1273

Gupta T, Seifoddini H (1990) Production data based similarity coefficient for machine-component grouping decisions in the design of a cellular manufacturing system. Int J Prod Res 28:1247–1269

Han C, Ham I (1986) Multiobjective cluster analysis for part family formations. J Manuf Syst 5:223–230

Ho YC, Lee C, Moodie CL (1993) Two sequence-pattern, matching-based, flow analysis methods for multi-flowlines layout design. Int J Prod Res 31:1557–1578

Ho YC, Moodie CL (1996) Solving cell formation problems in a manufacturing environment with flexible processing and routeing capabilities. Int J Prod Res 34:2901–2923

Holley JW, Guilford JP (1964) A note on the G index of agreement. Edu Psycho Measure 24:749–753

Hon KKB, Chi H (1994) A new approach of group technology part families optimization. Ann CIRP 43(1):425–428

Hsu CP (1990) Similarity coefficient approaches to machine-component cell formation in cellular manufacturing: a comparative study. PhD thesis. Department of Industrial and Manufacturing Engineering, University of Wisconsin-Milwaukee

Hwang H, Ree P (1996) Routes selection for the cell formation problem with alternative part process plans. Comput Ind Eng 30:423–431

Irani SA, Khator SK (1986) A microcomputer-based design of a cellular manufacturing system. In: Proceedings of the 8th Annual Conference on Computers and Industrial Engineering, vol 11, pp 68–72

Islam KMS, Sarker BR (2000) A similarity coefficient measure and machine-parts grouping in cellular manufacturing systems. Int J Prod Res 38:699–720

Jaccard P (1908) Novelles recgerches sur la distribution florale. Bull Soc Vaud Sci Nat 44:223–270

Jeon G, Broering M, Leep HR, Parsaei HR, Wong JP (1998a) Part family formation based on alternative routes during machine failure. Comput Ind Eng 35:73–76

Jeon G, Leep HR, Parsaei HR (1998b) A cellular manufacturing system based on new similarity coefficient which considers alternative routes during machine failure. Comput Ind Eng 34:21–36

Josien K, Liao TW (2000) Integrated use of fuzzy c-means and fuzzy KNN for GT part family and machine cell formation. Int J Prod Res 38:3513–3536

Kamrani AK, Parsaei HR, Chaudhry MA (1993) A survey of design methods for manufacturing cells. Comput Ind Eng 25:487–490

Kang SL, Wemmerlöv U (1993) A work load-oriented heuristic methodology for manufacturing cell formation allowing reallocation of operations. Eur J Oper Res 69:292–311

Kaparthi S, Suresh NC, Cerveny RP (1993) An improved neural network leader algorithm for part-machine grouping in group technology. Eur J Oper Res 69:342–356

King JR (1980) Machine-component grouping in production flow analysis: an approach using a rank order clustering algorithm. Int J Prod Res 18(2):213–232

King JR, Nakornchai V (1982) Machine component group formation in group technology: review and extension. Int J Prod Res 20:117–133

Kitaoka M, Nakamura R, Serizawa S, Usuki J (1999) Multivariate analysis model for machine-part cell formation problem in group technology. Int J Prod Econ 60–61:433–438

Kulkarni UR, Kiang MY (1995) Dynamic grouping of parts in flexible manufacturing systems – a self-organizing neural networks approach. Eur J Oper Res 84:192–212

Kumar CS, Chandrasekharan MP (1990) Grouping efficacy: a quantitative criterion for goodness of block diagonal forms of binary matrices in group technology. Int J Prod Res 28:233–243

Kumar KR, Kusiak A, Vannelli A (1986) Grouping of parts and components in flexible manufacturing systems. Eur J Oper Res 24:387–397

Kumar KR, Vannelli A (1987) Strategic subcontracting for efficient disaggregated manufacturing. Int J Prod Res 25:1715–1728

Kusiak A (1985) The part families problem in flexible manufacturing systems. Ann Oper Res 3:279–300

Kusiak A (1987) The generalized group technology concept. Int J Prod Res 25:561–569

Kusiak A, Boe WJ, Cheng C (1993) Designing cellular manufacturing systems: branch-and-bound and A* approaches. IIE Trans 25:46–56

Kusiak A, Cho M (1992) Similarity coefficient algorithms for solving the group technology problem. Int J Prod Res 30:2633–2646

Kusiak A, Chow WS (1987) Efficient solving of the group technology problem. J Manuf Syst 6:117–124

Kusiak A, Heragu SS (1987) The facility layout problem. Eur J Oper Res 29:229–251

Kusiak A, Vannelli A, Kumar KR (1986) Clustering analysis: models and algorithms. Control Cybernetics 15:139–154

Lashkari RS, Boparai R, Paulo J (2004) Towards an integrated model of operation allocation and material handling selection in cellular manufacturing systems. Int J Prod Econ 87:115–139

Lashkari RS, Gunasingh KR (1990) A Lagrangian relaxation approach to machine allocation in cellular manufacturing systems. Comput Ind Eng 19:442–446

Lee H, Garcia-Diaz A (1996) Network flow procedures for the analysis of cellular manufacturing systems. IIE Trans 28:333–345

Lee MK, Luong HS, Abhary K (1997) A genetic algorithm based cell design considering alternative routing. Comput Integr Manuf Syst 10:93–107

Leem CW, Chen JJG (1996) Fuzzy-set-based machine-cell formation in cellular manufacturing. J Intell Manuf 7:355–364

Lee-post A (2000) Part family identification using a simple genetic algorithm. Int J Prod Res 38:793–810

Liggett RS (2000) Automated facilities layout: past, present and future. Autom Constr 9:197–215

Lin TL, Dessouky MM, Kumar KR, Ng SM (1996) A heuristic-based procedure for the weighted production-cell formation problem. IIE Trans 28:579–589

Logendran R (1991) Effect of the identification of key machines in the cell formation problem of cellular manufacturing systems. Comput Ind Eng 20:439–449

Lozano S, Adenso-Diaz B, Eguia I, Onieva L (1999) A one-step tabu search algorithm for manufacturing cell design. J Oper Res Soc 50:509–516

Luong LHS (1993) A cellular similarity coefficient algorithm for the design of manufacturing cells. Int J Prod Res 31:1757–1766

Mansouri SA, Husseini SMM, Newman ST (2000) A review of the modern approaches to multi-criteria cell design. Int J Prod Res 38:1201–1218

Marcotorchino F (1987) Block seriation problems: a unified approach. Appl Stoch Models Data Analysis 3:73–91

Masnata A, Settineri L (1997) An application of fuzzy clustering to cellular manufacturing. Int J Prod Res 35:1077–1094

McAuley J (1972) Machine grouping for efficient production. Prod Eng 51:53–57

McCormick WT, Schweitzer PJ, White TW (1972) Problem decomposition and data reorganization by a clustering technique. Oper Res 20(5):993–1009

Mehrez A, Rabinowitz G, Reisman A (1988) A conceptual scheme of knowledge systems for MS/OR. Omega 16:421–428

Milligan GW, Cooper SC (1987) Methodology review: clustering methods. Appl Psycho Measure 11(4):329–354

Miltenburg J, Zhang W (1991) A comparative evaluation of nine well-known algorithms for solving the cell formation problem in group technology. J Oper Manage 10:44–72

Mitrofanov SP (1966) Scientific principles of group technology, part I. National Lending Library of Science and Technology, Boston

Mosier CT (1989) An experiment investigating the application of clustering procedures and similarity coefficients to the GT machine cell formation problem. Int J Prod Res 27:1811–1835

Mosier CT, Taube L (1985a) The facets of group technology and their impacts on implementation – a state of the art survey. Omega 13:381–391

Mosier CT, Taube L (1985b) Weighted similarity measure heuristics for the group technology machine clustering problem. Omega 13:577–583

Mosier CT, Yelle J, Walker G (1997) Survey of similarity coefficient based methods as applied to the group technology configuration problem. Omega 25:65–79

Murthy CVR, Srinivasan G (1995) Fractional cell formation in group technology. Int J Prod Res 33:1323–1337

Nair GJK, Narendran TT (1996) Grouping index: a new quantitative criterion for goodness of block-diagonal forms in group technology. Int J Prod Res 34(10):2767–2782

Nair GJK, Narendran TT (1998) CASE: A clustering algorithm for cell formation with sequence data. Int J Prod Res 36:157–179

Nair GJK, Narendran TT (1999) ACCORD: A bicriterion algorithm for cell formation using ordinal and ratio-level data. Int J Prod Res 37:539–556

Ng SM (1993) Worst-case analysis of an algorithm for cellular manufacturing. Eur J Oper Res 69:384–398

Offodile OF (1991) Application of similarity coefficient method to parts coding and classification analysis in group technology. J Manuf Syst 10:442–448

Offodile OF (1993) Machine grouping in cellular manufacturing. Omega 21:35–52

Offodile OF, Mehrez A, Grznar J (1994) Cellular manufacturing: a taxonomic review framework. J Manuf Syst 13:196–220

Offodile OF, Grznar J (1997) Part family formation for variety reduction in flexible manufacturing systems. Int J Oper Prod Manage 17:291–304

Onwubolu GC, Mlilo PT (1998) Manufacturing cell grouping using similarity coefficient-distance measure. Prod Plan Control 9:489–493

Optiz H, Eversheim W, Wienhal HP (1969) Work-piece classification and its industrial applications. Int J Mach Tool Des Res 9:39–50

Qiao LH, Yang ZB, Wang HP (1994) A computer-aided process planning methodology. Comput Ind 255:83–94

Rajagopalan R, Batra JL (1975) Design of cellular production system: a graph theoretic approach. Int J Prod Res 13:567–579

Reisman A, Kirshnick F (1995) Research strategies used by OR/MS workers as shown by an analysis of papers in flagship journals. Oper Res 43:731–739

Reisman A, Kumar A, Motwani J, Cheng CH (1997) Cellular manufacturing: a statistical review of the literature (1965–1995). Oper Res 45:508–520

Ribeiro JFF, Pradin B (1993) A methodology for cellular manufacturing design. Int J Prod Res 31:235–250

Rogers DJ, Tanimoto TT (1960) A computer program for classifying plants. Science 132:1115–1118

Romesburg HC (1984) Cluster analysis for researchers. Lifetime Learning, Wadsworth Inc., Belmont, CA

Samatova NF, Potok TE, Leuze MR (2001) Vector space model for the generalized parts grouping problem. Robot Comput Integr Manuf 17:73–80

Sarker BR (1996) The resemblance coefficients in group technology: a survey and comparative study of relational metrics. Comput Ind Eng 30:103–116

Sarker BR, Islam KMS (1999) Relative performances of similarity and dissimilarity measures. Comput Ind Eng 37:769–807

Sarker BR, Li Z (1998) Measuring matrix-based cell formation considering alternative routings. J Oper Res Soc 49(9):953–965

Sarker BR, Mondal S (1999) Grouping efficiency measures in cellular manufacturing: a survey and critical review. Int J Prod Res 37(2):285–314

Sarker BR, Xu Y (2000) Designing multi-product lines: job routing in cellular manufacturing systems. IIE Trans 32:219–235

Seifoddini H (1987) Incorporation of the production volume in machine cells formation in group technology applications. In: A Mital (ed) Recent developments in production research. Elsevier Science, Amsterdam, pp 562–570

Seifoddini H (1989a) Single linkage versus average linkage clustering in machine cells formation applications. Comput Ind Eng 16:419–426

Seifoddini H (1989b) A note on the similarity coefficient method and the problem of improper machine assignment in group technology applications. Int J Prod Res 27:1161–1165

Seifoddini H, Djassemi M (1995) Merits of the production volume based similarity coefficient in machine cell formation. J Manuf Syst 14:35–44

Seifoddini H, Djassemi M (1996) A new grouping measure for evaluation of machine-component matrices. Int J Prod Res 34(5):1179–1193

Seifoddini H, Hsu CP (1994) Comparative study of similarity coefficients and clustering algorithms in cellular manufacturing. J Manuf Syst 13:119–127

Seifoddini H, Tjahjana B (1999) Part-family formation for cellular manufacturing: a case study at Harnischfeger. Int J Prod Res 37:3263–3273

Seifoddini H, Wolfe PM (1986) Application of the similarity coefficient method in group technology. IIE Trans 18:271–277

Seifoddini H, Wolfe PM (1987) Selection of a threshold value based on material handling cost in machine-component grouping. IIE Trans 19:266–270

Selim HM, Askin RG, Vakharia AJ (1998) Cell formation in group technology: review, evaluation and directions for future research. Comput Ind Eng 34:3–20

Selvam RP, Balasubramanian KN (1985) Algorithmic grouping of operation sequences. Eng Cost Prod Econ 9:125–134

Sevier AJ (1992) Managing employee resistance to just-in-time: creating an atmosphere that facilitates implementation. Prod Inventory Manage J 33:83–87

Shafer SM, Meredith JR (1990) A comparison of selected manufacturing cell formation techniques. Int J Prod Res 28(4):661–673

Shafer SM, Meredith JR, Marsh RF (1995) A taxonomy for alternative equipment groupings in batch environments. Omega 23:361–376

Shafer SM, Rogers DF (1993a) Similarity and distance measures for cellular manufacturing. Part II. A survey. Int J Prod Res 31:1133–1142

Shafer SM, Rogers DF (1993b) Similarity and distance measures for cellular manufacturing. Part II. An extension and comparison. Int J Prod Res 31:1315–1326

Shambu G, Suresh NC (2000) Performance of hybrid cellular manufacturing systems: a computer simulation investigation. Eur J Oper Res 120:436–458

Shiko G (1992) A process planning-orientated approach to part family formation problem in group technology applications. Int J Prod Res 30:1739–1752

Silveira GD (1999) A methodology of implementation of cellular manufacturing. Int J Prod Res 37:467–479

Singh N (1993) Design of cellular manufacturing systems: an invited review. Eur J Oper Res 69:284–291

Singh N (1996) Systems approach to computer-integrated design and manufacturing. Wiley, New York

Singh N, Rajamani D (1996) Cellular manufacturing systems: design, planning and control. Chapman and Hall, London

Sneath PHA, Sokal RR (1973) Numerical taxonomy. Freeman, San Francisco

Sofianopoulou S (1997) Application of simulated annealing to a linear model for the formulation of machine cells in group technology. Int J Prod Res 35(2):501–511

Sokal RR, Michener CD (1958) A statistical method for evaluating systematic relationships. University of Kansas Sci Bull 38:1409–1438

Solimanpur M, Vrat P, Shankar R (2004) A heuristic to minimize makespan of cell scheduling problem. Int J Prod Econ 88:231–241

Srinivasan G (1994) A clustering algorithm for machine cell formation in group technology using minimum spanning trees. Int J Prod Res 32:2149–2158

Srinivasan G, Narendran TT, Mahadevan B (1990) An assignment model for the part-families problem in group technology. Int J Prod Res 28:145–152

Srinivasan G, Narendran TT (1991) GRAFICS – a nonhierarchical clustering algorithm for group technology. Int J Prod Res 29:463–478

Srinivasan G, Zimmers EW (1999) Fractional cell formation – issues and approaches. Int J Ind Eng 5:257–264

Steudel HJ, Ballakur A (1987) A dynamic programming based heuristic for machine grouping in manufacturing cell formation. Comput Ind Eng 12:215–222

Suer GA, Cedeno AA (1996) A configuration-based clustering algorithm for family formation. Comput Ind Eng 31:147–150

Tam KY (1990) An operation sequence based similarity coefficient for part families formations. J Manuf Syst 9:55–68

Tarsuslugil M, Bloor J (1979) The use of similarity coefficients and cluster analysis in production flow analysis. In: Proceedings 20th International Machine Tool Design and Research Conference, Birmingham, UK, September 1979, pp 525–532

Vakharia AJ, Kaku BK (1993) Redesigning a cellular manufacturing system to handle long-term demand changes: a methodology and investigation. Decis Sci 24:909–930

Vakharia AJ, Wemmerlöv U (1987) A new design method for cellular manufacturing systems. In: Proceedings of the 9th ICPR, Cincinnati, OH, pp 2357–2363

Vakharia AJ, Wemmerlöv U (1990) Designing a cellular manufacturing system: a materials flow approach based on operation sequences. IIE Trans 22:84–97

Vakharia AJ, Wemmerlöv U (1995) A comparative investigation of hierarchical clustering techniques and dissimilarity measures applied to the cell formation problem. J Oper Manage 13, 117–138.

Viswanathan S (1996) A new approach for solving the p-median problem in group technology. Int J Prod Res 34:2691–2700

Waghodekar PH, Sahu S (1984) Machine-component cell formation in group technology: MACE. Int J Prod Res 22:937–948

Wang J (1998) A linear assignment algorithm for formation of machine cells and part families in cellular manufacturing. Comput Ind Eng 35:81–84

Wang J, Roze C (1995) Formation of machine cells and part families in cellular manufacturing: an experimental study. Comput Ind Eng 29:567–571

Wang J, Roze C (1997) Formation of machine cells and part families: a modified p-median model and a comparative study. Int J Prod Res 35:1259–1286

Wei JC, Gaither N (1990) A capacity constrained multiobjective cell formation method. J Manuf Syst 9:222–232

Wei JC, Kern GM (1989) Commonality analysis: a linear cell clustering algorithm for group technology. Int J Prod Res 27:2053–2062

Wei JC, Kern GM (1991) Discussion: reply to "A note on a linear cell clustering algorithm". Int J Prod Res 29:217–218

Wemmerlöv U, Hyer NL (1986) Procedures for the part family/machine group identification problem in cellular manufacturing. J Oper Manage 6:125–147

Wemmerlöv U, Hyer NL (1987) Research issues in cellular manufacturing. Int J Prod Res 25:413–431

Wemmerlöv U, Johnson DJ (1997) Cellular manufacturing at 46 user plants: implementation experiences and performance improvements. Int J Prod Res 35:29–49

Wemmerlöv U, Johnson DJ (2000) Empirical findings on manufacturing cell design. Int J Prod Res 38:481–507

Won YK (2000a) New p-median approach to cell formation with alternative process plans. Int J Prod Res 38:229–240

Won YK (2000b) Two-phase approach to GT cell formation using efficient p-median formulation. Int J Prod Res 38:1601–1613

Won YK, Kim SH (1997) Multiple criteria clustering algorithm for solving the group technology problem with multiple process routings. Comput Ind Eng 32:207–220

Wu N, Salvendy G (1993) A modified network approach for the design of cellular manufacturing systems. Int J Prod Res 31:1409–1421

Yasuda K, Yin Y (2001) A dissimilarity measure for solving the cell formation problem in cellular manufacturing. Comput Ind Eng 39:1–17

Zhang C, Wang HP (1992) Concurrent formation of part families and machine cells based on the fuzzy set theory. J Manuf Syst 11:61–67

Chapter 10
Manufacturing Cells Design by Cluster Analysis

We introduce a cell formation problem that incorporates various real-life production factors, such as the alternative process routing, operation sequence, operation time, production volume of parts, machine capacity, machine investment cost, machine overload, multiple machines available for machine types and part process routing redesigning cost. None of the cell formation models in the literature has considered these factors simultaneously. We develop a similarity coefficient that incorporates alternative process routing, operation sequence, operation time and production volume factors. Although very few studies have considered the machine capacity violated issue under the alternative process routing environment, due to the difficulties of the issue discussed in this chapter, these studies fail to deal with this issue because they depend on some unrealistic assumptions. Five solutions have been proposed in this chapter and are used to cope with this difficulty. A heuristic algorithm that consists of two stages is developed. The developed similarity coefficient is used in stage 1 to obtain basic machine cells. Stage 2 solves the machine capacity violated issue, assigns parts to cells, selects process routing for each part and refines the final cell formation solution. Some numerical examples are used to compare with other related approaches in the literature and we also solve two large-sized problems to test the computational performance of the developed algorithm. The computational results suggest that the approach is reliable and efficient either in the quality or in the speed for solving cell formation problems.

10.1 Introduction

Over the last three decades, group technology (GT) has attracted a lot of attention from manufacturers because of its many applications and positive impacts in the batch type manufacturing system. GT is a manufacturing philosophy that attempts to increase production efficiency by processing part families within machine cells. The basic idea of GT is to identify and capitalize on the similar attributes of product design and manufacturing processes. Similar parts are grouped into a part family and manufactured by a cluster of dissimilar machines. Group technology takes full advantage of similarities to develop simplified and rationalized procedures in all stages of design and manufacture. The application of GT results in the mass production effect to multiproduct, small lot-sized production and leads to a lot of advantages such as reduction of material handling times and cost, reduction of labors and paper works, decrease of in-process inventories, shortening of production lead time, increase of machine utilization, and others (Ham *et al.* 1985).

One application of group technology to production is the cellular manufacturing (CM). Amongst the problems of CM, cell formation (CF) is considered to be the first and foremost problem in designing a CM system. The main objective of CF is to construct machine cells, to identify part families and to allocate part families to machine cells so as to minimize intercellular movements of parts.

A large number of approaches and practical reports have been published to identify machine cells and their associated part families. Many of them use a machine part incidence matrix which contains 0s and 1s elements to indicate the machine requirements of each part. In an incidence matrix, the "1" in the row and column represents the idea that the part needs an operation on the machine, and the "0" in the row and column represents the idea that the machine is not needed to process the part.

The most primitive method is to rearrange rows and columns of the incidence matrix on trial and error until a satisfactory solution is found. This method was used by Burbidge (1971) in his production flow analysis (PFA). PFA is basically an intuitive method and is relatively easy to implement. PFA may be suitable for small-sized problems, but it would definitely have difficulty dealing with large-scale problems when the machine part incidence matrix becomes more complex because of problem size.

Methodologies of clustering techniques in the literature can be divided into the following four groups (Yasuda and Yin 2001):

1. Descriptive methods: PFA proposed by Burbidge (1971), component flow analysis (CFA) by El-Essawy and Torrance (1972), and production flow synthesis (PFS) by De Beer and De Witte (1978).
2. Array-based methods: rank order clustering (ROC) algorithm developed by King (1980), ROC2 algorithm enhanced by King and Nakornchai (1982), and direct clustering algorithm (DCA) proposed by Chan and Milner (1982).
3. Similarity coefficient methods: clustering approach introduced by McAuley (1972), subsequently employed by Mosier and Taube (1985), Seifoddini and Wolfe (1987), and Gupta and Seifoddini (1990); also graph theoretic approach introduced by Rajagopalan and Batra (1975), subsequently employed by De Witte (1980), Chandrasekharan and Rajagopalan (1986), and Vannelli and Kumar (1986).
4. Other analytical methods: mathematical programming approach proposed by Purcheck (1975), Steudel and Ballakur (1987), Co and Araar (1988), and Shtub (1989), and also set-theoretic technique developed by Purcheck (1974).

An extension review of the various approaches for cell formation is available in the literature (Kumar and Vannelli 1983; Wemmerlöv and Hyer 1986; Chu and Pan 1988; Lashkari and Gunasingh 1990; Reisman *et al.* 1997; Selim *et al.* 1998). Wemmerlöv and Johnson (1997) employed a mail survey methodology and provided a study of implementation experiences and performance achievements at 46 user firms. Miltenburg and Zhang (1991) carried out a comparative study of nine well-known algorithms.

In the design of cellular manufacturing systems, many production factors should be involved when the cells are created, *e.g.*, machine requirement, machine set-up times, utilization, workload, alternative routings, machine capacities, operation sequences, setup cost and cell layout (Wu and Salvendy 1993). Due to the complexity of the cell formation problem, it is impossible to consider all the manufacturing factors in one method. A few approaches have been developed to incorporate different factors. In this research, we propose a new similarity coefficient to involve alternative process routings, operation sequences, operation times and production volumes. We also consider a heuristic that incorporates machine capacity, machine investment cost, machine overload, redesigning cost of part routing and multiple machines available for some machine types.

The remainder of this chapter is organized as follows. In Section 10.2, we discuss the background of several production factors, difficulties of capacity violated issue and objective of this study. Section 10.3 develops a new similarity coefficient that considers several production factors. This is followed, in Section 10.4 by a description of a two-stage heuristic algorithm. An important concept, key process routing is also introduced in this section. In Section 10.5, three numerical examples are presented to compare with other approaches in the literature; two large-sized problems are also used to test the computational performance. Finally, the conclusions from this study are given in Section 10.6.

10.2 Background, Difficulty and Objective of this Study

10.2.1 Background

10.2.1.1 Alternative Process Routings

Numerous cell formation methods have appeared in the literature. In most cell formation methods, parts are assumed to have a unique part process plan. However, it is well known that alternatives may exist in any level of a process plan. In some cases, there may be many alternative process plans for making a specific part, especially when the part is complex (Qiao *et al.* 1994). Explicit consideration of alternative process plans invoke changes in the composition of all manufacturing cells so that lower capital investment in machines, more independent manufacturing cells and higher machine utilization can be achieved (Hwang and Ree 1996).

The cell formation problem incorporating alternative process routings is called the generalized GT problem. Kusiak (1987) is the first person who described the cell formation problem in which alternative process routings are available. Kusiak (1987) presented an integer programming model (generalized p-median model) on the design of cells considering alternative process routings.

Hwang and Ree (1996) proposed a two-stage procedure for the cell formation problem with alternative process routings. At the first stage, the route selection problem is solved with the objective of maximizing the sum of compatibility coefficients among selected process routings. At the second stage, part families are formed based on the result of the first stage using the p-median problem.

Won and Kim (1997) considered a generalized machine similarity coefficient and used a multiple criteria clustering algorithm to obtain machine cells.

Sofianopoulou (1999) proposed a two-dimensional simulated annealing heuristic for the design of medium-sized cellular manufacturing systems with replicate machines and/or alternative process routings for some or all of the parts produced.

Zhao and Wu (2000) developed a genetic algorithm for manufacturing cell formation with multiple routes and multiple objectives. The multiple objectives include the minimization of the total within cell load variation, total intercellular/intracellular part movements and total exceptional elements.

10.2.1.2 Operation Sequences

Another important manufacturing factor in the design of a cellular manufacturing system is the operation sequences of parts. The operation sequence is defined as an ordering of the machines on which the part is sequentially processed (Vakharia and Wemmerlöv 1990).

Choobineh (1988) presented a two-stage procedure for the design of a cellular manufacturing system based on the operation sequences. The first stage uses a similarity coefficient to form part families. In the second stage, an integer programming model is developed to obtain machine cells.

Vakharia and Wemmerlöv (1990) proposed a similarity coefficient based on operation sequences to integrate the intracellular flow with the cell formation problem by using clustering methodology.

Logendran (1991) developed an algorithm to form the cells by evaluating the intercellular and intracellular moves with the operation sequences. He also indicated the impact of the sequence of operations and layout of cells in the cell formation problem.

Wu and Salvendy (1993) considered a network analysis method by using an undirected graph (network) to model the cell formation problem with taking into account the operation sequences factor.

Sarker and Xu (1998) presented a brief review of the methods of cell formation based on the operation sequences. A number of operation sequence-based similarity/dissimilarity coefficients are discussed in their research. They classified the methods of cell formation based on the operation sequences into four kinds: mathematical programming, network analysis, materials flow analysis method, and heuristics.

10.2.1.3 Part Production Volume and Machine Capacity

Gupta and Seifoddini (1990) depicted the merits of incorporating production volumes of parts into the cell formation procedure and developed a production data-based similarity coefficient. Among various production factors, operation time and machine capacity are two particularly relevant factors to production volume.

In the design of cellular manufacturing systems, available capacities of machines need to be sufficient to satisfy the production volume required by parts. Previous studies suggested that the number of machines for each machine type must be known *a priori*. For a specific machine type, if the exact number of machines required to process the parts has not been provided, it has to be determined before solving the cellular manufacturing problem (Heragu 1994; Heragu and Gupta 1994). This is realistic and it is easy to calculate these numbers under the traditional cell formation environment that does not incorporate alternative process routings. Interested readers may refer to the mathematical model presented in Heragu and Gupta (1994). However, it becomes difficult to calculate these numbers under the situation that considers alternative process routings. We will discuss this difficulty and propose solutions in the next section.

10.2.2 Objective of this Study and Drawbacks of Previous Research

The objective of this study is to formulate a new similarity coefficient that incorporates alternative process routings, operation sequences, operation times and production volumes. As for the machine capacity issue, under alternative process routings available, we also consider some realistic production factors such as multiple machines available for machine types, machine investment cost, and part routing redesigning cost to overcome the drawbacks of previous studies.

The importance of the above mentioned production factors has been emphasized constantly by previous studies. For example, the huge flexibility and lower machine capital investment brought by alternative process routings have been discussed by Kusiak (1987), Hwang and Ree (1996); Choobineh (1988), Sarker and Xu (1998) emphasized that the operation sequence is the most relevant attribute and ignoring this factor may erase the impact of material flow; Heragu (1994) indicated that it is obvious that machine capacity is more important than the other production factors and it is therefore necessary to first ensure that adequate capacity (in machine hours) is available to process all the parts.

Although a large number of cell formation methods have been developed thus far, most of them focus only on one or two production factors mentioned above.

Very few approaches have been proposed to cope with various production factors. To the best of our knowledge, the similarity coefficient developed by Gupta (1993) is the only coefficient that takes into account alternative process routings, operation sequences, operation times and production volumes. Gupta (1993) also considered the machine capacity by testing the process routing usage factors.

As for the machine capacity under the model that alternative process routings are available, it is impossible to calculate the number of machines for each machine type before solving the cell formation problem, because the routing selection for each part is unknown and different process routings of a part use different machines. Moreover, the operation time may be different for the same machine in different process routings. Due to these uncertain situations, it becomes complicated to ensure the capacity constraint under alternative process routings consideration. Previous research dealt with this problem based on some unrealistic hypothesis. Gupta (1993) ignored the situation that multiple machines are needed for a specific machine type in order to satisfy the capacity constraint. In his model, each machine type only contains a single machine and the capacities would be sufficient by adjusting routing usage factors of parts. A recent contribution by Caux *et al.* (2000) took into account the situation that multiple machines are available for machine types. However, they assumed that the number of machines for each machine type is known and they treated multiple machines of each machine type as different machine types. Furthermore, they assumed that at least one process routing for each part exists which has enough machine capacities to produce the required part quantities.

The assumptions mentioned above are not realistic. In this chapter, we are interested in finding solutions of cell formation problems, which respect the following real-life production situations.

For some machine types, multiple machines are available and should not be treated as different machine types. However, the number of machines needed for each type to ensure capacity constraints is unknown.

In the system designing stage, there is the possibility that none of the alternative process routings of a part can ensure the production of the required part quantity without exceeding available capacities of machines.

Based on the above analysis, it can be concluded that the capacity constraint violated issue is unavoidable in the system designing stage. Therefore, in order to ensure sufficient capacity to process all the parts, we should consider the following question: "If available machine capacity cannot guarantee the production of the required part quantity, what should we do under current production conditions?" Since the machine capacity insufficient issue emerges only after some process routing has been selected for producing the corresponding part, we propose several approaches in this chapter to cope with this problem when some routing is selected for a part whereas machine capacity in this routing is insufficient for producing the required part quantity. The approaches are described as follows:

1. Use multiple machines for the capacity violated machine types.
2. Use other alternative process routings.
3. Redesign part process routing.

4. Buy new/additional machines.
5. Overload.

We use these approaches in the cell formation procedure to guarantee the quantities of produced parts. Concrete steps will be discussed in stage 2 of the proposed algorithm.

10.3 Problem Formulation

10.3.1 Nomenclature

Indices:

i, k	machine ($i, k = 1, \ldots, M$)
j	part ($j = 1, \ldots, P$)
r	process routing ($r = 1, \ldots, R_j$)
o	operation ($o = 1, \ldots, n_i^{jr}$)
c, f, g	cell ($= 1, \ldots, C$)
l	key process routing
l'	first-key routing

Parameters:

$$a_i^{jr} = \begin{cases} 1 & \text{if machine } i \text{ is used in the process route } r \text{ of part } j\ ; \\ 0 & \text{otherwise .} \end{cases}$$

$$a_i^j = \begin{cases} 1 & \text{if } a_i^{jr} = 1 \text{ for some } r \in R_j\ ; \\ 0 & \text{otherwise .} \end{cases}$$

indicates whether part j is used by machine i or not.

$$a_{ik}^j = \begin{cases} 1 & \text{if } a_i^{jr} = a_k^{jr} = 1 \text{ for some } r \in R_j\ ,\quad i \neq k\ ; \\ 0 & \text{otherwise ,} \end{cases}$$

indicates whether part j is used by both machines i and k or not

$N_i = \sum_{j=1}^{P} a_i^j$ the number of parts processed by machine i

$N_{ik} = \sum_{j=1}^{P} a_{ik}^j$ the number of parts processed by both machines i and k

gs_{ik} generalized similarity coefficient, Equation 10.1. $0 \leq gs_{ik} \leq 1$
S_{ik} modified generalized similarity coefficient, Equation 10.2. $0 \leq S_{ik} \leq 1$
SR_{ik} sequence ratio, Equation 10.3. $0 \leq SR_{ik} \leq 1$
MLR_{ik} machine-load ratio, Equation 10.8. $0 \leq MLR_{ik} \leq 1$
X_{ik} number of actual movements of parts between machines i and k, Equation 10.4
D_{ik} number of possible movements of parts between machines i and k, Equation 10.5
x_{ik}^{j} number of times that part j moves between machines i and k
d_{ik}^{j} number of possible produced movements of part j between machines i and k
x_{ik}^{jr} number of times that part j moves between machines i and k in the routing r
d_{ik}^{jr} number of possible movements of part j between machines i and k in the routing r
sr_{ik}^{jr} sequence ratio between machines i and k in the routing r of part j, Equation 10.6
n_{i}^{jr} number of times that part j visits machine i in the process routing r

$$n_{ik}^{jr} = \text{Min}\left(n_{i}^{jr}, n_{k}^{jr}\right)$$

fl_{i}^{jr} whether or not the first or/and last operation in routing r of part j is performed by machine i, Equation 10.7
$i(k)$ indicates either machine i or k, determined by Table 10.1
Y_{ik} *min*-production volume factor between machines i and k, Equation 10.9
E_{ik} *max*-production volume factor between machines i and k, Equation 10.10
y_{ik}^{j} *min*-production volume factor of part j between machines i and k
e_{ik}^{j} *max*-production volume factor of part j between machines i and k
mlr_{ik}^{jr} machine-load ratio between machines i, k in the routing r of part j, Equation 10.11
y_{ik}^{jr} *min*-production volume factor between machines i and k in the routing r of part j, Equation 10.12
e_{ik}^{jr} *max*-production volume factor between machines i and k in the routing r of part j, Equation 10.12
v_{j} production volume for part j
t_{io}^{jr} operation time of the oth operation on machine i in routing r of part j

$t_{i}^{jr} = \sum_{o=1}^{n_{i}^{jr}} t_{io}^{jr}$ total operation times of part j in routing r on machine i

$vt_{i}^{jr} = v_{j}^{*} t_{i}^{jr}$ total operation times of part j in routing r on machine i during a production period

NM_f	number of machines in the machine cell f
KR_j	the set of key process routings of part j
T	total number of the intercellular movements in the system
T_j	number of intercellular movements in some key process routing of part j
rmc_{jl}	the required cost by purchasing new machines for routing l of part j to ensure adequate capacities for producing required quantity of part j, Equation 10.15
rmc_{jl}^m	the required cost by purchasing new machines for machine type m in routing l of part j to ensure adequate capacity of m to produce required quantity of part j, Equation 10.16
$nbnm_m$	the required number of new machines for machine type m in the current routing
mc_m	the investment cost of new machine m
amm_m	available number of multiple machines for machine type m
bnm_m	number of machine of type m that has newly bought
umc	the machine unit capacity = 8 h/day
rrc_j	the cost of redesigning a new process routing for part j
B	the set denotes the number of decreased intercellular movements
B_{mc}	an entry in B, denotes the number of decreased intercellular movements by moving machine
m	from current cell to cell c, $m \notin c$

$B_{mc}^h = \max_{m \in M, c \in C} B_{mc}$, the entry that has the maximum value in the set B.

10.3.2 Generalized Similarity Coefficient

Generalized machine similarity coefficient (Won and Kim 1997) which was proposed for considering alternative process routings is basically an extension of the Jaccard similarity coefficient. The deficiency of this coefficient is that it ignores the actual impacts of realistic production data such as material flows and part quantities. In this study, we extend the generalized machine similarity coefficient to incorporate the operation sequences, operation times and production volumes of parts.

According to Won and Kim (1997), the generalized similarity between machine i and k is given by the Equation 10.1:

$$gs_{ik} = \frac{N_{ik}}{N_i + N_k - N_{ik}} \quad (0 \leq gs_{ik} \leq 1) . \tag{10.1}$$

From the above definition, $a_i^j = 1$ indicates that if machine i is used by some process routing of part j the number of parts processed by machine i is counted as one for that part even if the remaining process routings of part j also use machine i. This idea follows from the basic assumption of the generalized GT problem that in the final solution only one process routing is selected for each part (Kusiak

1987; Won and Kim 1997). Similarly, if some process routing of part j uses both machines, then the remaining routings of part j are ignored in the definition.

10.3.3 Definition of the New Similarity Coefficient

In this section, we extend the generalized machine similarity coefficient to cope with cell formation problems, which consider alternative process routings, operation sequences, operation times and production volumes of parts simultaneously.

In order to reflect the impact of operation sequences in the generalized machine similarity coefficient, we add a sequence ratio SR_{ik} into Equation 10.1. We also add a machine-load ratio MLR_{ik} into Equation 10.1 to consider the totally required operation machine time. We define the similarity coefficient as follows:

$$S_{ik} = \frac{N_{ik}}{N_i + N_k - N_{ik}} \times SR_{ik} \times MLR_{ik} \quad (0 \le S_{ik} \le 1)\,. \tag{10.2}$$

1. Definition of the sequence ratio SR_{ik}

The value of the ratio varies from 0 to 1. The sequence ratio is defined as follows:

$$SR_{ik} = \frac{X_{ik}}{D_{ik}} \quad (0 \le SR_{ik} \le 1)\,. \tag{10.3}$$

The denominator D_{ik} indicates the number of possible produced movements of parts between machines i and k. The numerator X_{ik} is an indication of the number of actual movements of parts between machines i and k.

$$X_{ik} = \sum_{j=1}^{P} x_{ik}^{j} \tag{10.4}$$

$$D_{ik} = \sum_{j=1}^{P} d_{ik}^{j}\,. \tag{10.5}$$

In the generalized similarity coefficient, part j is regarded as using machines i and k if both machines are used by some process routing of part j; other remaining process routings of part j are ignored (Won and Kim 1997). Similarly, we define the intermachine movements of parts between a pair of machines by using some process routing of that part. In other words, the x_{ik}^{jr} and d_{ik}^{jr} of some process routing of part j are selected to represent the x_{ik}^{j} and d_{ik}^{j} of part j. Therefore, the problem is changed to find out the appropriate process routing r for each part that can facilitate the problem formulation. Two principles are introduced here in finding the process routing r.

Principle 10.1. The highest degree of sequence ratio between machines i and k can be achieved in process routing r for part j.

Principle 10.2. If Principle 10.1 cannot judge candidate, select the one that has minimal possible intermachine movements d_{ik}^{jr}. $d_{ik}^{jr} \neq 0$.

The idea of Principle 10.1 is that the existence of alternative process routings increases the chance of obtaining mutually independent machines cells. In other words, it increases the degree of similarity between two machines by selecting appropriate process routings of parts. Principle 10.2 is obvious since one of the important objectives of cellular manufacturing is to decrease the intermachine movements.

We define the similarity between machines i and k in the process routing r for part j as follows:

$$sr_{ik}^{jr} = \frac{x_{ik}^{jr}}{d_{ik}^{jr}} \quad (d_{ik}^{jr} \neq 0) \,. \tag{10.6}$$

Assume $r' \in R_j$ and $sr_{ik}^{jr'} = \underset{r \in R_j}{\text{Max}}(sr_{ik}^{jr})$, x_{ik}^{j} and d_{ik}^{j} for each part j are calculated as Table 10.2.

Table 10.1 The determination of machine $i(k)$

	if $n_k^{jr} < n_i^{jr}$	*if* $n_k^{jr} < n_i^{jr}$	*if* $n_i^{jr} = n_k^{jr} \neq 0$ *if* $fl_i^{jr} = 2$	*if* $fl_k^{jr} = 2$	Otherwise
$i(k)$	i	k	i	k	Either i or k

In Table 10.2, if $\sum_{r=1}^{R_j} d_{ik}^{jr} > 0$ and $\sum_{r=1}^{R_j} x_{ik}^{jr} > 0$, then Principle 10.3 is used to get x_{ik}^{j} and d_{ik}^{j}. On the other hand, if $\sum_{r=1}^{R_j} d_{ik}^{jr} > 0$ and $\sum_{r=1}^{R_j} x_{ik}^{jr} = 0$, then Principle 10.4 is performed to obtain x_{ik}^{j} and d_{ik}^{j} for part j.

The last problem is the definition of the possible produced intermachine movements d_{ik}^{jr}, which is established as follows:

$$fl_i^{jr} = \begin{cases} 2 & \text{if both the first and last operations in routing } r \\ & \text{of part } j \text{ are performed by machine } i \text{ ;} \\ 1 & \text{else if either first or last operation in routing } r \\ & \text{of part } j \text{ is performed by machine } i \text{ ;} \\ 0 & \text{otherwise .} \end{cases} \tag{10.7}$$

$$fl_{i(k)}^{jr} = \begin{cases} fl_i^{jr} & \text{if } i(k) = i \text{ ;} \\ fl_k^{jr} & \text{if } i(k) = k \text{ .} \end{cases}$$

Table 10.2 The calculation of x^j_{ik}, d^j_{ik}. * $\underset{r \in R_j}{\text{Min}}(d^{jr}_{ik}) > 0$

	if $\sum_{r=1}^{R_j} d^{jr}_{ik} > 0$		*if* $\sum_{r=1}^{R_j} d^{jr}_{ik} = 0$
	if $\sum_{r=1}^{R_j} x^{jr}_{ik} > 0$	*if* $\sum_{r=1}^{R_j} x^{jr}_{ik} = 0$	
$\left(x^j_{ik}, d^j_{ik}\right)$	$\left(x^{jr'}_{ik}, d^{jr'}_{ik}\right)$	$\left(0, \underset{r \in R_j}{\text{Min}}\,(d^{jr}_{ik})^*\right)$	$(0, 0)$

Table 10.3 The formulation of possible produced intermachine movements d^{jr}_{ik}

	if $n^{jr}_i \neq n^{jr}_k$			*if* $n^{jr}_i = n^{jr}_k \neq 0$		
	if $fl^{jr}_{i(k)} = 2$	*if* $fl^{jr}_{i(k)} = 1$	*if* $fl^{jr}_{i(k)} = 0$	*if* $fl^{jr}_{i(k)} = 2$	*if* $fl^{jr}_{i(k)} = 1$	*if* $fl^{jr}_{i(k)} = 0$
d^{jr}_{ik}	$2n^{jr}_{ik} - 2$	$2n^{jr}_{ik} - 1$	$2n^{jr}_{ik}$	$2n^{jr}_{ik} - 2$	$2n^{jr}_{ik} - 1$	$2n^{jr}_{ik} - 1$

Thus, d^{jr}_{ik} is formulated as in Table 10.3.

And if $n^{jr}_i = n^{jr}_k = 0$, then $d^{jr}_{ik} = 0$.

2. Definition of the machine-load ratio MLR_{ik}

$$MLR_{ik} = \frac{Y_{ik}}{E_{ik}} \quad (0 \leq MLR_{ik} \leq 1) \tag{10.8}$$

$$Y_{ik} = \sum_{j=1}^{P} y^j_{ik} \tag{10.9}$$

$$E_{ik} = \sum_{j=1}^{P} e^j_{ik} \,. \tag{10.10}$$

Similar with the definition of sequence ratio, we select some routing to calculate y^j_{ik} and e^j_{ik}. The two principles are modified as follows:

Principle 10.3. The highest degree of machine-load ratio between machines i and k can be achieved in process routing r for part j.

Principle 10.4. If Principle 10.3 cannot judge the candidate, select the one that has minimal production time.

$$mlr^{jr}_{ik} = \frac{y^{jr}_{ik}}{e^{jr}_{ik}} \quad \left(e^{jr}_{ik} \neq 0\right) \tag{10.11}$$

where

$$e_{ik}^{jr} = \max\left(vt_i^{jr}, vt_k^{jr}\right), \quad y_{ik}^{jr} = \min\left(vt_i^{jr}, vt_k^{jr}\right) \tag{10.12}$$

Assume $r' \in R_j$ and $mlr_{ik}^{jr'} = \underset{r \in R_j}{\text{Max}}(mlr_{ik}^{jr})$, y_{ik}^j and e_{ik}^j for each part j are calculated as in Table 10.4.

Table 10.4 The calculation of y_{ik}^j, e_{ik}^j. * $\underset{r \in R_j}{\text{Min}}(e_{ik}^{jr}) > 0$

	if $\sum_{r=1}^{R_j} e_{ik}^{jr} > 0$		if $\sum_{r=1}^{R_j} e_{ik}^{jr} = 0$
	if $\sum_{r=1}^{R_j} y_{ik}^{jr} > 0$	if $\sum_{r=1}^{R_j} y_{ik}^{jr} = 0$	
$\left(y_{ik}^j, e_{ik}^j\right)$	$\left(y_{ik}^{jr'}, e_{ik}^{jr'}\right)$	$\left(0, \underset{r \in R_j}{\text{Min}}\left(e_{ik}^{jr}\right)^*\right)$	$(0, 0)$

In Table 10.2, if $\sum_{r=1}^{R_j} e_{ik}^{jr} > 0$ and $\sum_{r=1}^{R_j} y_{ik}^{jr} > 0$, then Principle 10.3 is used to obtain y_{ik}^j and e_{ik}^j. On the other hand, if $\sum_{r=1}^{R_j} e_{ik}^{jr} > 0$ and $\sum_{r=1}^{R_j} y_{ik}^{jr} = 0$, then Principle 10.4 is performed to obtain y_{ik}^j and e_{ik}^j for part j.

10.3.4 Illustrative Example

We use a simple example to illustrate the definition of the proposed similarity coefficient. Assume there are four parts, for each part, the alternative process routings that include production data: operation sequence (denoted by machine number) and operation time (in parentheses, unit is minutes) are as follows:

Part 1 has three alternative process routings:
p1(r1): m4(3), m2(2), m1(3), m2(2), m1(3)
p1(r2): m1(3), m3(5), m4(4), m2(1), m5(3)
p1(r3): m5(5), m3(3), m4(4), m1(4)

Part 2 has two alternative process routings:
p2(r1): m1(3), m2(3), m1(2), m2(3), m1(2)
p2(r2): m1(2), m2(2), m1(1), m3(2), m1(1), m4(2), m2(1)

Part 3 has two alternative process routings:
p3(r1): m4(1), m1(2), m4(3), m5(1), m4(2)
p3(r2): m3(1), m5(3), m4(2), m2(2), m5(4)

Part 4 has two alternative process routings:

p4(r1): m4(3), m1(3), m4(5), m5(2), m2(2)
p4(r2): m3(1), m1(3), m4(2), m5(5)

We calculate the similarity between machines 1 and 2. From the above operational data, we construct the machine-part matrix as in the following Table 10.5. The elements in the matrix indicate the operation sequences of process routings

Table 10.5 A computing case

	p1			p2		p3		p4	
r	1	2	3	1	2	1	2	1	2
m 1	3, 5	1	4	1, 3, 5	1, 3, 5	2		2	
m 2	2, 4	4		2, 4	2, 7		4	5	1

For part 1, routing 1 includes both machines 1 and 2 twice. The last operation of routing 1 is processed on machine 1. Hence, the coefficient with routing 1 is computed as follows:

$n_{12}^{11} = n_1^{11} = n_2^{11} = 2\,;\quad fl_1^{11} = 1\,,\quad fl_2^{11} = 0\,;\quad i(k) = \text{either 1 or 2}\,.$ Finally, $d_{12}^{11} = 2n_{12}^{11} - 1 = 3\,,\quad x_{12}^{11} = 3\quad \text{and}\quad sr_{12}^{11} = 3/3 = 1\,.$

Similarly, for routing 2 and routing 3 of part 1, the coefficients are computed as follows:

$d_{12}^{12} = 1\,,\quad x_{12}^{12} = 0$ and $sr_{12}^{12} = 0;\ d_{12}^{13} = 0\,,\ x_{12}^{13} = 0$ and sr_{12}^{13} does not exist.

Therefore, Principle 10.3 is applied to obtain x_{12}^1 and d_{12}^1 as follows:

$$x_{12}^1 = x_{12}^{11} = 3 \quad \text{and} \quad d_{12}^1 = d_{12}^{11} = 3\,.$$

The same procedure is performed on part 2 and the results are given as follows:

$$d_{12}^{21} = 2 \times n_{12}^{21} = 4\,,\ x_{12}^{21} = 4 \quad \text{and} \quad sr_{12}^{21} = 1\,;$$
$$d_{12}^{22} = 2 \times n_{12}^{22} - 1 = 3\,,\ x_{12}^{22} = 2 \quad \text{and} \quad sr_{12}^{22} = 2/3\,;$$
$$x_{12}^2 = x_{12}^{21} = 4 \quad \text{and} \quad d_{12}^2 = d_{12}^{21} = 4\,.$$

Since $\sum\limits_{r \in R_3} d_{12}^{3r} = 0$, part 3 does not contribute to the sequence ration. For part 4, Principle 10.4 is applied and the results are given as follows:

$$\sum_{r \in R_4} d_{12}^{4r} > 0 \quad \left(d_{12}^{41} = 1\right) \quad \text{and} \quad \sum_{r \in R_4} x_{12}^{4r} = 0\,,\ \text{so}$$
$$x_{12}^4 = 0 \quad \text{and} \quad d_{12}^4 = 1\,.$$

Hence, $X_{12} = \sum\limits_{j=1}^{4} x_{12}^j = 7$, $D_{12} = \sum\limits_{j=1}^{4} d_{12}^j = 8$ and $SR_{12} = 7/8$.

As for MLR_{12}, the computing procedure is similar with SR_{12}, we assume the production volume of each part for this example is as $v_{1-4} = 10, 20, 10, 20$. Then, we get y^j_{ik}, e^j_{ik} and the final result as follows:

$$y^1_{12} = y^{11}_{12} = 5 \times 10\,, \quad e^1_{12} = e^{11}_{12} = 5 \times 10\,;$$
$$y^2_{12} = y^{21}_{12} = 6 \times 20\,, \quad e^2_{12} = e^{21}_{12} = 7 \times 20\,;$$
$$y^3_{12} = e^3_{12} = 0\,; \quad y^4_{12} = y^{41}_{12} = 2 \times 20\,, \quad e^4_{12} = e^{41}_{12} = 3 \times 20\,.$$

Hence, $Y_{12} = \sum_{j=1}^{4} y^j_{12} = 210$, $E_{12} = \sum_{j=1}^{4} e^j_{12} = 250$ and $MLR_{12} = 21/25$.

Since $gs_{12} = 3/5$, at last

$$S_{12} = gs_{12} \times SR_{12} \times MLR_{12} = (3/5) \times (7/8) \times (21/25) = 441/1000\,.$$

10.4 Solution Procedure

For solving cell formation problems by similarity coefficient methods, there exist two different solution methodologies: the optimal solution methodology and the heuristic approach. In heuristic approaches, suboptimal solutions are sought where it is expected that the optimal solution methodologies may not work well for large instances, although the grouping efficiency in suboptimal solutions may decrease. In solving cell formation problems, as the size (dimension) of a problem increases, the number of both variables and constraints increases and, at a certain point, the optimal solution methodology fails to solve larger instances of problem (Islam and Sarker 2000). In this section, we propose a two-stage heuristic algorithm for solving cell formation problems. Stage 1 applies the proposed similarity coefficient to obtain basic machine groups, and stage 2 refines the solution and solves the machine capacity problem by several approaches.

10.4.1 Stage 1

The objective of stage 1 is to obtain basic machine cells. At first, construct a similarity matrix by Equation 10.2, which is a symmetric matrix with M^*M entries. An entry s_{ik} in the matrix indicates the similarity between machine i and machine k. Then, group two machines into a machine cell and revise the similarity matrix. The procedure is iterated until the predefined number of machine cells has been obtained.

An average similarity coefficient is used for revising the similarity matrix. The coefficient is defined to evaluate the similarity between two machine cells f and g, and it is described as follows:

$$s_{fg} = \frac{\sum\limits_{i \in f} \sum\limits_{k \in g} s_{ik}}{NM^*_f NM_g}\,. \tag{10.13}$$

The general procedure of the proposed heuristic algorithm is presented as follows:

Step 1 Produce the similarity matrix s_{ik}.

Step 2 Join the two machines that have the highest value into a new machine cell.

Step 3 Check the constraint of the number of cells.
If (the predefined number of cells has been obtained)
stop;
else go to step 4.

Step 4 Revise the similarity coefficients between the new machine cell and other remainder machines (machine cells) in the similarity matrix by Equation 10.13. Go back to step 2.

After finishing stage 1 of the algorithm, we have obtained the basic machine cells that meet the cell number constraint. In order to finish the cell formation problem, we need to decide the part family for each machine cell and select the process routing for each part. We also need to ensure that the machine capacities of the selected routing can guarantee the production volume of the part. We solve these problems in the stage 2.

10.4.2 Stage 2

Before describing the procedure in detail, we define the concept of key process routings as follows: if part j can be processed in the process routing l with minimum number of intercellular movements, then we call the process routing l as the key process routing of part j. A part can have several key process routings that form a key routing set KR.

We define the total number of the intercellular movements in the system as follows:

$$T = \sum_{j=1}^{P} T_j \times v_j \, . \tag{10.14}$$

In stage 2 of the algorithm, step 1 forms the set of key process routings KR for each part. Since the members in KR need a minimum number of intercellular movements, they are the candidates for the finally selected routing of the part.

Step 2 invokes add-in-step which reassigns each machine to other machine cells to check the possibility of reducing intercellular movements T. If the reassignment of some machine reduces intercellular movements, then reassign the machine to the machine cell that will reduce intercellular movements maximally. The add-in-step is iterated until no more reduction of intercellular movements can be produced by reassignments of machines.

For each part j, steps 3 and 4 ensure the required quantity to be produced. Firstly, a key routing l' that contains the smallest number of operations is selected. We call l' as the first-key routing of part j. Based on the first-key routing, a lot of objectives such as minimization of intercellular movements, minimization of the intracellular movements are achieved.

Step 3.1 checks the capacity constraints of used machines in the first-key routing l', if the capacity constraints are violated, then other second-key routings in KR are utilized in step 3.2 to ensure the machine capacities. If machine capacities in all key routings of part j are insufficient for producing the required part quantity, the approaches introduced in Section 10.2.2 are employed in step 4 to ensure the machine capacity constraint. The details are as follows:

Approach 1 For first and second-key routings: If multiple machines are available for the capacity violated machine types, add machines into the manufacturing system until the required machine capacity has been satisfied (step 4.1). Else, use approach 2.

Approach 2 Use non-key routings that need more intercellular movements than keys (step 4.2).

If both approach 1 and approach 2 are not viable, which indicates that multiple machines are not available or the number of multiple machines is not sufficient for all routings of part j. Hence, other approaches need to be considered for coping with the insufficient machine capacity issue.

Approach 3 Buy new machines (step 4.3.1).

Approach 4 Redesign new process routing in which machine capacities can ensure the required part quantity (step 4.3.2).

Approach 5 If the cost of a new machine is very high and the capacity of the machine is only very little exceeded, consider machine overload. This approach can be used as a supplement for other approaches. Since overload is not normal for machines, it should be used only for special cases. We do not use this approach in this chapter.

In the selection of the above approaches, a balance model among several costs needs to be considered. These costs include machine purchase cost, process routing redesigning cost, and machine overload cost. Since different approaches may select different process routing, intercellular and intracellular movements costs also need to be considered. Finally, the fixed usage cost of newly added multiple machines also needs to be considered. We suggest the model as a further study topic and will discuss it in conjunction with approach 5 in other research.

For applying approach 3, some parameters are calculated as follows:

$$rmc_{jl} = \sum_{m=1}^{M} rmc_{jl}^{m} \tag{10.15}$$

$$rmc_{jl}^{m} = mc_m \times nbnm_m \tag{10.16}$$

$$nbnm_m = \left(\sum_{i=1}^{M} a_i^{jr} * vt_i^{jr} - (amm_m + bnm_m) \times umc \right) / umc, nbnm_m \geq 0 \tag{10.17}$$

and if $nbnm_m$ contains a decimal fraction, $nbnm_m = nbnm_m + 1$.

After finishing step 4, the process routing that satisfies the capacity constraint has been selected for each part. Step 5 re-executes the add-in-step to improve the cell formation solution. Step 6 assigns multiple or newly purchased machines to cells.

The details of stage 2 is presented as follows:

Step 1 For each part j, form KR.

Step 2 Improvement of the solution obtained in stage 1.
Invoke add-in-step (at the end of the steps). Go to next step.
Check machine capacity constraint. Initialize: set part number $j = 1$.

Step 3 For part j, find the current first-key routing l' that contains the smallest number of operations.

Step 3.1 If the capacity constraints of used machines in routing l' are satisfied, select l'. Part number $j = j + 1$. If $j = P + 1$, go to step 5; else, go back to the top of step 3.

Step 3.2 If the capacity constraints of used machines in routing l' are violated, invalidate l' in set KR. If there is no valid routing in the KR, go to step 4; else, go back to the top of step 3.

Step 4 Validate all members in KR of part j.

Step 4.1 For part j, find the first-key routing l' that contains the smallest number of operations.

For the capacity constraint violated machine types, if the number of multiple machines is sufficient to ensure the needed capacity, go to step 4.1.1; else, go to step 4.1.2.

Step 4.1.1 Add required machines into system. Select routing l'. $j = j + 1$, if $j = P + 1$, go to step 5; else, go back to the top of step 3.

Step 4.1.2 Invalidate l' in set KR. If there is no valid routing in the KR, go to step 4.2; else, go back to the top of step 4.1.

Step 4.2 Remove all key routings of part j. If there are no other routings for part j, go to step 4.3; else form KR, go back to the top of step 3.

Step 4.3 Find the routing l'' in which $rmc_{jl''} = \min_{\forall l} rmc_{jl}$. If $rrc_j > rmc_{jl''}$, go to step 4.3.1; else, go to step 4.3.2.

Step 4.3.1 Select l''. For the capacity violated machine types, add available multiple machines into system. For the machine types whose number of multiple machines is not sufficient, buy new machines until capacities are satisfied. $j = j + 1$, if $j = P + 1$, go to step 5; else, go back to step 3.

Step 4.3.2 Redesign a new process routing for part j in which machine capacity constraints can be ensured by current machines or adding multiple machines that are available into the system. $j = j + 1$, if $j = P + 1$, go to step 5; else, go back to the top of step 3.
Initialize: Remove all unselected routings for each part.

Step 5 Reimprovement of the solution.
If step 4 has been invoked by any part, invoke add-in-step.
Assign each part to the cell that processes the first operation of the part. Go to next step.

Step 6 For each added multiple machines or newly purchased machines that are identified in step 4, assign it to the cell that maximize the utilization of the machine.

Add-in-step Initialize: calculate the number of intercellular movements T in the system by Equation 10.14.

Add-in 1 Create a new matrix $B(B_{mc})$. $m = 1, \ldots, M; c = 1, \ldots, C$.
Initialize: set $B_{mc} = 0, \forall m, \forall c$; set $m = 1$.
loop 1{ (loop 1 begins here)
Initialize: set $c = 1$
loop 2{ (loop 2 begins here)
move m from current cell c' to cell c $(c' \neq c)$. For each part, reform KR.
calculate the number of intercellular movements (T') by expression (14)
set $B_{mc} = T - T'$
set $c = c + 1$
if $(c \leq C)$ return to the top of loop 2;
else exit loop 2. (loop 2 ends here)
}
set $m = m + 1$
if $(m \leq M)$ return to the top of loop 1;
else exit loop 1. (loop 1 ends here)
}

Add-in 2 Find the element that bears the highest value B^h_{mc} in the matrix B.
if(B^h_{mc} >0)
reassign machine m to cell c
go back to the top of the step that invoked this add-in-step
else end.

10.5 Comparative Study and Computational Performance

In this section, we use our method to solve three problems and compare the results with the one given by Gupta (1993). The reason for choosing Gupta's is that his similarity coefficient is the only coefficient that considers alternative process routings, operation sequences, operation times and production volumes simultaneously in the literature. Furthermore, the machine capacity issue is also discussed in his research. Therefore, it can be concluded that his model is the most relevant one that compares to ours. In order to test the computational performance of the proposed heuristic approach, we also solve two big size problems by judging the performance of the approach. The algorithm has been coded in C++ and implemented on a Pentium II based IBM-PC compatible.

10.5.1 Problem 1

There are six machine types and eight parts in this problem. The operation sequences (shown by machine number) and operation times (in parentheses, unit is in minutes) for the alternative process routings of each part are as follows:

p1 r1: m1(2), m4(4), m2(2);
r2: m2(2), m3(2), m5(5), m6(6);
r3: m3(3), m2(2), m5(4), m6(6).
p2 r1: m3(3), m6(5), m5(5).
p3 r1: m3(3), m5(5), m6(5).
p4 r1: m1(1), m4(4);
r2: m2(2), m1(3), m4(4).
p5 r1: m6(6), m3(5), m2(2), m5(5);
r2: m3(4), m6(6).
p6 r1: m1(1), m2(2), m3(2);
r2: m1(1), m2(2), m6(5).
p7 r1: m5(5), m6(6), m2(2);
r2: m5(3), m6(6), m3(3);
r3: m2(2), m6(5).
p8 r1: m4(3), m2(2).

For each part, the production volume for a day is 50, 30, 20, 30, 20, 10, 15, 40, respectively. By using the approach proposed by Gupta (1993), the machine groups (MG), part families (PF) and finally selected routing for each part are given as follows:

MG-1: m1, m4; MG-2: m2, m3, m5, m6.
PF-1: p4(r1), p8(r1); PF-2: p1(r2), p6(r1), p3(r1), p2(r1), p5(r2), p7(r3).

The result is shown in Table 10.6. A total of 50 intercellular movements is produced by Gupta's model.

Table 10.6 Final solution by Gupta

Machine/Part									
		4	8	1	6	3	2	5	7
	r	1	1	2	1	1	1	2	3
m1		1			1				
m4		2	1						
m2			2	1	2			1	
m3				2	3	1	1	1	
m5				3		2	3		
m6				4		3	2	2	2

However, by applying our model, the result is given as following MGs, PFs and shown in Table 10.7.

MG-1: m1, m4, m2; MG-2: m3, m5, m6.
PF-1: p1(r1), p4(r1), p8(r1), p6(r1); PF-2: p3(r1), p2(r1), p5(r2), p7(r2).

Table 10.7 Final solution by our model

Machine/Part									
		1	4	8	6	3	2	5	7
	r	1	1	1	1	1	1	2	2
m1		1	1		1				
m4		2	2	1					
m2		3		2	2				
m3					3	1	1	1	3
m5						2	3		1
m6						3	2	2	2

Our model produces only 10 intercellular movements, which is absolutely better than the solution provided by Gupta.

10.5.2 Problem 2

This problem includes five machine types and seven parts. The operation sequences and operation times for the alternative process routings of each part are as follows:

p1 r1: m4(4), m1(2); r2: m1(2), m3(2), m4(4).
p2 r1: m1(3), m4(4).
p3 r1: m5(5), m2(4); r2: m2(3), m5(5), m3(3).
p4 r1: m2(2), m1(4), m5(5); r2: m2(3), m5(4), m3(3).
p5 r1: m1(3), m4(4); r2: m1(5), m3(3), m4(4).
p6 r1: m2(3), m5(5).
p7 r1: m5(3), m2(2).

For each part, the production volume for a day is 50, 5, 20, 30, 40, 10, 35, respectively. The results by Gupta's is given as follows:

MG-1: m1, m4, m3; MG-2: m2, m5.
PF-1: p1(r1), p2(r1), p5(r1); PF-2: p3(r1), p4(r2), p6(r1), p7(r1).

Thirty intercellular movements are produced in Gupta's model. However, the result by our model gets perfect partition for this problem, where the intercellular movement has been totally eliminated by our proposed approach. This problem again shows the efficiency of our model, and the result is as follows:

MG-1: m1, m4; MG-2: m2, m5, m3.
PF-1: p1(r1), p2(r1), p5(r1); PF-2: p3(r1), p4(r2), p6(r1), p7(r1).

10.5.3 Problem 3

This problem includes more important production data such as multiple machines of machine types, machine investment cost, *etc.* The model developed by Gupta (1993) fails to cope with these real-life production factors. The input data are given as follows:

Operation sequence of each routing:

p 1 r1: m1, m3, m4, m5, m6; r2: m1, m2, m3, m4, m3.
p 2 r1: m4, m1, m2, m5, m7; r2: m1, m4, m3, m5, m3.
p 3 r1: m6, m3, m2, m5; r2: m6, m2, m4, m1.
p 4 r1: m2, m3, m5, m9, m1; r2: m1, m2, m4, m3, m10.
p 5 r1: m5, m3, m5; r2: m4, m5, m2.
p 6 r1: m7, m5, m3, m1, m4; r2: m1, m8, m2, m4, m5.
p 7 r1: m2, m6, m1, m4, m5; r2: m8, m1, m2, m3, m5.
p 8 r1: m5, m4, m3, m2, m1; r2: m5, m4, m1, m2, m3.
p 9 r1: m2, m6, m9, m8, m9; r2: m7, m5, m9, m6, m9.
p10 r1: m6, m8, m7, m9, m10; r2: m6, m4, m7, m10, m9.
p11 r1: m6, m7, m8, m9; r2: m6, m8, m10, m1.
p12 r1: m6, m9, m10, m9; r2: m5, m6, m8, m5.
p13 r1: m6, m8, m7, m9; r2: m6, m7, m10, m9.
p14 r1: m8, m6, m9, m7; r2: m8, m6, m7, m10.
p15 r1: m6, m7, m8, m9, m10; r2: m7, m4, m6, m9, m10.
p16 r1: m6, m7, m9, m10; r2: m10, m7, m8, m9.

We assume the operation time for each operation is 3 min. The production volume for each part is as follows: p1 = 15, p2 = 30, p3 = 30, p4 = 20, p5 = 30, p6 = 20, p7 = 20, p8 = 20, p9 = 40, p10 = 20, p11 = 25, p12 = 50, p13 = 20, p14 = 30, p15 = 20, p16 = 30. The available number of multiple machines for each machine type is 2. The investment cost for purchasing new machine for each machine type is assumed as 2. The cost for redesigning a new process routing for each part is 3.

Based on above input data, the machine groups and part families of the heuristic are as follows:

MG-1: m1, m2, m4, m5, m3; MG-2: m6, m8, m9, m7, m10.

PF-1: p1(r2), p2(r2), p4(r2), p5(r2), p8(r1), p9(r1);

PF-2: p3(r1), p6(r1), p7(r2), p10(r1), p11(r1), p12(r1), p13(r1), p14(r1), p15(r1), p16(r1).

Since the machine capacities for some machine types cannot satisfy the requirements of production volumes of parts, step 4 of stage 2 added available multiple machines into the designing system. The number of added machines for each machine type is as follows:

m2 = 1, m3 = 1, m6 = 1, m7 = 1, m8 = 1, m9 = 1.

For machine type 9, even if the all available multiple machines have been added into the system, its capacity still cannot ensure the required part quantities. Thus, step 4 takes approach 3 to solve this capacity violated issue, and a new machine for machine type 9 is bought based on the comparison of the cost between purchase and redesign.

Finally, step 6 assigns all these machines to the cells, and the result is as follows (the number of machines for each machine type is in parentheses):

MG-1: m1(1), m2(1), m4(1), m5(1), m3(1), m8(1), m9(1);
MG-2: m2(1), m3(1), m6(2), m8(1), m9(2), m7(2), m10(1).

10.5.4 Computational Performance

Finally, in order to test the computational performance of the developed heuristic approach, two large-sized problems are solved on a Pentium II based IBM-PC compatible to check the CPU time. Problem 3 in Section 10.5.3 is used as the base to generate these two problems.

The first large problem includes 60 machines (30 machine types) and 96 alternative process routings (48 parts). The final solution is obtained within 1 min (44 s). The second big problem includes 80 machines (40 machine types) and 128 alternative process routings (64 parts), and it only cost 4 min to obtain the final solution.

From the test results, it can be concluded that the proposed heuristic is efficient either in the quality of the solutions or in the speed that leads to the solutions.

10.6 Conclusions

Various production factors have been discussed in this chapter. Due to the complexity of the cell formation problem, most approaches in the literature only involved a few factors mentioned in this chapter. The difficulty of the machine capacity violated issue is discussed and previous studies cannot cope with this issue accurately. We introduce five approaches to deal with this issue and apply these approaches in the developed heuristic algorithm. A new similarity coefficient that incorporates several important production factors is developed. The similarity coefficient extends the generalized similarity coefficient by using a sequence ratio and a machine-load ratio. A two-stage heuristic algorithm is developed to apply the proposed similarity coefficient. In stage 1, the basic machine cells are obtained. In stage 2, the part families are formed and appropriate process routing for each part is selected. The machine capacity violated problem is also solved in stage 2. Five numerical examples are solved to demonstrate the effectiveness of the proposed similarity coefficient and the solution procedure of the heuristic algorithm. The computational results show that the approach provides feasible solutions rapidly.

References

Burbidge JL (1971) Production flow analysis. Prod Eng 50:139–152

Caux C, Bruniaux R, Pierreval H (2000) Cell formation with alternative process plans and machine capacity constraints: a new combined approach. Int J Prod Econ 64:279–284

Chan HM, Milner DA (1982) Direct clustering algorithm for group formation in cellular manufacture. J Manuf Syst 1(1):65–75

Chandrasekharan MP, Rajagopalan R (1986a) An ideal seed non-hierarchical clustering algorithm for cellular manufacturing. Int J Prod Res 24:451–464

Choobineh F (1988) A framework for the design of cellular manufacturing systems. Int J Prod Res 26:1161–1172

Chu CH, Pan P (1988) The use of clustering techniques in manufacturing cellular formation. In: Proceedings of the International Industrial Engineering Conference, Orlando, Florida, pp 495–500

Co HC, Araar A (1988) Configuring cellular manufacturing systems. Int J Prod Res 26:1511–1522

De Beer C, De Witte J (1978) Production flow synthesis. Ann CIRP 27:389–392

De Witte J (1980) The use of similarity coefficients in production flow analysis. Int J Prod Res 18:503–514

El-Essawy IGK, Torrance J (1972) Component flow analysis – an effective approach to production systems' design. Prod Eng 51:165–170

Gupta T (1993) Design of manufacturing cells for flexible environment considering alternative routeing. Int J Prod Res 31:1259–1273

Gupta T, Seifoddini H (1990) Production data based similarity coefficient for machine-component grouping decisions in the design of a cellular manufacturing system. Int J Prod Res 28:1247–1269

Heragu SS (1994) Group technology and cellular manufacturing. IEEE Trans Syst Man Cybernetics 24:203–215

Heragu SS, Gupta YP (1994) A heuristic for designing cellular manufacturing facilities. Int J Prod Res 32:125–140

Ham I, Hitomi K, Yoshida T (1985) Group technology: applications to production management Kluwer-Nijhoff, Boston

Hwang H, Ree P (1996) Routes selection for the cell formation problem with alternative part process plans. Comput Ind Eng 30:423–431

Islam KMS, Sarker BR (2000) A similarity coefficient measure and machine-parts grouping in cellular manufacturing systems. Int J Prod Res 38:699–720

King JR (1980) Machine-component grouping in production flow analysis: an approach using a rank order clustering algorithm. Int J Prod Res 18(2):213–232

King JR, Nakornchai V (1982) Machine component group formation in group technology: review and extension. Int J Prod Res 20:117–133

Kumar KR, Vannelli A (1987) Strategic subcontracting for efficient disaggregated manufacturing. Int J Prod Res 25:1715–1728

Kusiak A (1987) The generalized group technology concept. Int J Prod Res 25:561–569

Lashkari RS, Gunasingh KR (1990) A Lagrangian relaxation approach to machine allocation in cellular manufacturing systems. Comput Ind Eng 19:442–446

Logendran R (1991) Impact of sequence of operations and layout of cells in cellular manufacturing. Int J Prod Res 29:375–390

McAuley J (1972) Machine grouping for efficient production. Prod Eng 51:53–57

Miltenburg J, Zhang W (1991) A comparative evaluation of nine well-known algorithms for solving the cell formation problem in group technology. J Oper Manage 10:44–72

Mosier CT, Taube L (1985) Weighted similarity measure heuristics for the group technology machine clustering problem. Omega 13:577–583

Purcheck GFK (1974) Combinatorial grouping: a Lattice-theoretic method for the design of manufacturing systems. J Cybernetics 4:27–60

Purcheck GFK (1975) A mathematical classification as a basis for the design of group-technology production cells. Prod Eng 54:35–48
Qiao LH, Yang ZB, Wang HP (1994) A computer-aided process planning methodology. Comput Ind 255:83–94
Rajagopalan R, Batra JL (1975) Design of cellular production system: a graph theoretic approach. Int J Prod Res 13:567–579
Reisman A, Kumar A, Motwani J, Cheng CH (1997) Cellular manufacturing: a statistical review of the literature (1965–1995). Oper Res 45:508–520
Sarker BR, Xu Y (1998) Operation sequences-based cell formation methods: a critical survey. Prod Plan Control 9:771–783
Seifoddini H, Wolfe PM (1987) Selection of a threshold value based on material handling cost in machine-component grouping. IIE Trans 19:266–270
Selim HM, Askin RG, Vakharia AJ (1998) Cell formation in group technology: review, evaluation and directions for future research. Comput Ind Eng 34:3–20
Shtub A (1989) Modeling group technology cell formation as a generalized assignment problem. Int J Prod Res 27:775–782
Sofianopoulou S (1999) Manufacturing cells design with alternative process plans and/or replicate machines. Int J Prod Res 37:707–720
Steudel HJ, Ballakur A (1987) A dynamic programming based heuristic for machine grouping in manufacturing cell formation. Comput Ind Eng 12:215–222
Vakharia AJ, Wemmerlöv U (1990) Designing a cellular manufacturing system: a materials flow approach based on operation sequences. IIE Trans 22:84–97
Vannelli A, Kumar KR (1986) A method for finding minimal bottle-neck cells for grouping part-machine families. Int J Prod Res 24:387–400
Wemmerlöv U, Hyer NL (1986) Procedures for the part family/machine group identification problem in cellular manufacturing. J Oper Manage 6:125–147
WemmerlÃ¶v U, Johnson DJ (1997) Cellular manufacturing at 46 user plants: implementation experiences and performance improvements. Int J Prod Res 35:29–49
Won YK, Kim SH (1997) Multiple criteria clustering algorithm for solving the group technology problem with multiple process routings. Comput Ind Eng 32:207–220
Wu N, Salvendy G (1993) A modified network approach for the design of cellular manufacturing systems. Int J Prod Res 31:1409–1421
Yasuda K, Yin Y (2001) A dissimilarity measure for solving the cell formation problem in cellular manufacturing. Comput Ind Eng 39:1–17
Zhao CW, Wu ZM (2000) A genetic algorithm for manufacturing cell formation with multiple routes and multiple objectives. Int J Prod Res 38:385–395

Chapter 11
Fuzzy Approach to Quality Function Deployment-based Product Planning

This chapter presents a fuzzy modeling approach and a genetic-based interactive approach to QFD planning taking the financial factor and design uncertainties into consideration. Before formulating the optimization model, a QFD-based integrated product development process model is presented firstly. By introducing some new concepts of planned degree, actual achieved degree, actual primary costs required and actual planned costs, two types of fuzzy nonlinear optimization models are introduced here. These models not only consider the overall customer satisfaction, but also the enterprise satisfaction with the costs committed to the product. With the interactive approach, the best balance between enterprise satisfaction and customer satisfaction can be obtained, and the preferred solutions under different criteria can be achieved by means of the human–computer interaction.

11.1 Introduction

In a fiercely competitive global market, being able to develop new products with a shorter lead time, cheaper prices and better quality has become a key success factor of manufacturing enterprises. New product development (NPD) is a complex informational, technical and business process, and it must be managed efficiently and effectively in a company as well as throughout its supply network.

Product quality design is an important function in NPD to ensure higher quality, lower cost and shorter development time. Quality function deployment (QFD), which originated in the 1960s in Japan, is an overall concept that provides an efficient means to plan product design activities from customer requirements to product through the phases of product planning, parts planning, process planning and production planning (Akao 1990). Since then, QFD has been widely adopted as a customer-driven approach to plan product quality and improve customer satisfaction, particularly in a concurrent engineering (CE) environment. It has been used as a customer-oriented approach and tool to product development including new product development and product improvement in a structured way on the basis of assessment of customer requirements. It describes the interrelationships between customer requirements (CRs) and technical attributes (TAs) of a product and the co-relationship of TAs, which have to be considered in order to achieve higher overall customer satisfaction.

The complex relationships between customer requirements and technical attributes, and the correlation between different TAs, can be illustrated in a typical "house of quality" (HoQ) (Akao 1990; Bode and Fung 1998), and the development and formulation of these relationships and correlation are important procedures in the QFD process. There are seven steps in the product development project using QFD, among which the establishment of the relationship matrix of the customer requirements/attributes (CRs) and the technical attributes (TAs), with its correlation matrix among TAs are the important starting points.

However, overall customer satisfaction can be achieved through meeting individual customer requirements, which may well conflict with one another. The prioritization of the customer requirements can be formulated according to their relative importance, and it reflects their individual contributions towards the overall performance of a product (Fung *et al.* 1996, 1998). On the other hand, a given CRs may be related to a number of TAs. Therefore, the overall customer satisfaction has to be formulated by mapping the CRs onto the TAs.

The determination of degree of attainment (target levels) for the TAs of a product with a view to achieve higher level of overall customer satisfaction is usually the focus on the QFD process planning (Fung *et al.* 1996, 1998a,b, 2003). Traditional methods for setting the degree of attainment is mostly accomplished in a subjective, *ad hoc* manner (Fung *et al.* 2002) or a heuristic way, such as prioritized-based (Fung *et al.* 1998b; Hauser and Clausing 1988), both of which aim at arriving at a feasible design, rather than an optimal one. These prioritization-based methods could not achieve global optimization, and most of these models take little consideration of the correlation between TAs.

Moreover, these models and methods are technically one-sided without considering the limited design budget; however, they are unreasonable in QFD planning in practice. In fact, the resources and cost budget for target level of TAs for a product are not infinite, but limited. Therefore, the financial factor is also an important consideration and should not be neglected in QFD planning.

Owing to the fact that these methods seldom consider the correlation among TAs, they can not explore the relationship between degrees of attainment of two different TAs, resulting in a linear formulation of the cost of achieving the degree of attainment of TAs. In addition, it is assumed that the costs committed for fully attaining the target of TAs under the condition that there are no other costs for other TAs is a deterministic value. In practice, however the primary cost required may be expressed as a fuzzy number with imprecision in order to cope with the uncertainties in a design process. These uncertainties include ill-defined or incomplete understanding of the relationship between TAs and CRs, and the degree of dependence among TAs, as well as the subjective nature and preference in the decision process. Under these circumstances, a fuzzy modeling approach based on fuzzy set theory may be more suitable and efficient for integrating the financial factors into a QFD planning process.

This chapter describes a fuzzy modeling approach and a genetic-based interactive approach to QFD planning taking the financial factor and design uncertainties into consideration. Before formulating the optimization model, a QFD-based integrated product development process model is presented firstly. By introducing some new concepts of planned degree, actual achieved degree, actual primary costs required and actual planned costs, two types of fuzzy nonlinear optimization models are introduced. These models not only consider the overall customer satisfaction, but also the enterprise satisfaction with the costs committed to the product. With the interactive approach, the best balance between enterprise satisfaction and customer satisfaction can be obtained, and the preferred solutions under different criteria can be achieved by means of the human–computer interaction.

11.2 QFD-based Integration Model for New Product Development

11.2.1 Relationship Between QFD Planning Process and Product Development Process

In general, the processes of developing mechanic and electronic products consist of the major stages of customer requirements capture, conceptual design, engineering design, process design, parts manufacturing and assembly. As indicated in the Section 10.1, being a customer-driven approach, QFD is in effect a systematic methodology that provides an efficient planning tool to deploy customer requirements (CRs) in a hierarchical way and to plan the whole process through the phases of product planning, part planning, process planning and production planning via the house of quality (HoQ). Starting with customer requirements (CRs), the QFD deployment and planning processes in product development are conducted in four planning phases including product planning, parts deployment planning, process planning and production planning, each corresponding to individual activity of product development processes, *i.e.*, conceptual design, engineering design, process planning and manufacturing, respectively. Thus there exist corresponding relations between QFD planning processes and product development processes. During the QFD planning phases, the HoQ provides an effective means of deployment and planning from one phase to the next.

11.2.2 QFD-based Integrated Product Development Process Model

Basing on the QFD planning approach and taking into account customer participation in product conceptual design and earlier supplier involvement in parts and process deployment in new product development, an integrated product development process model (IPDP/QFD) based on QFD is proposed in this section. The basic idea and features lie in the following aspects:

1. The whole product development processes are decomposed into five major phases, *i.e.*, customer requirement synthesis, product concept design, engineering design, process design and manufacturing, each corresponding to an individual task and objective.
2. QFD is taken as the basic deployment and planning tools, by which HoQ_1, HoQ_2, HoQ_3 and HoQ_4 correspond to concept design, engineering design, process design and manufacturing, respectively.
3. Fuzzy formulation and quantitative analysis, artificial neural network, expert system, case-based reasoning and optimization theory are embedded into the deployment and planning processes. In particular, some optimization models and algorithms are developed as planning optimizer to support the planning processes.
4. Customer requirements are obtained in a new way in the global market area via Internet technology and synthesized and evaluated in an automatic and intelligent way.
5. Earlier supplier involvement (ESI) is considered in the part deployment and planning process.
6. Enterprise knowledge including product knowledge, process knowledge and resources knowledge, as well as the design team's knowledge and experience are collected and developed as knowledge base and combined into the model.

Combining the above features, the QFD-based integrated product development process model (IPDP/QFD) is presented in Figure 11.1.

The integrated product development process model (IPDP/QFD) is a conceptual model, and it consists of six modules. They are user interface to the Internet, customer requirement (CR) capture and synthesis, knowledge express and fuzzy information process, QFD deployment and planning, QFD planning optimizer, development tool and platform. QFD planning optimizer is composed of product planning optimizer, part planning optimizer, process planning optimizer and production planning optimizer, where each is supported by several optimization models and solution algorithms that are stored in model base (MB) and arithmetic base (AB), respectively. Development tool and platform is a support module for other components of the IPDP/QFD model and it is not only composed of development tools and development platforms, but also of databases, knowledge bases, and rule and method bases. The database is mainly composed of product-related database, customer information database, resource database and customer requirements information database. The rule and method bases consist of algorithms for solving the planning models and methods for fuzzy quantities rules, fuzzy rules and cases. Knowledge base includes the marketing engineer's knowledge of CRs, for example predefined keywords on CRs; it also includes the designer's knowledge of product bills of materials (BOM) and product specification, as well as knowledge on the relationship between CRs.

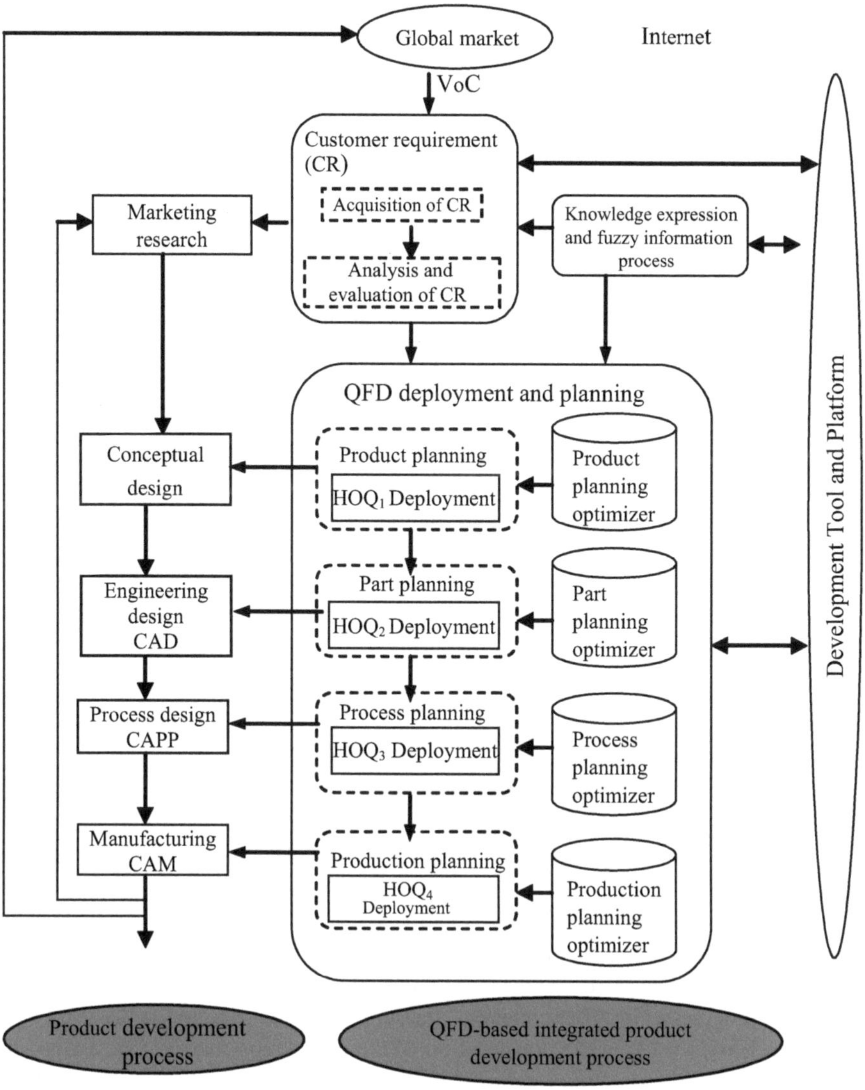

Figure 11.1 Integrated product development process model

11.3 Problem Formulation of Product Planning

The basic concept of QFD in the product design is to translate the desires of the customer into the determination of TAs, which is illustrated by a HoQ (Akao 1990; Bode and Fung 1998). Assuming for a product design, there are m CRs denoted by CR_i, $i = 1, 2, \ldots, m$ and n TAs denoted by TA_i, $i = 1, 2, \ldots, n$. The overall customer satisfaction is gauged in terms of the satisfaction with all individual customer requirements, which are generally conflicting with one another. Let d_i $(i = 1, 2, \ldots, m)$ be the weight of importance of the ith CR indicating the relative importance of ith CR towards the overall customer satisfaction, whereas $w_j(j = 1, 2, \ldots, n)$ denoting the relative weight of importance of TAs, is determined from the relationship between CRs and TAs. Let R be the relationship matrix between CRs and TAs, the elements R_{ij} of which indicates the strength of impact of the jth TA on the satisfaction of the ith CR. The value of R_{ij} may be assigned by the scale of 1-3-9 or 1-9-15 to denote weak, medium and strong relationships, respectively, or using quantified HoQ method 2. R^*_{ij}, normalized R_{ij}, may be interpreted as the contribution quota of the jth TA towards complete fulfillment of the ith CR when the target of the jth TA is met. With this interpretation, the weights w_j $(j = 1, 2, \ldots, n)$ may be derived in terms of the following equation:

$$w_j = \sum_{i=1}^{m} d_i R^*_{ij}, j = 1, 2, \ldots, n \,. \tag{11.1}$$

Let the decision variable x_j be the degree of attainment of the jth TA. The traditional formulation of a QFD planning problem may be expressed as (Fung *et al.* 2002):

$$\text{(P1)} \qquad \text{Max } Z = \sum_{i=1}^{m} d_i v_i \tag{11.2}$$

$$s.t.\ v_i = \sum_{j=1}^{n} R_{ij} x_j \,, \quad i = 1, 2, \ldots, m \tag{11.3}$$

$$0 < x_j = f_j(x_1, x_2, \ldots, x_n) \leq 1 \,, \quad j = 1, 2, \ldots, n \tag{11.4}$$

where v_i, $f_j(x)$ are the perception degree of customer satisfaction with the ith CR and the functional relationship between the jth TA and other TAs.

In general, correlation exists between different types of TAs. Let T be the correlation matrix of the TAs; the element T_{ij} denotes the correlation factor between ith and jth TAs. If there is no dependence between ith and jth TAs, $T_{ij} = 0$; else it represents the degree/scale of dependence. Of course, the TA is defined as the strongest dependence with itself in the construction of the correlation matrix, *i.e.*, T_{ii} is defined as the maximum degree/scale. When it is normalized $T_{ij} \in (-1, 1]$ with this formulation, we not only can formulate the dependence or independence between TAs, but also formulate the strong degree of dependence. In addition, $T_{ij} < 0$ implies that there is a conflict between the ith and jth TAs, *i.e.*, the ith TA has a negative impact on the jth TA, and *vice versa*. Similarly, they have a positive effect on each other if $T_{ij} > 0$.

In constructing the correlation matrix/element, the degree/scale of dependence of a TA on itself is defined to be much larger than that between two different TAs. For example, the correlation between two TAs is scaled by way of 1-3-5-15, the point 1, 3, 5, are scaled to formulate the degree of weak, medium, and strong dependency between two different TAs, whereas point 15 is defined to be the scale of dependence of one TA on itself. After normalization $T_{ii} = 1$, while $T_{ij} \in (-1, 1)$ for $i \neq j$.

In this case, the normalized correlation element T_{ij} may be interpreted as incremental changes of degree of the attainment of the jth TA when the degree of attainment of the ith TA is increased by one unit.

11.4 Actual Achieved Degree and Planned Degree

On account of their correlation between two different types of TAs, it is difficult to formulate the relation between x_i and x_j. As indicated above, the normalized correlation element T_{ij} may be interpreted as incremental changes of the degree of attainment of the jth TA when the degree of attainment of the ith TA is increased by one unit. With this interpretation, it is necessary to introduce the concept of planned degree of attainment of TA in order to distinguish it from the actual achieved degree of attainment of TA.

Let the planned degree of attainment of the jth TA be y_j, which is the decision variable we focus on. Owing to the correlation between TAs, x_j, the actual achieved degree of attainment of the jth TA, may be formulated as:

$$x_j = y_j + \sum_{k \neq j} T_{kj} y_k = \sum_{k=1}^{n} T_{kj} y_k \, . \tag{11.5}$$

In the case that $1 - \sum_{k \neq j} T_{kj} y_k = 0$, it implies that other TAs have strongly positive correlation with a certain TA, say the jth TA, such that the jth TA is fully achieved with no planned degree of attainment, *i.e.*, $y_j = 0$. It also implies that when the planned degree of attainment of some other TAs reaches some certain level, the target of the jth TA is attained completely owing to its correlation with other TAs; that is to say, the planned degree of attainment of jth TA is zero.

In some extreme cases, for some planned degree of attainment $y_j (j = 1, 2, \ldots, n)$, the factor $y_j + \sum_{k \neq j}^{n} T_{kj} y_k$ may be greater than 1 owing to the positive effects on a certain TA, or less than zero when there are some negative effects on the TA, both of which are infeasible. In order to guarantee the feasibility, the following constraints should be imposed:

$$0 \leq y_j + \sum_{k \neq j}^{n} T_{kj} y_k \leq 1 \, , \quad j = 1, 2, \ldots, n \tag{11.6}$$

In addition, for a given jth TA, the planned degree of attainment is guaranteed to meet $0 \leq y_j \leq 1$.

11.5 Formulation of Costs and Budget Constraint

Assume that there are multiple resources required to support the design of a product, including technical engineers, advanced facilities, tools or any other facilities. At the level of strategy planning, the above types of resources can be aggregated in financial terms.

According to the past experiences, the costs committed for the full degree of attainment of the jth TA under the condition that no other costs for other TAs is assumed to be c_j, which is defined as the primary cost required, *i.e.*, the cost needed to reach the highest level of the jth TA without considering other TA contributions to the attainment of the jth TA.

For simplicity, assuming that the cost function C_j for achieving the degree of attainment of the jth TA is scaled linearly to the degree of attainment x_j results in $C_j(x_j) = c_j x_j$.

Due to the correlation of TAs, the actual primary costs required c_j^* is defined as the costs required for the fulfillment of the target of the jth TA when there are other costs for other TAs. It may be formulated as follows:

$$c_j^* = c_j \left(1 - \sum_{i \neq j} T_{ij} y_i\right) . \tag{11.7}$$

On account of the fact that the degree of attainment of the jth TA x_j comes from two parts, one is the planned degree which is directly from the committed costs, and the other one is indirectly from the correlation of other TAs, so the costs $C_j(x_j)$ for achieving the degree of attainment x_j is just the actual planned costs for the planned degree of attainment. It may be formulated as:

$$C_j(x_j) = c_j^* x_j = c_j \left(1 - \sum_{k \neq j} T_{kj} y_k\right)\left(y_j + \sum_{k \neq j} T_{kj} y_k\right) . \tag{11.8}$$

In the case that there is no correlation between the jth TA and other TAs, *i.e.*, the jth TA is independent from other TAs, $T_{kj} = 0, (k \neq j)$. Under this circumstance, $c_j^* = c_j$ and $C_j(x_j) = c_j y_j = c_j x_j$, which coincides with the actual scenario. As indicated above, when $1 - \sum_{k \neq j} T_{kj} y_k = 0$, it implies that the full attainment of the jth TA is achieved completely due to the correlation of other TAs, *i.e.*, the planned degree of attainment of the jth TA is zero. This results in that the actual planned costs needed for the jth TA is zero. It is consistent with Equation 11.8.

Assume that the design costs for product development or product improvement are constrained to a budget B. Under this circumstance, the design budget constraints are formulated as:

$$C(y) = \sum_{j=1}^{n} C_j(y_j) = \sum_{j=1}^{n} c_j \left(1 - \sum_{k \neq j} T_{kj} y_k\right) \sum_{k=1}^{n} T_{kj} y_k \leq B . \tag{11.9}$$

Owing to the uncertainties in the design process, such as ill-defined or incomplete understanding of the relationship between TAs and CRs, as well as the human subjective judgment on the dependence between TAs, c_j can be expressed as a fuzzy number. Let c_j be a triangular fuzzy number denoted by $\tilde{c}_j = (c_j^p, c_j, c_j^o)$, where

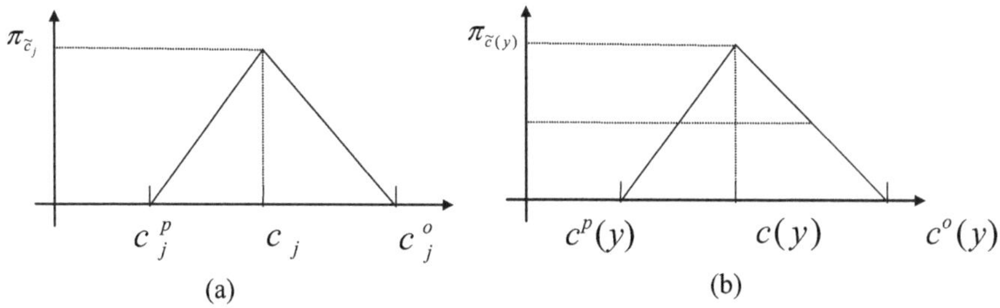

Figure 11.2 The possibility formulation of triangular fuzzy number $\tilde{c}_j$: (a) possibility distribution function $\pi_{\tilde{c}_j}(t)$, and (b) objective function $\tilde{C}(y)$

c_j^p, c_j, c_j^o are most pessimistic, most likely and most optimistic values, respectively. The possibility distribution function $\pi_{\tilde{c}_j}(t)$ is shown in Figure 11.2 (a). In light of the fuzzy arithmetic, for any $y = (y_1, y_2, \ldots, y_n) \in [0, 1]^n$, let

$$\tilde{C}(y) = \sum_{j=1}^{n} \tilde{c}_j \left(1 - \sum_{k \neq j} T_{kj} y_k\right) \sum_{k=1}^{n} T_{kj} y_k \tag{11.10}$$

The objective function $\tilde{C}(y)$ denoted by $(c^p(y), c(y), c^o(y))$, is also a triangular fuzzy number with possibility distribution, as shown in Figure 11.2 (b), where $c^p(y), c(y), c^o(y)$ are most pessimistic, most likely and most optimistic values of the objective function $\tilde{C}(y)$, and can be determined as follows:

$$c^p(y) = \sum_{j=1}^{n} c_j^p \left(1 - \sum_{k \neq j} T_{kj} y_k\right) \sum_{k=1}^{n} T_{kj} y_k \tag{11.11}$$

$$c(y) = \sum_{j=1}^{n} c_j \left(1 - \sum_{k \neq j} T_{kj} y_k\right) \sum_{k=1}^{n} T_{kj} y_k \tag{11.12}$$

$$c^o(y) = \sum_{j=1}^{n} c_j^o \left(1 - \sum_{k \neq j} T_{kj} y_k\right) \sum_{k=1}^{n} T_{kj} y_k \,. \tag{11.13}$$

11.6 Maximizing Overall Customer Satisfaction Model

The level of overall customer satisfaction with a product is denoted by S. It is achieved through the aggregation of degrees of customer satisfaction s with individual CRs, *i.e.*,

$$S = \sum_{i=1}^{m} d_i s_i \,. \tag{11.14}$$

On the other hand, the degrees of customer satisfaction with CRs are derived from the *actual achieved degrees* of attainment of TAs through normalizing the relationship matrix between CRs and TAs. With normalized correlation elements, the relationship between s_i and x_j can be given as follows:

$$s_i = \sum_{j=1}^{n} R_{ij}^* x_j \,, \quad i = 1, 2, \ldots, m \,. \tag{11.15}$$

In terms of the actual achieved degree of attainment of TAs, the overall customer satisfaction is stated as follows:

$$S = \sum_{j=1}^{n} w_j, x_j \tag{11.16}$$

Let

$$w_k^* = \sum_{j=1}^{n} w_j T_{kj} \,, \quad k = 1, 2, \ldots, n \tag{11.17}$$

where w_k^* represents the contribution of kth TA to overall customer satisfaction due to their correlation of TAs when one unit of planned degree of attainment of the TA is committed. It is noted that $\sum_{k=1}^{n} w_k^* \geq 1$.

With the planned degree of attainment of TAs, the overall customer satisfaction level can be expressed as follows:

$$S = \sum_{j=1}^{n} w_j \sum_{k=1}^{n} T_{kj} y_k = \sum_{k=1}^{n} \sum_{j=1}^{n} w_j T_{jk} y_k = \sum_{k=1}^{n} w_k^* y_k \,. \tag{11.18}$$

Considering the imprecision of the primary cost required, limited design budget and other technical constraints, the QFD planning problem can be formulated as *FP1*[12]:

$$\text{Max } S = \sum_{j=1}^{n} w_j \sum_{k=1}^{n} T_{kj} y_k = \sum_{k=1}^{n} \sum_{j=1}^{n} w_j T_{kj} y_k = \sum_{k=1}^{n} w_k^* y_k \tag{11.19}$$

$$\text{s.t.} \sum_{k=1}^{n} T_{kj} y_k \geq \theta_0 \,, \quad j = 1, 2, \ldots, n \tag{11.20}$$

$$C(y) = \sum_{j=1}^{n} \tilde{c}_j \left(1 - \sum_{k \neq j} T_{kj} y_k \right) \sum_{k=1}^{n} T_{kj} y_k \tilde{\leq} B \tag{11.21}$$

$$\sum_{k=1}^{n} T_{kj} y_k \leq 1 \,, \quad j = 1, 2, \ldots, n \tag{11.22}$$

$$0 \leq y_j \leq 1 \,, \quad j = 1, 2, \ldots, n \tag{11.23}$$

where θ_0 and B are the preferred acceptable degree of attainment of TA and total limited budget respectively, y_j are the planned degree of attainment of TAs and decision variables in this model.

The preferred acceptable degree θ_0 reflects the decision maker's (DM) preference and subjectivity; its specification depends on the DM preference on technical requirements. Additionally, different levels of θ_0 may be applied to different TAs in practical scenarios.

FP1 is a fuzzy nonlinear programming model with linear objective and nonlinear constraints. Fung *et al.* (2002) developed a fuzzy solution to *FP1* and made a comparison with the traditional prioritization-based QFD planning methods.

11.7 Minimizing the Total Costs for Preferred Customer Satisfaction

Model *FP1* aims at maximizing the overall customer satisfaction by considering the planned degree of attainment of TAs under a limited budget and specific level of TA constraints. In some cases, the enterprise hopes to achieve a preferred acceptable customer satisfaction at the smallest design cost, which leads to the financial model *FP2* formulated as follows:

$$\text{Min}\, \tilde{C}(y) = \sum_{j=1}^{n} C_j(y_j) = \sum_{j=1}^{n} \tilde{c}_j \left(1 - \sum_{k \neq j} T_{kj} y_k\right) \sum_{k=1}^{n} T_{kj} y_k \tag{11.24}$$

$$s.t. \sum_{j=1}^{n} w_j \sum_{k=1}^{n} T_{kj} y_k = \sum_{k=1}^{n} w_k^* y_k \geq \beta_0 \tag{11.25}$$

$$\sum_{k=1}^{n} T_{kj} y_k \leq 1\,, \quad j = 1, 2, \ldots, n \tag{11.26}$$

$$\sum_{k=1}^{n} T_{kj} y_k \geq \beta_j\,, \quad j = 1, 2, \ldots, n \tag{11.27}$$

$$0 \leq y_j \leq 1\,, \quad j = 1, 2, \ldots, n \tag{11.28}$$

where β_0, and β_j are the preferred acceptable customer satisfaction and preferred requirements for the actual achieved degree of attainment of TAs, respectively.

In model *FP2*, Equation 11.24 is the objective of minimizing total design costs, and $\tilde{C}(y)$ is a fuzzy objective function with triangular fuzzy number defined in Equations 11.11–11.13. Equation 11.25 is the constraint of the overall customer satisfaction with a preferred level, while Equations 11.26 and 11.27 are the preferred requirements of the actual achieved degree of attainment of TAs.

11.8 Genetic Algorithm-based Interactive Approach

FP2 is a fuzzy quadratic programming asymmetric model with fuzzy objective coefficients and crisp constraints. Unlike crisp mathematical programming, the solutions to fuzzy mathematical programming have a different meaning, depending on the DM understanding of the problem and the interpretation of the optimal solution. There are various approaches to fuzzy mathematical programming (Lai and Hwang 1992), including symmetric and asymmetrical approaches. In this section we will develop overall procedures to solve *FP2* by introducing the concept of enterprise satisfaction and solving its equivalent crisp model with a genetic algorithm.

11.8.1 Formulation of Fuzzy Objective Function by Enterprise Satisfaction Level

Since the objective function in model *FP2* is a fuzzy number with a possibility distribution, we cannot directly solve it by traditional optimization methods. The commonly used method is to translate it into the solution to an equivalent crisp model, including multiple-objective model and single-objective model, depending on the understanding and interpretation of minimizing objective in the fuzzy sense. On the other hand, if the optimal objective of *FP2* is larger than the limited budget B, then this implies that it is impossible to achieve the preferred customer satisfaction under the limited budget. An alternative way is either to degrade the preferred customer satisfaction or to increase the design budget. Therefore, there is a trade-off between overall customer satisfaction and the enterprise satisfaction with the costs committed to the product, which motivate one to formulate a fuzzy objective function by the concept of enterprise satisfaction level.

In fact, on one hand, the enterprise hopes to obtain maximum overall customer satisfaction, while on the other hand, the minimized design cost is also a main concern of the enterprise. Therefore, there needs to be a trade-off between overall customer satisfaction with the product and enterprise satisfaction with the costs committed to the product. Let the enterprise expected cost or aspiration level of cost be C_0, the maximum budget be B, *i.e.*, the enterprise is fully satisfied if the costs are no more than C_0, and partially satisfied when the total cost is more than C_0, but less than the budget B, and not satisfied when the total cost are equivalent or larger than B. The enterprise satisfaction level with the costs denoted by ES may be formulated as a linear or nonlinear membership function as shown in Figure 11.3 (a). The problem may be described as how to organize the planned degrees of attainment of TAs in order to obtain the highest possible customer satisfaction at least possible cost, *i.e.*, how to find the best balance between customer satisfaction and enterprise satisfaction.

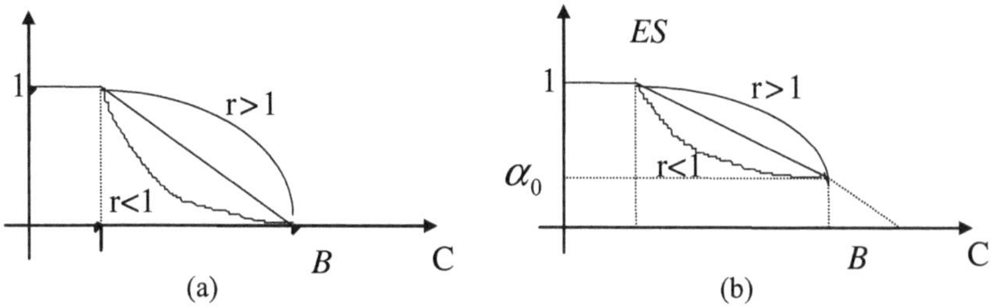

Figure 11.3 The formulation of enterprise satisfaction with costs: (a) ES as linear or nonlinear membership function, and (b) ES formulated as non-zero when the cost committed for the product reaches B

11.8.2 Transforming FP2 into a Crisp Model

With formulation of enterprise satisfaction with costs, the *FP2* model is transferred to the equivalent trade-off problem between overall customer satisfaction and enterprise satisfaction. In this sense, the QFD planning problem *FP2* is equivalent to the following crisp model CP:

$$\text{Max}\,\alpha \tag{11.29}$$

$$s.t.\; ES \geq \alpha \tag{11.30}$$

$$S = \sum_{k=1}^{n} w_k^* y_k \geq \alpha \tag{11.31}$$

$$\theta_j \leq \sum_{k=1}^{n} T_{kj} y_k \leq 1\,, \quad j = 1, 2, \ldots, n \tag{11.32}$$

$$0 \leq y_j \leq 1\,, \quad j = 1, 2, \ldots, n \tag{11.33}$$

$$0 \leq \alpha \leq 1 \tag{11.34}$$

$$ES = \begin{cases} 1 & C \leq C_0 \\ 1 - \left(\dfrac{C - C_0}{B - C_0} \right)^r & C_0 < C < B \\ 0 & C \geq B \end{cases} \tag{11.35}$$

where $r > 0$. C is a kind of most likely value of fuzzy costs $\tilde{C}(y)$, and it can be determined in the form (Lai and Hwang 1992) as follows:

$$C = [4c(y) + c^p(y) + c^m(y)]/6\,. \tag{11.36}$$

Of course other forms of fuzzy costs, such as weighted sum of optimistic and pessimistic values may be applied to C. In the model CP, Equation 11.29 is the objective of maximizing best balance between customer satisfaction and enterprise satisfaction, Equation 11.30 is the enterprise satisfaction constraint, and Equation 11.31 is the overall customer satisfaction constraint.

In some practical situations, it may be more reasonable that the enterprise satisfaction level ES may be formulated as non-zero, *i.e.*, $\alpha_0 > 0$, when the cost committed for the product reaches the budget B. This scenario can be illustrated in Figure 11.3 (b). Under this circumstance, Equation 11.35 may be replaced by the following formula:

$$ES = \begin{cases} 1 & C \le C_0 \\ (1-\alpha_0)\left(\dfrac{B-C}{B-C_0}\right)^r + \alpha_0 & C_0 < C \le B \\ 0 & C > B \end{cases} \tag{11.37}$$

In general, Equation 11.37 is adopted to express the enterprise satisfaction level with the costs committed to the product, where α_0 is named as the acceptable enterprise satisfaction with the cost of the budget. Under this circumstance, ES is a nonlinear discontinuous function of planned degree of attainment of TAs; therefore, CP is also a nonlinear optimization model.

In any case, model CP is equivalent to model CP-1 as follows:

$$\text{(CP-1)} \qquad \text{Max Min}\left\{ES, \sum_{k=1}^{n} w_k^* y_k\right\} \tag{11.38}$$

$$S.t. \quad (32)\text{–}(33)$$

$$\alpha_0 \le ES \le 1 \tag{11.39}$$

$$\alpha_0 \le \sum_{k=1}^{n} w_k^* y_k \le 1\,. \tag{11.40}$$

11.8.3 Genetic Algorithm-based Interactive Approach

Model CP-1 is a nonlinear optimization model. Moreover, the enterprise satisfaction level in Equation 11.30 with Equation 11.37 is a discontinuous function, and thus cannot be solved by traditional optimization methods. In this chapter a hybrid genetic algorithm (Tang *et al.* 1998) with mutation along the weighted gradient direction is suggested as a solution. The basic idea is described as follows: first, an initial population is randomly produced; each individual is selected to reproduce children by means of mutation along the incremental direction of enterprise satisfaction level and overall customer satisfaction level according to the selection probability depending on the fitness function value. For the individuals with both enterprise and overall customer satisfaction level less than α_0, give it a less fitness value by means of penalty so that it may have less of a chance than others to be selected to reproduce children in the later generation. With the increase of generations, both the enterprise and overall customer satisfaction level corresponding to each individual in the generation is not only greater than α_0 but also closer to the optimal solution. Moreover, these individuals will form a neighbor domain including the exact optimal solution.

In order for the decision maker (DM) to select the preferred solution from the neighbor domain of the optimal solution, a human–computer interactive procedure is designed as follows (Tang and Wang 1997). Firstly, the DM is asked to prefer an acceptable satisfaction level α_0 by both customer and enterprise. Secondly, by means of human–computer interaction, the preferred enterprise satisfaction is constructed and elicits the DM to point out which criteria are most important. These criteria include overall customer satisfaction level, enterprise satisfaction level and best balance between customer and enterprise satisfaction, reflecting the DM's preference. Thirdly, by way of a hybrid genetic algorithm (Tang *et al.* 1998) with mutation along the weighted gradient direction, the near optimal solutions under different criteria will be obtained and updated in each generation. When the generation terminates, these solutions will be presented for the DM to select through human–computer interaction.

11.9 Illustrated Example and Simulation Results

To clarify the performance of the model and solution approach, a simple revised example (Wassermann 1993; Fung 1998b) is introduced and some results are illustrated in this section.

A corporation is undergoing a new product development project. According to the survey in the market place, there are four major customer attributes (CRs), *i.e.*, easy to handle (CR1), does not smear (CR2), point lasts (CR3) and does not roll (CR4), and five technical attributes (TAs), *i.e.*, length of pencil (TA1), time between sharpening (TA2), lead dust generated (TA3), hexagonally (TA4) and minimal erasure residue (TA5). The relationship between CRs and TAs, and the correlation between TAs are illustrated in a HoQ as shown in Figure 11.4. The weights of the above CRs is as follows:

$$d_1 = 0.15, d_2 = 0.25, d_3 = 0.45, d_4 = 0.15.$$

Assuming that the budget B is 15.0 (1000) unit, C_0 is 10.0 (1000) units, and primary costs required for each TAs are expressed as follows:

$$\tilde{c}_1 = (9.8, 10.0, 10.2)\,,\ \tilde{c}_2 = (3.8, 4.0, 4.2)\,,\ \tilde{c}_3 = (4.9, 5.0, 5.1)\,,$$

The normalized relationship matrix R^*_{ij} by scale of 1-3-9 and correlation matrix T_{ij} by scale of 1-3-9-18 are given as follows, where $1, 3, 9$ represent the week, medium and strong relationships between CRs and TAs, as well as TAs and TAs, respectively, and 18 denotes the strongest dependency between the TA and itself.

$$R^*_{ij} = \begin{bmatrix} 0.25 & 0.00 & 0.00 & 0.75 & 0.00 \\ 0.00 & 0.19 & 0.405 & 0.00 & 0.405 \\ 0.023 & 0.185 & 0.396 & 0.00 & 0.396 \\ 0.10 & 0.00 & 0.00 & 0.90 & 0.00 \end{bmatrix}$$

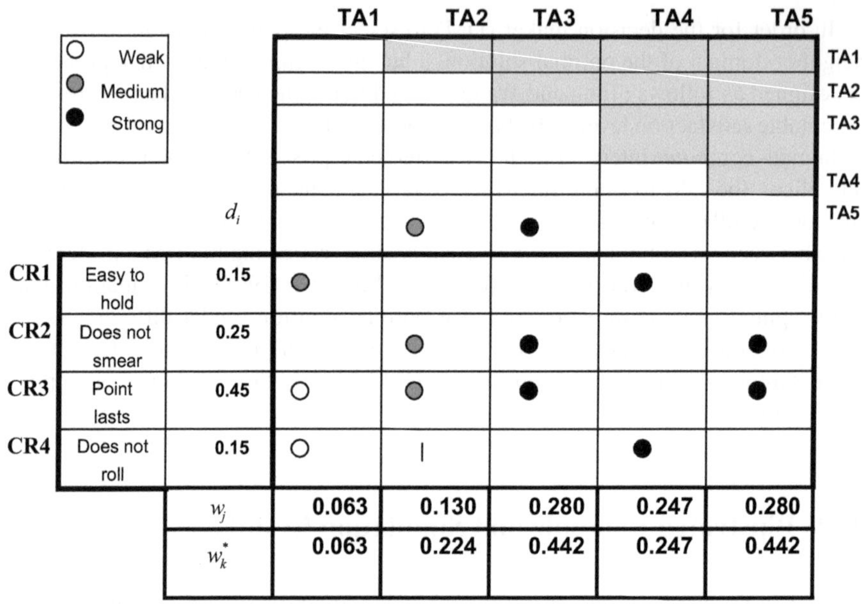

Figure 11.4 House of quality (HoQ) for illustrated example

Table 11.1 Simulation results for the example by way of genetic algorithm-based interactive approach

Solution	Criteria	Best balance (α^*)	Maximum overall customer satisfaction	Minimum total costs
Planned degree of TAs	TA1	0.46914	0.4549	0.5049
	TA2	0.47105	0.4830	0.7003
	TA3	0.39555	0.5648	0.2406
	TA4	0.59740	0.6160	0.4793
(y_j)	TA5	0.39932	0.5364	0.2849
Actual achieved degree of TAs (x_j)	TA1	0.46914	0.4549	0.5049
	TA2	0.60379	0.6669	0.7881
	TA3	0.67387	0.9137	0.4999
	TA4	0.59740	0.6159	0.4793
	TA5	0.67576	0.8995	0.5221
Actual planned costs ($c_j(y)$) (1000)	TA1	4.69141	4.5488	5.04974
	TA2	2.09458	2.1771	2.87574
	TA3	2.43158	2.9746	1.85148
	TA4	1.79220	1.8479	1.43801
	TA5	2.93360	3.4375	2.38956
Total actual costs $\sum c_j(y)$ (1000)		13.94348	14.9859	13.60454
Satisfaction level (α)		0.56622	0.77574	/

$$T_{ij} = \begin{bmatrix} 1.0 & 0.0 & 0.0 & 0.0 & 0.0 \\ 0.0 & 1.0 & 0.167 & 0.0 & 0.167 \\ 0.0 & 0.167 & 1.0 & 0.0 & 0.5 \\ 0.0 & 0.0 & 0.0 & 1.0 & 0.0 \\ 0.0 & 0.167 & 0.5 & 0.0 & 1.0 \end{bmatrix} .$$

In light of Equation 11.1 and 11.13, w_j and w_k^* are employed as follows:

$$w_1 = 0.063, w_2 = 0.130, w_3 = 0.280, w_4 = 0.247, w_5 = 0.280$$
$$w_1^* = 0.063, w_2^* = 0.224, w_3^* = 0.442, w_4^* = 0.247, w_5^* = 0.442 .$$

Assuming that the acceptable degree of attainment of all TAs should not be less than $\beta_j = 0.45$, and both acceptable overall customer satisfaction and enterprise satisfaction level can be given by $\alpha_0 = 0.45$. The simulation results as $r = 1$ using a genetic algorithm-based interactive approach are shown in Table 11.1.

From Table 11.1, we can see that the overall customer satisfaction, minimized design costs, as well as the best balance between overall customer satisfaction and enterprise satisfaction can be achieved by the interactive approach. It can offer the DM more choice to support his work under different criteria, especially under a fuzzy environment.

References

Akao Y (1990) Quality function deployment: integrating customer requirements into product design. Productivity, Cambridge, MA

Bode J, Fung RYK (1998) Cost engineering with quality function deployment. Comput Ind Eng 35(3/4):587–590

Fung RYK, Popplewell K, Xie J (1998a) An intelligent hybrid system for customer requirements analysis and product attribute targets determination. Int J Prod Res 36(1):13–34

Fung RYK, Ren S, Xie J (1996) The prioritisation of attributes in customer requirement management. In: Proceedings of the 1996 IEEE International Conference on Systems, Man and Cybernetics, Beijing, P. R. China, pp 953–958

Fung RYK, Tang JF, Tu YL (2003) Modeling of quality function deployment planning with resource allocation. Res Eng Design Appl 14(4):247–255

Fung RYK, Tang JF, Tu Y, Wang D (2002) Product design resources optimization using a nonlinear fuzzy quality function deployment model. Int J Prod Res 40(3):585–599

Fung RYK, Tang J, Wang D (1998b) Fuzzy financial optimisation in product design using quality function deployment. Int J Prod Res 40(3):585–599

Lai YJ, Hwang CL (1992) Fuzzy mathematical programming: methods and applications. Springer, Berlin Heidelberg New York

Tang J, Wang D (1997) An interactive approach based on a genetic algorithm for a type of quadratic programming problems with fuzzy objective and resources. Comput Oper Res 24(5):413–422

Tang J, Wang D, Ip A, Fung RYK (1998) A hybrid genetic algorithm for a type of non-linear programming problem. Comput Math Appl 36(5):11–21

Wassermann GS (1993) On how to prioritise design requirements during the QFD planning process. IIE Trans 25(3):59–65

Chapter 12
Decision Making with Consideration of Association in Supply Chains

In many supply chain systems, inventory control is a key decision-making problem. ABC classification is usually used for inventory items aggregation because the number of inventory items is so large that it is not computationally feasible to set stock and service control guidelines for each individual item. Then different managing methods have been applied to control the inventory level of the items. However, because of the complexity of inter-relationships among items, there is a small amount of research that treats decision making with correlation of inventory items. Therefore, how to treat the correlation is a challenge when developing inventory strategies. This chapter firstly establishes a new algorithm of inventory classification based on the association rules; by using the support-confidence framework the consideration of the cross-selling effect is introduced to generate a new criteria, which is then used to rank inventory items. Then, a numerical example is used to explain the new algorithm and empirical experiments are implemented to evaluate its effectiveness and utility, comparing with traditional ABC classification.

This chapter is organized as follows. Section 12.1 introduces the importance of the consideration of association. Section 12.2 provides an overview of relational research. Section 12.3 outlines our approach and the issues to be addressed, and provides the detailed descriptions of the new algorithm. Section 12.4 and Section 12.5 present a numerical example and the empirical experiments, respectively. Section 12.6 presents conclusions and outlines future research.

12.1 Introduction

In many inventory control systems, it has been considered that the number of items (usually called stock-keeping units (SKU)) is so large that it is not computationally feasible to set stock and service control guidelines for each individual item. As a result, items are often grouped together and generic inventory control policies (such as service level/order quantity/safety stock coverage) are applied to each item in a group. Such grouping methods provide management with more effective means for specifying, monitoring, and controlling system performance, since strategy objectives and organization factors can often be represented more naturally in the terms of item groups.

In practice, the ABC classification scheme is the most frequently used method for items grouping. It groups items based on the fact that a small fraction of items account for a high percentage of total dollar usage. The principles of ABC classification have been around for a long time, at least since Pareto made his famous observations on the inequality of the distribution of incomes (Silver *et al.* 1988). Astute managers have continued to apply the principle by concentrating on the "significant few" (the A items) and spending less time on the "trivial many" (the C items). Moreover, to classify the items into the A, B, and C categories, one criterion had to be on the basis, just as Pareto did. For inventory items, such a criterion is often the dollar usage (price multiplied by annual usage) of the item, although it is sometimes just the item's cost.

For many items, however, ABC classification is not suitable for inventory control. Managers have to shift some items among categories for a number of reasons. Several researchers considered there may be other criteria that represent important considerations for management. The certainty of supply, the rate of obsolescence, and the impact of a stock out of the item are all examples of such considerations. Some of these may even weigh more heavily than dollar usage in the management of the items. Several criteria have been identified as important in the management of maintenance inventories (Chase *et al.* 1998). Also, several researchers suggested that multiple criteria should be used in the classification of inventories (Flores and Whybark 1987; Cohen and Ernst 1988; Lenard and Roy 1995).

On the other hand, we consider a fundamental problem of classification of inventories. For some inventory items, evaluating the importance of one item comes not only from its own value, but also from its influence on the other items, *i.e.*, the "cross-selling effect" (Anand *et al.* 1997). In such a situation, it should be explained clearly whether the cross-selling effects would influence the ranking of items or not, and how to group the items if such effects existed, not concerning what and how many criteria could be used. However, it could not be solved in the traditional classification of inventories in the past. We reconsider the traditional and fundamental inventory classification problem here for two reasons. Firstly, current gradual availability of cheaper and better information technology has enabled us to process huge amounts of inventory data, which was considered impossible before. Secondly, popular techniques in knowledge discovery on databases (KDD) had been developed remarkably and it can lead us to build a new approach for classification of inventories.

By using the association rule, which is one of the most popular techniques in KDD, a new approach of ranking items with the consideration of the cross-selling effect has been presented (Kaku and Xiao 2008). This chapter summarizes the consideration of their ranking approach, which gives the meaning of the new approach as a new suggestion to turn data-mining technologies into inventory control (Cooper and Giuffrida 2000), and outlines several future possibilities.

12.2 Related Research

In this section, we review several topics: the classification of inventories, association rules and evaluating criteria.

12.2.1 ABC Classification

As Silver *et al.* (1988) described, the classification of inventories can be developed by two step procedures: firstly, the annual dollar usage of each item is calculated and ranked in descending order starting with the largest value of dollar usage. Secondly, the ranked list is divided into three groups by the value: A (most important), B (intermediate in importance), and C (least important). The number of groups appropriate for a particular company depends on its circumstances and the degree to which it wishes to differentiate the amount of effort allocated to various groups. Generally, a minimum of three categories is almost always used.

Sometimes, an item may be critical to a system if its absence will create a sizable loss. In this case, regardless of the item's classification, sufficiently large stock should be kept on hand to prevent from running out. One way to ensure closer control is to designate this item an A or a B, forcing it into the category even if its dollar usage does not warrant such inclusion. That means the ABC classification sometimes is not appropriate on the basis of the dollar usages alone. For solving this problem multiple criteria have been suggested in the classification of inventories. However, almost all of the approaches using the multiple criteria were just applied in special cases, and basically had no differences with single criterion approaches. Therefore, even though multiple criteria were used, the fundamental problem of the cross-selling effect among inventory items is not yet satisfactorily solved. The purpose of this section is to introduce the consideration of the cross-selling effect into the classification of inventories, especially to deal with the problem of how to rank the items when the cross-selling effects are considered.

12.2.2 Association Rule

When the cross-selling effects are considered, an item sometimes does not generate a large dollar usage itself, but plays some role for other items to generate a good dollar usage. Such a relationship can be handled by using an association rule, which has a similar intention of capturing association between items. Association rules, which are introduced by Agrawal *et al.* (1993), can be described by the following support-confidence framework:

Let $I = \{i_1, i_2, \ldots, i_k\}$ be a set of items. Let D be a set of transactions, where each transaction T is a set of items such that $T \subseteq I$. A transaction T contains X, a set of some items in I, if $X \subseteq T$. An association rule is an implication of the form $X \Rightarrow Y$, where $X \subset I$, $Y \subset I$, and $X \cap Y = \phi$. The rule $X \Rightarrow Y$ has support s in the transaction set D if $s\%$ of transactions in D contain $X \cup Y$. The rule $X \Rightarrow Y$ holds in the transaction set D with confidence c if $c\%$ of transactions in D that contain X also contain Y. Given a set of transactions D, the problem of mining association rules is to generate all association rules that have support and confidence both greater than user-specified minimum support (*min_sp*) and minimum confidence (*min_cf*).

Generating association rules involves looking for so-called frequent itemsets in the data. Indeed, the support of the rule $X \Rightarrow Y$ equals the frequency of the itemset $\{X, Y\}$, and the frequency of itemset X in D equals the fraction of transactions in D that contain X. Moreover, an itemset X is frequent in D if and only if its support is larger than the *min_sp*. Then we can determine whether an itemset is frequent or not just by calculating the support and comparing it with *min_sp*.

A typical approach (Agrawal and Srikant 1994) to discover all frequent itemsets is to use the knowledge that all subsets of a frequent itemset are also frequent. This insight simplifies the discovery of all frequent itemsets considerably. Once all frequent itemsets are known, finding association rules is easy. Namely, for each frequent set X and each $Y \in X$ verify whether the rule $X \backslash \{Y\} \Rightarrow Y$ has sufficiently high confidence. There are many algorithms used to discover frequent itemsets because developing a more effective algorithm is a main topic of research in data-mining technology. The detail of the algorithms can be found in Han and Micheline (2001), and Zhang and Zhang (2002).

In this chapter, we use the association rules to solve the ranking problem in classification of inventories.

12.2.3 Evaluating Index

When the cross-selling effect is introduced into inventory classification, the dollar usage is not appropriate as an evaluating index. We try to develop an index related to association rules in order to evaluate the inventory items.

Recently, Brijs *et al.* (1999, 2000a,b) firstly presented an index to evaluate the value of item in a store. They presented a PROFSET model to calculate the effects of cross-selling among items and a 0-1 programming algorithm to select which frequent itemset can reflect the "purchase intentions." The largest contribution of the PROFSET model is that it shows how to calculate the profit of a frequent itemset. However, because PROFSET model does not consider the strength of the relationship between items, PROFSET provides no relative ranking of selected items, which is important in classification of inventories. Moreover, to calculate the profit of a frequent itemset they used the maximal frequent itemsets. Unfortunately, the maximal frequent itemsets often do not reflect purchase intentions because they do not occur as frequently as their subsets. Therefore the PROFSET model cannot be used to classify inventory items because not only frequent items but also all inventory items should be ranked. That means the criterion of dollar usages of a few frequent itemsets can only be used to select some special items and is not appropriate for the classification of all inventory items.

To represent the strength of the relationship between items, Wang and Su (2002) presented a profit ranking approach of items based on a "hub-authority" analogy. Such an analogy exploits the relationship of hubs and authorities adopted in the web-pages ranking algorithm HITS (Kleinberg 1998), which is used in the well-known search engine Google. They determined the potential cross-selling links between frequent items by using the confidence of a pair of frequent items. The strength of the links is defined by (profit × confidence), and then the items in a frequent itemset can be ranked. However, the hub-authority ranking method is just used in the frequent items. It is also not clear how to deal with the confidence over two frequent items. Moreover, the ranking problem in classification of inventories is not only for frequent items but also for infrequent items; those items that belong to frequent itemsets were only subsets of all items.

12.3 Consideration and the Algorithm

It should be considered that the relationship among items might be important in the process of classifying inventory items. However, there were no methods or algorithms to solve such problems until now. Although several researchers had considered the shortcomings of traditional ABC classification and proposed some solutions for it, all of them had not treated the cross-selling effects among inventory items. Here, the cross-selling effects among inventory items are firstly considered in the classification of inventories. The basic idea is to treat a frequent itemset as a single item in the ranking approach. Based on this consideration, two questions must be answered. The first is what criterion should be used to evaluate the importance of items instead of the dollar usage. The second question is how to deal with the relationship among inventory items.

12.3.1 Expected Dollar Usage of Item(s)

When evaluating an inventory item, we should not only look at the individual dollar usage generated by the item, but also take into account the usage with other items in the ranking due to cross-selling effects. Therefore, it is more essential to look at frequent sets rather than at individual items since the former represents frequently co-occurring item combinations in inventory transactions. The importance of individual items must also be regarded as equal to that of the frequent set it belongs to as well, because any element's absence may prevent the frequent set from co-occurring in transactions. Therefore, it is clear that the traditional dollar usage of a single item used to evaluate its importance should not be good enough to fit for such a purpose. In this section, we suggest a new criterion of expected dollar usage

(EDU) to judge the importance of item(s), and serve as the evaluating index to rank inventory items in inventory classification. The EDU is just a new criterion that we are trying to look for to revaluate the importance of an item with the consideration of cross-selling effects.

To calculate the EDU of an item, we first need to deal with a frequent itemset as a special item, the dollar usage of which can be calculated out like a normal item, *i.e.*, the total dollar usage of the items co-occurring in all transactions, or can be simply taken as the set's price times the set's support. Then, all frequent itemsets will be considered in the ranking process together with individual items. Because one-item frequent itemsets, *i.e.*, the one-large itemset, equals the item itself, such special items just involve the frequent itemsets, which include over two items.

Suppose an itemset X in an inventory transaction database D has a support, denoted as sp(X).

$$\mathrm{sp}\,(X) = \frac{|X(t)|}{|D|}$$

where $X(t) = \{t \text{ in } D/t \text{ contains } X\}$ and t is a transaction.

The itemset X is called frequent if its support is equal to, or greater than, the threshold minimal support (*min_sp*) given by users. Note that even if an itemset is not frequent, its support can still be calculated.

Now consider an itemset, whether it is frequent or not. Its EDU can be calculated as follows:

$$C_X = \mathrm{sp}\,(X) \sum_{i \in X} p_i$$

where C_X is the EDU of an itemset X, p_i is the price of single item in X and $\sum_{i \in X} p_i$ reflects the set's price. It is easy to see that when X contains only one item its EDU is equal to the individual item's dollar usage. That means all of the inventory items can be evaluated by using this index.

By the definition above, the EDU of each frequent itemset and each item that is not frequent are then able to be calculated and ranked in descending order starting with the largest value of EDU. If a frequent itemset is ranked more ahead than all of its element items, this means the cross-selling effect in the itemset is greater than each of its element items. Then, such an itemset should be treated as a single item.

12.3.2 Further Analysis on EDU

From another viewpoint, the cross-selling effect might act as a role of changing the weight of the evaluating index of items appearing in all frequent itemsets. The rule is that if the change makes the weight greater than the old one, *i.e.*, the EDU of an individual item, the change should be accepted; otherwise, it should be ignored. However, an item is not always exclusively belonging to only one frequent itemset. It may be the element of many frequent itemsets with different lengths and different supports. The item's index weight can be any EDUs of frequent itemsets that contain it. In this case, the maximum EDU of itemsets should be taken as the final index weight because it represents the largest influence of individual item on dollar usage.

It is noticeable that the frequent itemset with longest length may not have the maximum EDU due to the diminishment of support of frequent itemset when its length becomes longer. Therefore, for example the EDU of {A, B, C} may not be greater than that of {A, B} because the support of the latter may be much greater than that of the former and correspondingly results in a greater EDU. That means the maximal frequent itemsets often do not reflect real purchase intentions because they may not occur as frequently as their subsets. As a mater of fact, when the length of frequent itemset increases, the support of it always drops.

To better understand the EDU of item, we present the notion of family set. All the frequent itemsets that contain an ith item build up a family set of ith item, the elements of which are also itemsets (*i.e.*, the frequent itemsets), and the length and the EDU of the elements can be plotted and curved in a rectangle plane. Figure 12.1 shows such a relationship between the length and the EDU of the frequent itemsets in a family set, which can be obtained by tracing one individual item, *i.e.*, the ith item, when the length of frequent itemsets that contain the ith item increases in the process of finding frequent itemsets by the Apriori algorithm. The vertex of the curve in Figure 12.1, *i.e.*, the p point, has some interesting meanings; the p point marks out the largest EDU related to the ith item, and indicates the maximum cross-selling value of ith item, which is of much concern to the inventory manager. The corresponding frequent itemset at p point is defined as the key frequent itemset of the family set of the ith item, which reflects the most valuable group of items related to the ith item. We can just think that the key frequent itemset generally reflects a large possibility of the strongest purchase intentions of customers who buy the ith item.

Summarizing, the cross-selling effect influenced by association rules is to use the key frequent itemset of each ith item to revaluate the weightiness of items. If the EDU of the key frequent itemset is greater than the traditional dollar usage of individual item, the EDU will be accepted as a new index weight for classification; otherwise, the ith item does not have a key frequent itemset, the traditional index weight will remain.

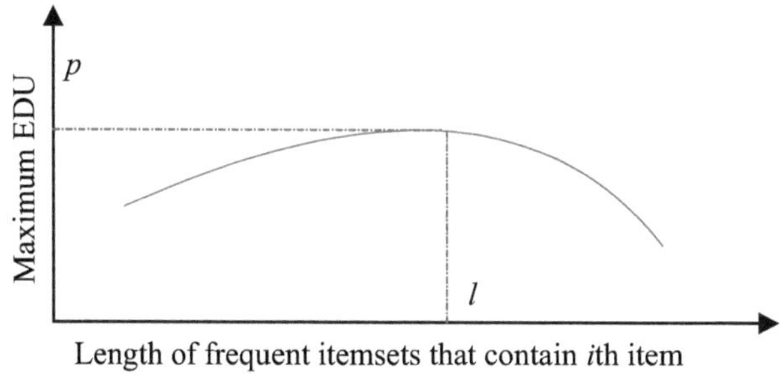

Figure 12.1 The relationship between length and EDU of an item

12.3.3 New Algorithm of Inventory Classification

Based on the consideration above, we present a new algorithm of inventory classification with the effect of association rules; the main steps of the algorithm are shown as follows:

Step 1 Calculation of frequent itemsets.
Calculate all frequent itemsets in a transaction database by using a frequent itemsets searching algorithm. An enhanced Apriori algorithm will be detailed in the next section, which is based on the well-known Apriori algorithm but improved to have the ability of ranking the items in the searching process.

Step 2 Calculate the EDU and rank the items.
Calculate the EDU of all frequent itemsets and of all single items; rank them in descending order starting with the largest value. The frequent itemsets are a special item ranked in the list. The key frequent itemset, according to its definition above, should have the largest EDU, and will be ranked ahead of other frequent itemsets in the same family set.

Step 3 Withdraw the repeated items in the ranking list.
Replace all frequent itemsets in the ranking list with their contained items, the interior order of which must remain. After that, scan the ranking list from the beginning to the end and withdraw the repeated items that have appeared for the second time to make each of the items unique in the list.

Step 4 Make the new ABC classification of inventory.
Make the ABC classification of inventory based on the new ranking list. The new A group is obtained by selecting items of the new ranking list from the beginning to the end till the total EDU of selected items reaches 80% of total dollar usage occupancy.

12.3.4 Enhanced Apriori Algorithm for Association Rules

In this section, the well-known Apriori algorithm was applied in order to find the frequent itemsets for the first step of the new algorithm of inventory classification. A little change had been made to this algorithm to make it rank the items of the large itemsets according to the subset's EDU during the searching process. To meet this feature and quicken the algorithm as well, we designed a special form for storing the large itemsets and the candidates, which has the following:

$$\langle \text{id}, \{\text{items}\}, \text{length}, \{\text{TIDs}\}, \text{support}, \text{EDU} \rangle$$

where the id is a sequence number used as identifier of the itemset, {items} is the items list, length is the count of items, {TIDs} is the set of transaction identifiers that contain {items}, support is the count of TIDs, and EDU is the expected dollar usage of an itemset.

Two special operators of $\cdot\vec{\vee}\cdot$ and $|\cdot|$ are specially used in the algorithm. The $\cdot\vec{\vee}\cdot$ is a union operator to calculate the union of two sets with a fixed sequence order of elements. By this operator, $\mathbf{A}\,\vec{\vee}\,\mathbf{B}$ is to add those elements of **B** that are not in **A** to the right side of **A**. The sequence of elements in **A** and **B** will remain in the new intersection. For example, $\{A, C, D\}\,\vec{\vee}\,\{B, C\} = \{A, C, D, B\}$ and $\{B, C\}\,\vec{\vee}\,\{A, C, D\} = \{B, C, A, D\}$. The other operator $|\cdot|$ is used to calculate the length of a set. For example, $3 = |\{A, B, C\}|$ and therefore support $= |\{\text{TIDs}\}|$.

Figure 12.2 gives the main steps of the enhanced algorithm we designed for searching frequent itemsets based on the Apriori algorithm (Agrawal and Srikant 1994). The first pass of the algorithm simply counts item occurrences to determine the large one-itemsets, *i.e.*, the L_1, as usual. The subsequent pass consists of two phases. In the first phase, the large itemsets L_{k-1} found in the $(k-1)$th pass are used to generate the candidate itemsets C_k by using the *apriori_gen* function described in Figure 12.3, where the support list of TIDs is also generated at the same

Algorithm Finding_FIS(D, *min_sp*)
Input: D, *min_sp*
Output: Answer
Begin
$L_1 = \{\text{large one-itemsets}\}$;
for ($\text{k} = 2; \text{L}_{\text{k}-1} \neq \theta; \text{k}++$) **do** {
    C_k = *apriori-gen*(L_{k-1}); // New candidates
    $L_k = \{c \in C_k | c.\text{ support} \geq min\_sp\}$
}
Answer $= \cup L_k$;
End

Figure 12.2 Main program of enhanced Apriori algorithm

Function *apriori_gen* (L_{k-1})
Input: L_{k-1}
Output: C_k
Begin
for (each pair $p, q \in L_{k-1}$) **do** {
if p.EDU$\geq$ q.EDU **then**
    Cand.{items}= p.{items}$\vec{\vee}$$q$.{items}
else
    Cand.{items}= q.{items}$\vec{\vee}$$p$.{items}
Cand.length= | Cand.{items}|
if *Cand*.length$\neq$ k **then** continue loop
Cand.{TIDs}= p.{TIDs}$\vee$$q$.{TIDs}
Cand.support =|*Cand*.{TIDs}|
Cand.EDU=get_EDU(*Cand*.{items})
if *Cand*$\notin$ C_k **then** put *Cand* into C_k
}
End

Figure 12.3 Algorithm of function *apriori_gen*($L_{k.-1}$)

time and is attached to the candidate, which will accelerate the program greatly by avoiding scanning the total transaction database to obtain the support of candidates. Next, count the support of all candidates in C_k. By comparing the support of all candidates in C_k with the threshold of *min_sp*, the large itemset L_k is then determined.

The *apriori_gen* function is crucial in the algorithm because the sequence of elements in candidate C_k must be managed correctly according to the EDU of two subsets in L_{k-1} that formed it, which is what we are mainly concerned with. From pass 2 to pass 4, the new candidate *Cand* is created as the intersection of each two sets of L_{k-1} under the rule of operator $\cdot\vec{\vee}\cdot$. Pass 7 ignored those new candidates whose length are greater than k. Pass 8 and pass 9 calculate the TIDs list of new candidates, as well as its support. Because of the special storing form of the candidate designed above, the itemsets and corresponding support is calculated out with high efficiency. The transactions in D that contain $p.\{\text{items}\}$ have been already saved as a subset attached to p, *i.e.*, $p.\ \{\text{TIDs}\}$. The intersection of $p.\ \{\text{TIDs}\}$ and $q.\{\text{TIDs}\}$, *i.e.*, $p.\{\text{TIDs}\}\vee q.\{\text{TIDs}\}$, include all the transactions that contain the items of both p and q. Pass 10 calculates the EDU of the new candidate. Pass 11 is to ensure that the candidates in C_k are unique.

12.3.5 Other Considerations of Correlation

Wong *et al.* (2003, 2005) proposed a loss rule, which described that total profit of an item (set) not only comes from itself but also from another item (set) that has a cross-selling effect associated with it. Therefore, the reduction in profits could not be ignored if the confidence of the loss rule is reasonable large. Note that the loss rule can apply to all items, even several items might have very small confidences, which is an important characteristic of inventory classification, so that we can use it as a criterion to evaluate the importance of an item. However, because the reduction of profit is not convenient to operate the items, the same consideration but a reverse version of loss rule can be used to contribute a criterion to classify inventory items. For example, we can use the concept of total loss profit (TLP) of an item (set) I to evaluate the importance of an item as follows:

$$\mathrm{TLP}(I) = \sum_{i=1}^{m} \left[prof(I, t_i) + \sum_{S \in t_i'} prof(S, t_i) \cdot conf(S \to I) \right]$$

where, $prof(I, t_i)$ is the profit of item I in transaction t_i. $prof(S, t_i)$ is the profit of item S in transaction t_i while S is not selected. $conf(S \to I)$ is the confidence of the loss rule and can be calculated as the number of transactions containing I and any element in S divided by the number of transactions containing I. Therefore, $\mathrm{TLP}(I)$ presents the total loss profit of item I if item I was lost. The more and the stronger the related items, the larger the TLP. We can rank the items by TLP when the TLP is considerably serious in the purchasing behaviors.

12.4 Numerical Example and Discussion

The consideration of the new classification algorithm is illustrated by using an example adopted from Zhang and Zhang (2002). Let an inventory itemset be $I = \{A, B, C, D, E\}$ and inventory transactions set be $D = \{1, 2, 3, 4\}$, and let the prices of items be A = 5, B = 4, C = 2, D = 2, E = 1. By using tridational ABC classification, let the dollar usage of items be A = 10, B = 12, C = 6, D = 2, E = 3. Then rank items in descending order starting with the largest value of dollar usage; the ranking list is (BACED).

As shown in Table 12.1, $TID = \{1, 2, 3, 4\}$ are the unique identifies of the four inventory transactions. Each row in Table 12.1 can be taken as an inventory transaction. The frequent itemsets can be calculated easily by using the enhanced Apriori algorithm.

Table 12.1 An inventory transaction database

TID	Items				
1	A		C	D	
2		B	C		E
3	A	B	C		E
4		B			E

Let *min_sp* = 2 (to be frequent, an itemset must occur in at least two inventory transactions). According to step 1 and step 2 of the new algorithm of inventory classification, the enhanced Apriori algorithm in Figure 12.2 is employed to generate the generations of all large frequent itemsets. The results are shown in the form of ⟨id, {items}, length, {TIDs}, support, EDU⟩ in Table 12.2. By ranking these individual items and frequent itemsets of Table 12.2 in descending order starting with the largest value of EDU, the following list of itemsets can be obtained:

$$\{BE\}, \{BEC\}, \{BC\}, \{B\}, \{A\}, \{AC\}, \{C\}, \{CE\}, \{E\}, \{D\} .$$

According to step 3 of the new algorithm, replace the frequent itemsets with their elements to get the ranking list of items as follows:

$$\{BEBECBCBAACCCEED\} .$$

Then scan the ranking list from the beginning to the end and withdraw the repeated items. Then, we obtain the final ranking list as below:

$$\{BECAD\} .$$

It is a very different order compared with the result {BACED} of the traditional ABC classification. To explain such a new order, we can see in traditional ABC

Table 12.2 The progress of finding frequent itemsets by enhanced Apriori algorithm

id	{Items}	*Length*	{TIDs}	*Support*	EDU	Is it frequent?
	One-large itemset					
1	{A}	1	{1, 3}	2	$2 * 5 = 10$	Y
2	{B}	1	{2, 3, 4}	3	$3 * 4 = 12$	Y
3	{C}	1	{1, 2, 3}	3	$3 * 2 = 6$	Y
4	{D}	1	{1}	1	$1 * 2 = 2$	N
5	{E}	1	{2, 3, 4}	3	$3 * 1 = 3$	Y
	Two-large itemset					
6	$\{B\} \vec{\vee} \{A\} = \{BA\}$	2	$\{2,3,4\} \vee \{1,3\} = \{3\}$	1		N
7	$\{B\} \vec{\vee} \{C\} = \{BC\}$	2	$\{2,3,4\} \vee \{1,2,3\} = \{2,3\}$	2	$2 * (4 + 2) = 12$	Y
8	$\{B\} \vec{\vee} \{E\} = \{BE\}$	2	$\{2,3,4\} \vee \{2,3,4\} = \{2,3,4\}$	3	$3 * (4 + 1) = 15$	Y
9	$\{A\} \vec{\vee} \{C\} = \{AC\}$	2	$\{1,3\} \vee \{1,2,3\} = \{1,3\}$	2	$2 * (5 + 2) = 10$	Y
10	$\{A\} \vec{\vee} \{E\} = \{AE\}$	2	$\{1,3\} \vee \{2,3,3\} = \{3\}$	1		N
11	$\{C\} \vec{\vee} \{E\} = \{CE\}$	2	$\{1,2,3\} \vee \{2,3,4\} = \{2,3\}$	2	$2 * (2 + 1) = 6$	Y
	Three-large itemset					
12	$\{BE\} \vec{\vee} \{BC\} = \{BEC\}$	3	$\{2,3,4\} \vee \{2,3\} = \{2,3\}$	2	$2 * (4 + 2 + 1) = 14$	Y
13	$\{BE\} \vec{\vee} \{AC\} = \{BEAC\}$	4		1		Length > 3, ignored
14	$\{BE\} \vec{\vee} \{CE\} = \{BEC\}$					Already exist
15	$\{BC\} \vec{\vee} \{AC\} = \{BCA\}$	3	$\{2,3\} \vee \{1,3\} = \{3\}$	1		N
16	$\{BC\} \vec{\vee} \{CE\} = \{BCE\}$					Already exist
17	$\{AC\} \vec{\vee} \{CE\} = \{ACE\}$	3	$\{1,3\} \vee \{2,3\} = \{3\}$	1		N

classification, the dollar usage of item E is small so that it was ranked at the latter of the list. However, in the new classification, item E has a cross-selling effect influence over other items, such as B, so that its ranking position is moved forward, even surpassing item A, which has a higher dollar usage.

Furthermore, for item E, its family set of frequent itemsets includes {E}, {BE}, {CE} and {BEC}. Although the {BEC} has a longer length, according to the definition, the {BE} should be regarded as the key frequent itemset of item E because it has the highest EDU.

12.5 Empirical Study

To evaluate the new algorithm and find out how much impact considering association rules will have on the inventory policy, we used two datasets to compare the results when traditional ABC classification and the new algorithm are applied.

12.5.1 Datasets

The first dataset is acquired from a Japanese convenient store in the countryside over one month. This dataset carries thousands of different items and 15,000 sales transactions. Each item in a transaction is associated with a Japanese Yen usage.

We also generated a synthetic database of transactions with stronger cross-selling effect to see more differences through comparing. This database is downloaded from a known benchmark for association rule data mining from the website: http://cucis.ece.northwestern.edu/projects/DMS/MineBench.html.

There are 96,554 transactions made by 1000 items on file:

data.ntrans_100.tlen_5.nitems_1.npats_2000.patlen_2 .

However, this database is not associated with the dollar usage. Because the purpose of the experiment data is to represent the difference of the algorithms, we randomly initiated the price of all items between 1 and 100 (dollar) to make the items follow the common rule that nearly 20% of items count for 80% of total dollar usage. The EDU of an item is then equal to (price $\times$ support).

12.5.2 Experimental Results

Firstly, we calculated the inventory classification by both the traditional algorithms and the new algorithm presented here to find out how much the difference of the two results will be. The comparison of the results when applied to the Japanese

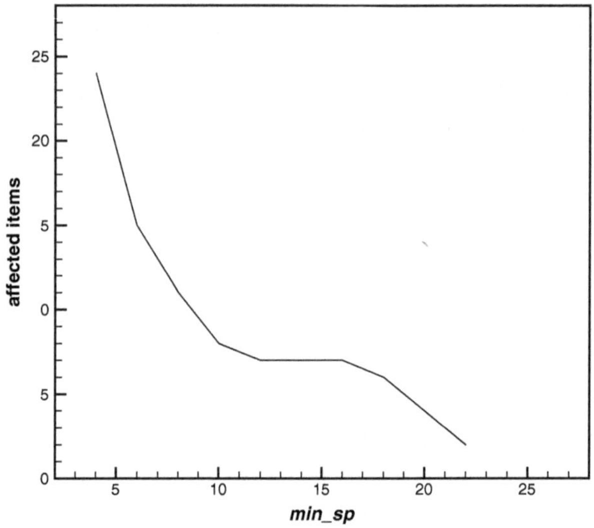

Figure 12.4 The number of items affected by the new algorithm

convenient store database is shown in Figure 12.4. The curve shows the variety of the number of affected items, *i.e.*, the items that have a larger expected Japanese Yen usage than when the association rule effect is considered, and should move forward in its position in the ranking list of ABC classification when the *min_sp* is changed. The curve was obtained by many iterations of calculation with increasing *min_sp*. There should be no doubt that some items will have a change in position, and would be moved forward when the association rule influences are taken into account. However, the main question is how much would this influence be? Is it enough for the manager to reconfigure his inventory policy? Generally, we found that about 0.5 to 2% of the key items that make up the 80% of total Japanese Yen usage will be changed in our experiences.

To what extent the difference will be depends on the strength of the association rules of the database and the variable *min_sp*, on which the association rules depend. Though the curve in Figure 12.4 shows a monotonically decreasing trend of the number of affected items when the *min_sp* varies, it should not be assumed to always being that way. According to our analysis, there is an antimonotonic rule that when the *min_sp* grows, the length of the key frequent itemset of individual items will always decrease; the EDU of individual items may have a vertex in this process.

Secondly, to know more about characteristics of the influence by association rules when classifying inventory items, the same comparisons had been made on a larger dataset downloaded from a known website, which is a benchmark dataset of transactions for data mining and has stronger association rules. The similar comparisons are shown in Figure 12.5, the curve in which shows the number of affected items by the new algorithm in percentage, which presents a similar result but with more items that are affected. Generally, about 10 to 40%, when the *min_sp* varies, of the

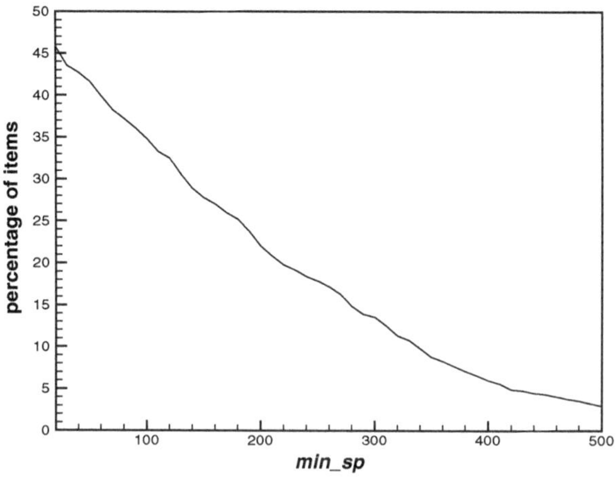

Figure 12.5 The percentage of items affected by the new algorithm

1000 total items have to be moved forward in the ranking list of the new classification because of their increased weights by the influences of association rules. It is very obvious that the manager should not ignore such a considerable influence when he makes the inventory policies of individual items. Because nearly half of the items have changed their weights in the new algorithm, it is also noticeable that the variable of *min_sp* is very crucial in the algorithm. The influence of association rule will be gradually reduced when the *min_sp* becomes large. How to determine an optimal *min_sp* will be discussed in the last part of this section.

Thirdly, We have made more calculations to find out how many items, with the consideration of association rules, that used to be assigned to B or C by traditional ABC classification will be forced into category A even if their dollar usage does not reach the traditional criterion. We also discover how many items that were to be assigned to group A will lose their qualification and be moved out from A. The dataset applied is the benchmark transactions. The result is shown in Figure 12.6, the two curves in which show the number of items that are forced into or out of the A group by the association rules, respectively. Curve 1 represents the items forced into A; curve 2 is the number of items moved out of A. Both are varied along with *min_sp*. The two curves are obtained by many cycles of calculations with different *min_sp*.

Finally, we focus on discussing the variable *min_sp*, because all the calculation results are of high concern. Generally, *min_sp* is the threshold criteria for judging an itemset to be frequent or not, and differing values of *min_sp* will lead to different EDUs of individual items as well. Therefore, there must be questions from managers: is there an optimal *min_sp* for all items or will each individual item have its own optimal one? How does one find this out? Liu *et al.* (1999) and Lee *et al.* (2005) thought that different items might have different criteria to judge its importance, and the minimum support requirements should then vary with different items. Lee *et al.*

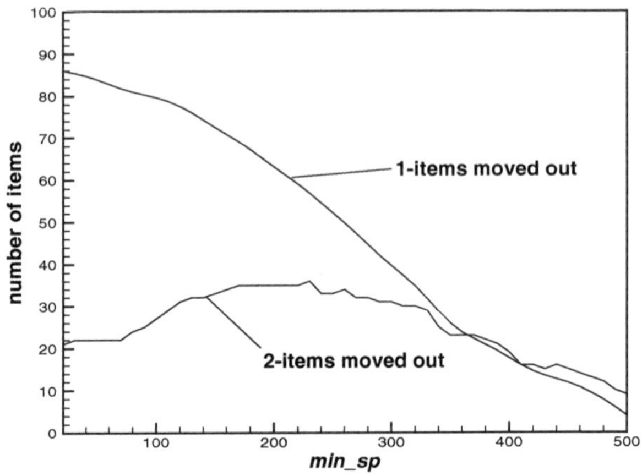

Figure 12.6 The number of items forced into and out of category A by association rules

(2005) gave a method of maximum constraint to defining the minimum supports of itemsets when items have different *min_sp*, and presented an improved Apriori algorithm as well. We are not intending to discuss the algorithm of finding association rules with multiple *min_sp* of items here. Instead, we are more concerned about how to determine the optimal *min_sp* of individual items. As mentioned before, the EDU of the ith item is related to its *min_sp*, which can be denoted by a function format $\text{EDU}_i = f_i(min\_sp)$.

Therefore, we can simply define the *min_sp* with the maximum EDU to be the optimal *min_sp* of an item. Every item has its own optimal *min_sp*. The following two definitions are accurate for an individual item and itemset:

1. The optimal *min_sp* of the ith item is the *min_sp* with maximum EDU_i.
2. The optimal *min_sp* of an itemset is the maximum optimal *min_sp* of the item in the set.

Based on the discussions above, we performed exhaustive calculations to search for the maximum EDU of all items by letting the *min_sp* increase gradually in the algorithm, and ranked the items by their maximum EDU starting with the largest value. After that, by comparing with the ranking list of traditional ABC classification, we found that 458 items that make up the 45.8% of total 1000 items have a greater EDU than traditional dollar usage and their positions in the new ranking list will be promoted consequently. About 86 items that make up the 8.6% of total 1000 items are added into A groups by the association rules effect. Here the A group is defined as 80% of dollar usage occupancy. Therefore, by the definitions of optimal *min_sp* of an item and itemset above, it is no longer necessary for the manager to spend time specifying the *min_sp* of individual items, because the optimal *min_sp* of each item are "hidden" in the transaction database and the computer program can automatically find and use them to determine the frequent itemsets.

12.6 Concluding Remarks

In this chapter, a new approach of ranking inventory items is presented. By using the association rules the cross-selling effect of items is introduced into classification of inventories. We presented a new algorithm for ranking all inventory items. A numerical example was presented to illustrate the utility of the new approach. Empirical experiments using a database from a benchmark and practice were conducted to evaluate the new algorithm's performance and effectiveness, which indicated that a considerable part of inventory items should be revaluated for their importance and therefore change positions in the ranking list of ABC classification. Many items that traditionally belonged to the B or C group were moved into the A group by the cross-selling effect.

The subsequent studies on this topic should be carried out in two respects. The first is to find out a more efficient algorithm of searching the maximum EDU of items with different *min_sp*. The second is to evaluate how much opportunity will be lost when the cross-selling effects are ignored. This insight can lead to the development of a new method to estimate opportunity cost (lost sales) quantitatively, which is an important decision factor in inventory control. Moreover, how to determine the inventory policies in the frequent itemsets is also an important decision problem because the items in a frequent itemset may be from different classes and correlate with each other.

Acknowledgements This research work is a cooperation including Dr. Yiyong Xiao and Professor Ikou Kaku. Their contribution is very much appreciated.

References

Agrawal R, Imilienski T, Swami A (1993) Mining association rules between sets of items in large database. In: Proceedings of SIGMOD, pp 207–216

Agrawal R, Srikant R (1994) Fast algorithm for mining association rules. In: Proceedings of VLDB, pp 487–499

Anand SS, Hughes JG, Bell DA, Patrick AR (1997) Tackling the cross-sales problem using data mining. In: Proceedings of PAKDD'97, pp 331–343

Brijs T, Swinnen G, Vanhoof K, Wets G (1999) Using association rules for product assortment decisions: a case study. In: Proceedings of KDD, pp 254–260

Brijs T, Goethals B, Swinnen G, Vanhoof K, Wets G (2000a) A data mining framework for optimal product selection in retail supermarket data: the generalized PROFSET model. In: Proceedings of ACMSIGKDD, pp 300–304

Brijs T, Vanhoof K, Wets G (2000b) Reducing redundancy in characteristic rule discovery by using integer programming techniques. Intell Data Analys 4:229–240

Chase RB, Aquilano NJ, Jacobs FR (1998) Production and operations management manufacturing and services, 8th edn. McGraw-Hill, Boston, pp 607–608

Cohen MA, Ernst R (1988) Multi-item classification and generic inventory stock control policies. Prod Inventory Manage J 3rd Quarter:6–8

Cooper LG, Giuffrida G (2000) Turning data mining into a management science tool: new algorithms and empirical results. Manage Sci 46(2):249–264

Flores BE, Whybark DC (1987) Implementing multiple criteria ABC analysis. J Oper Manage 7(1/2):79–85

Han J, Micheline K (2001) Data mining: concepts and techniques, chap 6. Morgan Kaufmann, San Fransisco, CA, pp 226–269

Kleinberg JM (1998) Authoritative Sources in a Hyperlink Environment, Proceedings of 9th ACM-SIAM Symposium on Discrete Algorithms, pp 668–677

Kaku I, Xiao Y (2008) A new algorithm of inventory classification based on the association rules. Int J Serv Sci 1(2):148–163

Lee YC, Hong TP, Lin WY (2005) Mining association rules with multiple minimum supports using maximum constraints. Int J Approx Reason 40:44–54

Lenard JD, Roy B (1995) Multi-item inventory control: a multicriteria View. Eur J Oper Res 87:685–692

Liu B, Hsu W, Ma Y (1999) Mining association rules with multiple minimum supports. In: Proceedings of 1999 International Conference on Knowledge Discovery and Data Mining, pp 337–341

Silver EA, Pyke DF, Peterson R (1988) Inventory Management and Production Planning and Scheduling (3rd Edition), pp 30–35

Wang K, Su MT (2002) Item selection by "hub-authority" profit ranking. In: Proceedings of SIGKDD'02, pp 652–657

Wong RC, Fu AW, Wang K (2003) MPIS: Maximal-profit item selection with cross-selling consideration. In: Proceedings of the 3rd IEEE International Conference on Data Mining, pp 371–378

Wong RC, Fu AW, Wang K (2005) Data mining for inventory item selection with cross-selling consideration. Data Minin Know Discov 11(1):81–112

Zhang C, Zhang S (2002) Association Rule Mining: Models and Algorithms, 25–36

Chapter 13
Applying Self-organizing Maps to Master Data Making in Automatic Exterior Inspection

In modern electronics and the electronic device industry, the manufacturing process has been changed tremendously by introducing surface mountain technology (SMT). Many automatic machines for inspecting exteriors have been added into the assembly line, in order to find automatically those products with exterior defects. Usually image processing technology and equipment are used in automatic exterior inspection due to the requirement of high inspection speed and accuracy. The pattern-matching method is the most frequently used method for image processing in exterior inspection, in which, a reference must be made as a representative image of the object to be inspected, the so-called master data. How the master data should be generated is a very important issue for increasing the inspection accuracy. In this chapter, we propose a method of making master data by using the self-organizing maps (SOM) learning algorithm and prove that such a method is effective not only in judgement accuracy but also in computational feasibility. We first apply the SOM learning algorithm to learn the image's feature from the input of samples. Secondly, we discuss theoretically the learning parameters of SOM used in the new master data making process. Thirdly, we propose an indicator, called continuous weight, as an evaluative criterion of learning effects in order to analyze and design the learning parameters. Empirical experiments are conducted to demonstrate the performance of the indicator. Consequently, the continuous weight is shown to be effective for learning evaluation in the process of making the master data.

This chapter is organized as follows. Section 13.1 introduces our motivation for the research. Section 13.2 presents how to make the master data. In Section 13.3 the sample selection methods are defined in detail. In Section 13.4 comparison experiments are presented and discussed. Concluding remarks are given in Section 13.5.

13.1 Introduction

In modern electronics and the electronic device industry, the exterior inspection equipment using image processing technologies has been dramatically developed (a survey can be found in Sakaue 1997). Unlike visual inspections previously performed by humans, the new modern equipment is very useful and effective. Employing not only a hardware unit (such as an advanced optical system) but also an image processing system (such as filters, judgment methods, and so on) leads to these highly effective imaging systems being used in exterior inspection in the electronics and electronic device industry. Such automatic exterior inspection systems dominate when compared to the human visual inspection speed; however, several weak points on the sides of cost and accuracy still exist. For example, exterior inspection is often operated to inspect some defects in the surface of the product, such as chips, projections, foreign material, spots, and especially letters and characters printed on the surface of product. Inspection of exterior letters and characters usually contains two requests, *i.e.*, "no good" products (NG, meaning defective products) must be removed and good products must be passed. In practice, these requirements sometimes are conflicting because that which determines if the product is good or not depends on the inspection method and judgment measure used. When the same inspection method and judgment measure are used the NG products often mingle with the good products. The accuracy of exterior inspection is usually defined to evaluate such methods and judgment measures. Since there is a strong enterprise principle that the NG must be removed, the accuracy of exterior inspection can be defined as how many good products may be misjudged as NG in one production cycle.

There are many judgment measures in the image processing system; the most useful one is pattern matching, which compares the inspecting image to a good image registered beforehand (master data) to determine whether the inspecting image is accurate (Iijima 1973). Although there are various ways to compare these images, the block-matching algorithm was proved to be very effective in the practical inspection process (Sakusabe 1991). However, even when the block-matching method has been used, the accuracy of exterior inspection is still a serious problem. According to our experiences, when exterior inspections are operated in a production line, the accuracy of exterior inspection varies between 1 and 5% of product outputs in the electronic device industry (Sakusabe 1991).

There are a number of reasons for the fluctuation of the accuracy of exterior inspection. One of the reasons to consider is the fact that the master data is made from one image, and is not representative of the whole population of inspecting images, since the master data is usually made from the first good product at the beginning of a production cycle. To overcome such a problem, the question of how to make the master data is an important one. We have made a series of studies by using the self-organizing maps (SOM) learning algorithm to solve this problem, in which the master data can be built automatically from one or several samples of inspecting images (Fujiwara *et al.* 2001, 2002a,b; Kaku *et al.* 2003). In this chapter several contributions of the studies are presented as follows. Firstly, the SOM learning algorithm is applied to learn the image's features from the input of samples. How should the input of samples be made and communicated to the algorithm is very important because that can influence the learning result directly. Also, sample selection methods from several samples will be discussed, which can contribute to a more effective accuracy in the exterior inspection. Secondly, we discuss theoretically the learning parameters of SOM used in the new master data making process, which can guarantee the inspection accuracy in the learning process. Thirdly, we propose an indicator, the continuous weight, as an evaluative criterion of learning effect in order to analyze and construct the learning parameters. Finally, empirical experiments are performed for proving the performance of the indicator. Consequently, the continuous weight is effective for learning evaluation in the process of making the master data.

13.2 Applying SOM to Make Master Data

Simply, SOM, which was developed by Kohonen (1997), is an unsupervised neural network approach, which is widely used in many research fields. The SOM is inspired by the functioning of the human brain in the group which does not require explicit tutoring by input-output correlations and which spontaneously self-organizes upon internal presentation of input patterns. We have given a theoretical illustration of the convergence of SOM in Chapter 5. Therefore, Using the SOM approach, the features of an input pattern can be detected automatically. Figure 13.1 shows the structure of SOM.

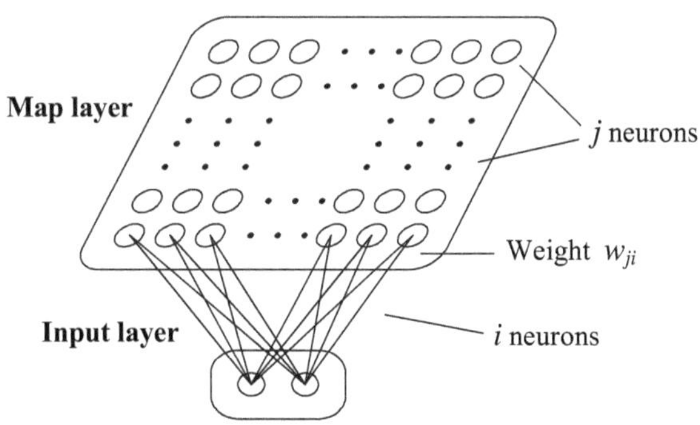

Figure 13.1 The self-organizing maps (SOM)

As shown in Figure 13.1, the SOM is formed by a two-layer competitive learning network, which consists of an input layer and a two-dimensional map layer. There are two i neurons on the input layer and there are j neurons with two dimensions on the map layer. Weights w_{ji} represent the competitive learning relation between a j neuron on the map layer and an i neuron on the input layer. The size of the map layer can be defined by the user, and in our master data-making problem it is defined as the same as the input image size. The SOM learning algorithm is shown as follows:

Step 1 Initialize the network.

All weights w_{ji} are initialized uniformly or located following a normal distribution, which has a mean in the center of the learning image.

Step 2 Samples selection.

Theoretically, the master data must be representative of the population of inspection images, so that some statistical concepts can be used to build a new sample selection method. Additionally, we have investigated several sample selection methods (Kaku *et al.* 2003). Four methods of sample selection have been considered. The first method is called random selection, by which the sample is selected randomly and will be changed at each learning time. Second method is called index image, by which the mean and the standard deviation of each coordinate will be estimated from all samples. Then, a statistical image can be generated as an input of the sample. This generation will be performed at each learning time. The third method is similar to the second method. The difference between them is just on the timing of generation. In the third method, the mean and the standard deviation of each coordinate is generated only by (the number of learning/the number of samples). Therefore, the third method is called simplified index image. Typically, traditional sample selection method is employed by input of samples, which is just changed by (the number of samples) times, where each sample will be learned continuously by (the number of learning/the number of samples) times. It is never used again if the sample of image has been learned. The result of comparing experiments shows the method of simplified index image has an advantage over other methods.

The simplified index image method is illustrated in Figure 13.2, where $x(\leq X)$ and $y(\leq Y)$ correspond to the coordinates of the horizontal axis and vertical axis, respectively, in the image. Assume $A(x, y)$ (the illumination value located at coordinate (x, y) with a gray-scale of $[0, 255]$) has a normal distribution, the mean and the standard deviation of $A(x, y)$ can be estimated from all samples. Then, a new image can be built which has the estimated mean and the estimated standard deviation of each coordinate. This generation will be done at each T/N time, where T is the number of total learning times and N is the number of selected samples.

Step 3 Generate an input vector.

An input vector $\boldsymbol{v} = [v_1, v_2]$ is generated from the description of an image sample with an occurrence probability $P_A(x, y)$, which is defined by

$$P_A(x, y) = \frac{A(x, y)}{\sum_{i=1,j=1}^{X,Y} A(i, j)} . \tag{13.1}$$

Using the occurrence probability $P_A(x, y)$, the input vector $\boldsymbol{v}$ can be generated with the following equation:

$$\boldsymbol{v} = \min \left\{ \begin{pmatrix} x \\ y \end{pmatrix} \middle| \sum_{i=1,j=1}^{x,y} A(i, j) > k \right\} , \tag{13.2}$$

where k is an integer generated randomly, and $\sum_{x=1,y=1}^{X,Y} A(x, y)$ is with lexicographic order.

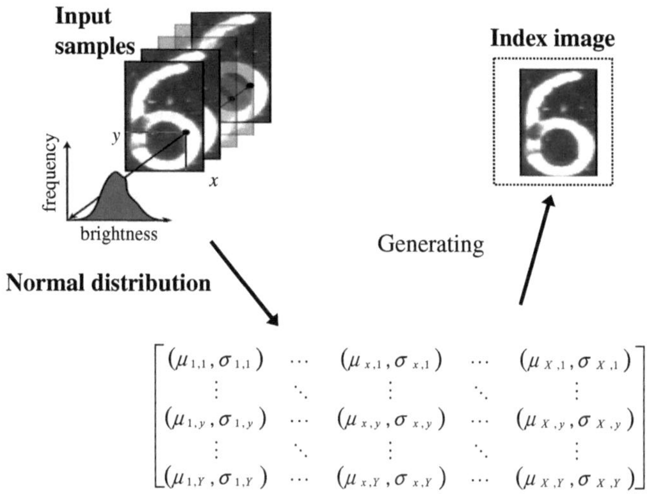

Figure 13.2 Simplified index image

Step 4 Calculate distances.

Distance is defined as a Euclidean distance between all weights and input vector. The distance d_j of the weight w_{ji} and an input vector $\boldsymbol{v}$ is denoted by

$$d_j = \left\| \boldsymbol{v}, \boldsymbol{w}_{j\bullet} \right\| . \tag{13.3}$$

Step 5 Competitive selections.

Minimize the calculated distance by

$$d_{j*} = \min\left(d_j\right) , \tag{13.4}$$

where j^* is the selected neuron.

Step 6 Update of weight.

The weights corresponding to the j^* neuron and its neighbors are updated following:

$$\Delta w_{ji} = \eta_t h_t\left(j, j^*\right) \cdot \left(v_i - w_{ji}\right) , \tag{13.5}$$

where t is the number of learning times, η_t is a learning multiplier and $h_t\left(j, j^*\right)$ is the neighbor multiplier, which are defined by

$$\eta_t = \begin{cases} \eta_0 \times \left(1 - \dfrac{t}{T_\eta}\right) & \text{if } t \leq T_\eta , \\ 0 & \text{otherwise} \end{cases} , \tag{13.6}$$

$$h_t\left(j, j^*\right) = \begin{cases} 1 - \dfrac{d}{r_t} & \text{if } d \leq r_t , \\ 0 & \text{otherwise} \end{cases} \tag{13.7}$$

and

$$r_t = \begin{cases} r_0 \times \left(1 - \frac{t}{T_r}\right) & \text{if } t < T_r \ , \\ 0 & \text{otherwise} \ . \end{cases} \tag{13.8}$$

In Equation 13.6, η_0 and T_η are the default value of learning multiplier and the coefficient of learning multiplier. In Equations 13.7 to 13.9, d is the Euclidean norm on the map layer between j and j^* as shown in Figure 13.3, r_t is the neighbor value, r_0 is the default value of the neighbor value and T_r is the coefficient of neighbor value. Equations 13.6 and 13.7 contribute a mechanism in which the neighbor value r_t of neuron j^* will be decreased to zero when $t = T_r$, so that learned neighborhood will become only neuron j^*. Also, the learning rate will be decreased when the numbers of learning increase.

The algorithm above will be repeated T times, defined as the number of learning times.

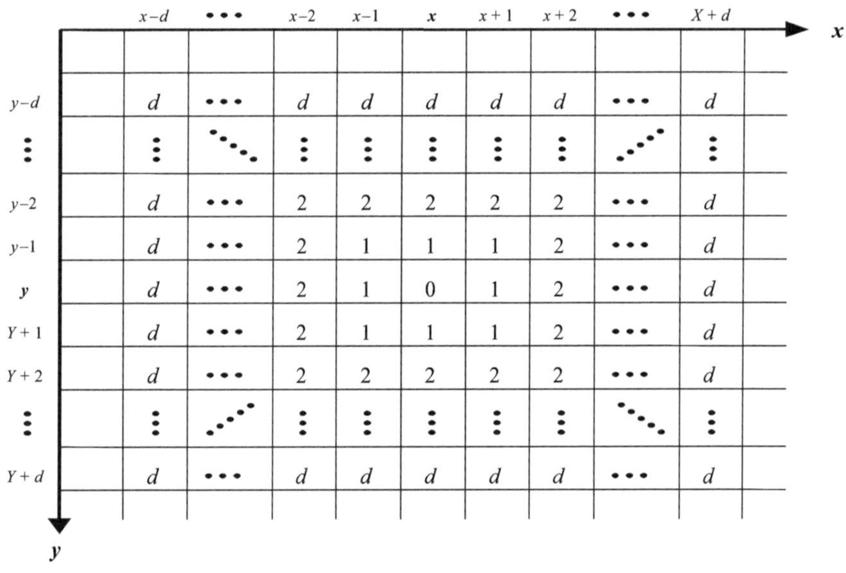

Figure 13.3 Euclidean distance on map layer

Figure 13.4 shows the flowchart of this algorithm and Figure 13.5 shows the learning process of the master data making by SOM algorithm. As shown in Figure 13.5, when the learning time is increasing, the feature of the original input image is learned perfectly.

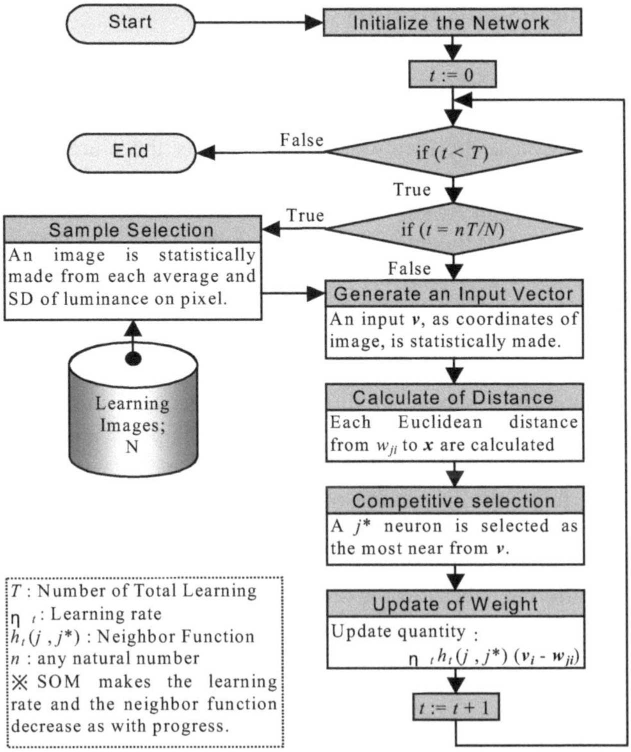

Figure 13.4 The flow chart of making master data

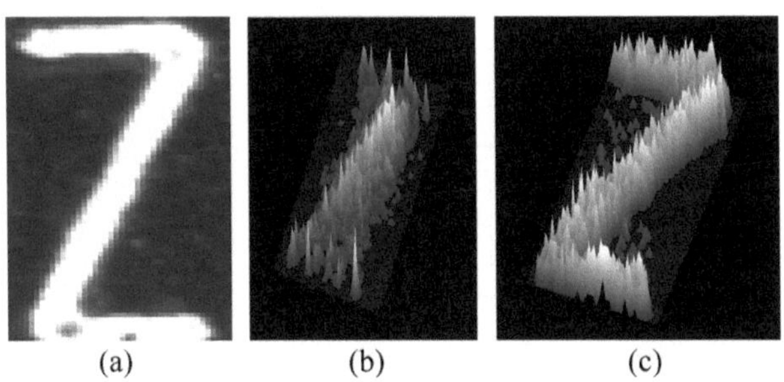

Figure 13.5 The learning process of SOM algorithm: (a) original, (b) after 1000 times, and (c) after 200,000 times

13.3 Experiments and Results

Usually, the Latin alphabet, Arabic numerals, and some special symbols (such as $+, -$) are used for practical purposes in electronic device products. The main task of exterior inspection is the identification of whether these are printed correctly on the surface of the products. In this chapter, several character samples (including the Latin alphabet and Arabic numerals) that are cut out from electronic device products are prepared for the empirical experiments (Table 13.1). Twenty samples of each character were prepared, which are good images but have been misjudged once by the common master data. Additionally, the material of the samples is plastic mould and the character is sealed by a YAG laser. The size of each character is uniformly set to 42×80 pixels. For comparing the learning effect of the criteria, we use the parameters shown in Table 13.2 throughout all the experiments. All programs were coded in Microsoft Visual C++ version 6 and implemented on a PC (CPU: Pentium IV 2 GHz, OS: Microsoft Windows XP).

Figure 13.6 shows the inspection accuracy of the selected products. In Figure 13.6, the horizontal axis represents the character samples (*i.e.*, 1: sample of 1, 2: sample of 6, 3: sample of N, 4: sample of Z) and vertical axis represents the inspection accuracy. Different bars show the different sample selection methods as defined below the graph. Comparing the inspection accuracy of different sample selection

Table 13.1 The samples of an image

	Type 1	Type 2	Type 3	Type 4
Example image				

Table 13.2 The parameter list of SOM

Factor	Symbol	Value
Initial value of learning multiplier	η_0	0.9
Coefficient of learning multiplier	T_η	600,000
Initial value of neighbor value	r_0	4
Coefficient of neighbor value	T_r	220,000
Number of total learning times	T	200,000

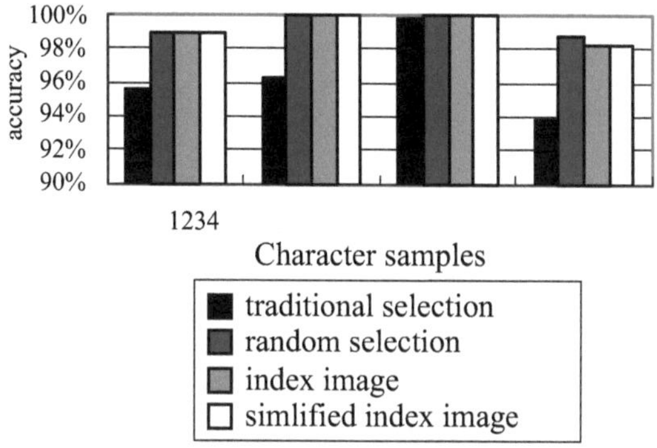

Figure 13.6 The inspective accuracy of the selected samples

methods, all of the sample selection methods except traditional selection can obtain a high level of inspection accuracy (note that these samples have been misjudged before). That means the SOM approach is very effective for such a misjudgement problem in exterior inspection. Therefore, these new sample selection methods are superior in inspection accuracy to the traditional one. Moreover, because the index image method takes many times in the master data making process (about 40 s), it can be considered that the index image method is imaginary.

13.4 The Evaluative Criteria of the Learning Effect

The above master data-making algorithm is proved be effective on inspection accuracy. However, several issues remain. One of them is how to evaluate the learning effect of the new algorithm during the learning process, and to decide when the learning process should be ended. This work is straightly related with the determination of the learning parameters (includes initial weight, number of total-learning times, learning rate, neighbor function), and how to update these parameters. In fact, whether the new master data can be used in practice is based on its computational feasibility. There are several criteria to evaluate the learning effect. For example, we have used the convergence rate of Euclidean distance between the input vector $\boldsymbol{v}$ and a vector whose elements are weights w_{ji} corresponding to neuron j^* for evaluating the learning effect. However, as shown in Figure 13.7, there is little difference between the convergence rate of successful learning and the failed one. The failed master data contains a lot of noise so it cannot be used in practice.

For evaluating the learning effect, we consider an indicator which is called continuous weight. Basically in the learning process like SOM, all weights are ordered continuously corresponding to the positions of neurons on the map layer when the learning times are increased. We define such a continuous order as continuous weight using a color system shown in Figure 13.8, *i.e.*, the learning process is successful if there is the continuous weight in the map layer. Otherwise, the learning process is a failure. Figure 13.9 shows the examples. In Figure 13.9 (a), the image

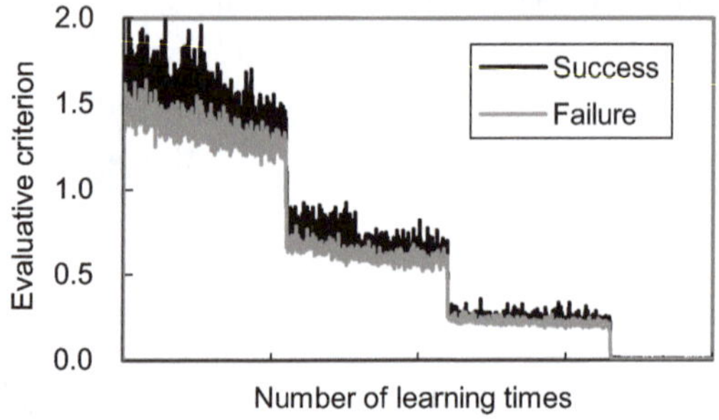

Figure 13.7 The ineffective criterion

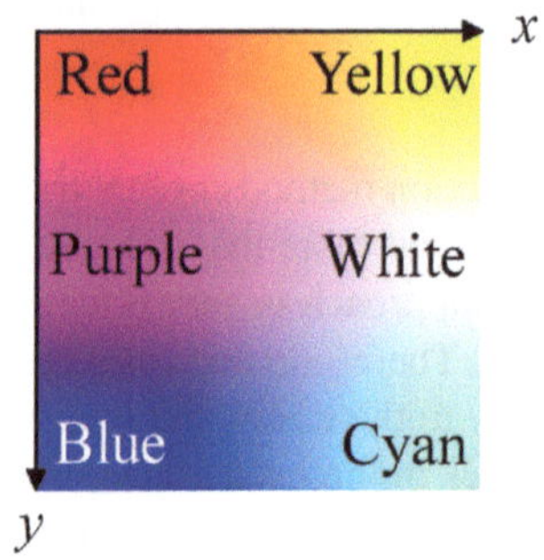

Figure 13.8 Color system of continuous weight

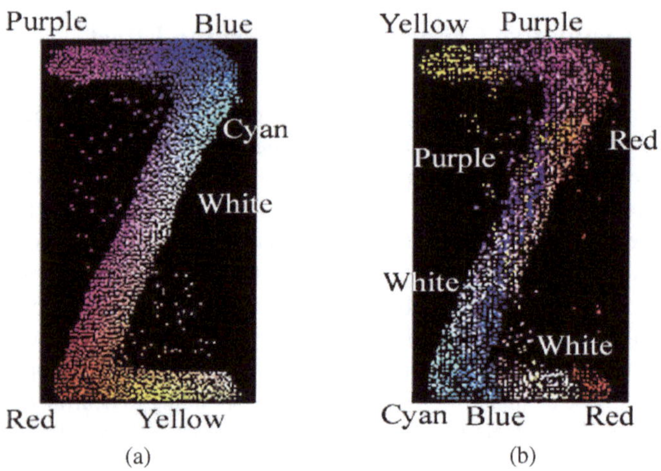

Figure 13.9 The examples of continuous weight: (a) success and (b) failure

is learned successfully because the continuous weight is in the color order defined by Figure 13.8. According to the same consideration, in Figure 13.9 (b), the image is learned unsuccessfully because the continuous weight is not in the correct color order. There are some discontinuous segments that can be found in Figure 13.9 (b).

Therefore, using the continuous weight, the learning effect of different criteria can be compared. Here, four evaluative criteria are proposed:

- Chi-squared test;
- Square measure of close loops;
- Distance between adjacent neurons;
- Monotony of close loops.

13.4.1 Chi-squared Test

First, we define the classical evaluative criterion of the chi-squared test as E_{chi} following:

$$E_{\text{chi}} = \sum_{m=1,n=1}^{4,4} \frac{\left(C_w(t,m,n) - A_D(m,n) \cdot \frac{S_A}{S_C}\right)^2}{A_D(m,n) \cdot \frac{S_A}{S_C}}, \tag{13.9}$$

where $A_D(m,n)$ is the sum of the illumination values located on each area (m,n) that is divided into 16 blocks as in Figure 13.10, $C_w(t,m,n)$ is the number of weight w_{ji} in the map layer corresponding to area (m,n) when the learning time is t, S_A is $\sum_{m=1,n=1}^{4,4} A_D(m,n)$ and S_C is the number of all neurons on the map layer.

It can be considered that the continuous weight is satisfied if E_{chi} is small enough.

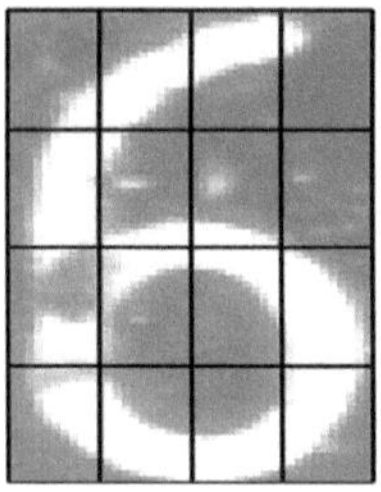

Figure 13.10 The chi-squared test

13.4.2 Square Measure of Close Loops

Second, we define the evaluative criterion of square measure of two close loops on the input image as E_{sqr} following:

$$E_{\text{sqr}} = \frac{\sum\limits_{j} (|a_1 \cdot b_2 - a_2 \cdot b_1| + |c_1 \cdot d_2 - c_2 \cdot d_1|)}{X \cdot Y}, \tag{13.10}$$

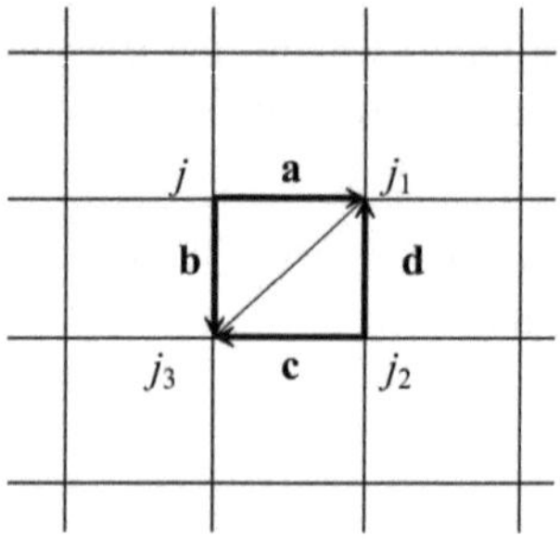

Figure 13.11 The positional relationship of neurons

where parameters $\{a_i, b_i, c_i, d_i | i = 1, 2\}$ are elements of each vectors $\boldsymbol{a}$ from j to j_1, $\boldsymbol{b}$ from j to j_3, $\boldsymbol{c}$ from j_2 to j_3, and $\boldsymbol{d}$ from j_2 to j_1, where j_1, j_2 and j_3 are adjacent neurons to a neuron j on the map layer as shown in Figure 13.11.

It can be considered that the continuous weight is satisfied if E_{sqr} diminishes to a value proportional to the total square of lines of the learned image.

13.4.3 Distance Between Adjacent Neurons

Third, we define the evaluative criterion of distance between adjacent neurons, using the Euclidean distances between j neuron and two neurons that are adjacent to it on the map layer in Figure 13.11, as E_{adj} following:

$$E_{\text{adj}} = \frac{\sum_j \left(\left\| \boldsymbol{w}_{j\bullet}, \boldsymbol{w}_{j_1\bullet} \right\| + \left\| \boldsymbol{w}_{j\bullet}, \boldsymbol{w}_{j_3\bullet} \right\| \right)}{X \cdot Y} . \tag{13.11}$$

It can be considered that E_{adj} also will be diminished if the learned map layer satisfies the continuous weight.

13.4.4 Monotony of Close Loops

Finally, we define the evaluative criterion of monotony as E_{mon} in the following:

$$E_{\text{mon}} = \frac{\left| \sum_j^p - \sum_j^q \right|}{X \cdot Y - \sum_j^r} , \tag{13.12}$$

where p, q and r are defined by

$$p = \begin{cases} 1 \text{ if } a_1 \cdot b_2 - a_2 \cdot b_1 > 0 , \\ 0 \text{ otherwise} , \end{cases} \tag{13.13}$$

$$q = \begin{cases} 1 \text{ if } a_1 \cdot b_2 - a_2 \cdot b_1 < 0 , \\ 0 \text{ otherwise} \end{cases} \tag{13.14}$$

and

$$r = \begin{cases} 1 \text{ if } a_1 \cdot b_2 - a_2 \cdot b_1 = 0 , \\ 2 \text{ otherwise} \end{cases} , \tag{13.15}$$

where parameters $\{a_i, b_i | i = 1, 2\}$ are defined above.

It can be considered that the continuous weight is satisfied if E_{mon} is close to 1.

13.5 The Experimental Results of Comparing the Criteria

The results of experiments are shown respectively in Figures 13.12 to 13.15 with different criteria. In these figures, the horizontal axis shows the number of learning times and the vertical axis shows the value of the evaluative criterion; the dark curves show success learning and light curves show failure learning. Each curve has been represented by each sample.

Figure 13.12 shows the chi-squared test criterion. From this figure, it can be considered that the chi-squared test criterion is not appropriate for evaluating the

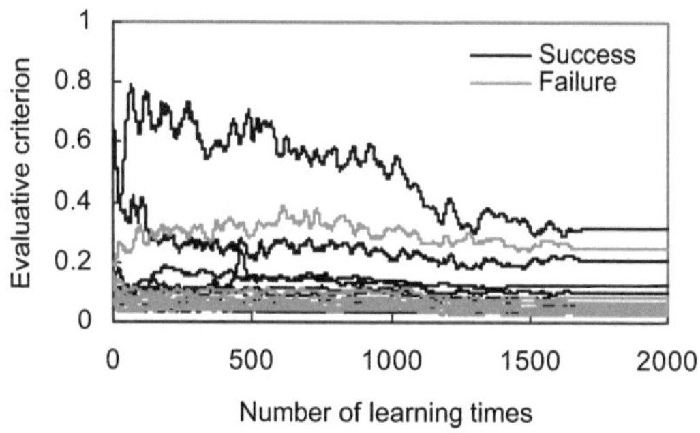

Figure 13.12 The Chi-squared test method

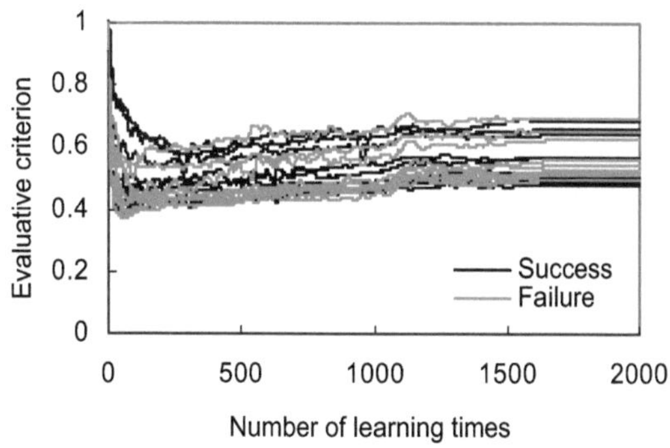

Figure 13.13 The square measure method

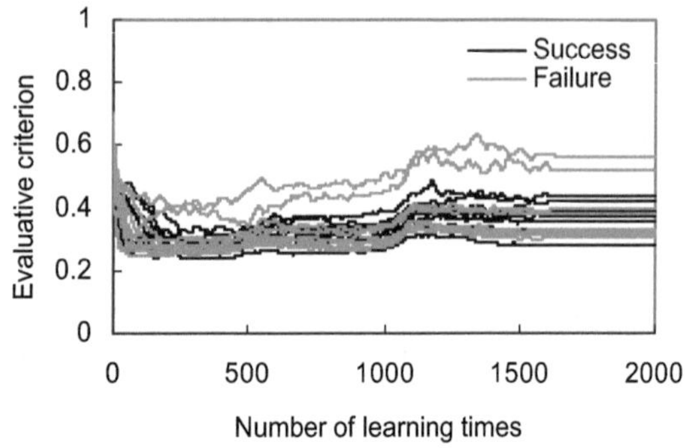

Figure 13.14 The adjacent distance method

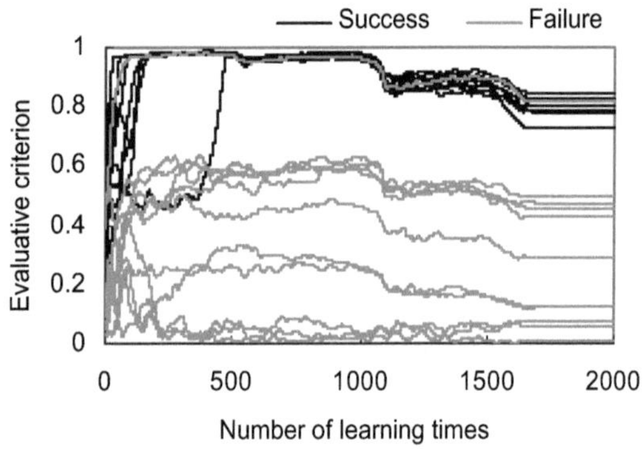

Figure 13.15 The monotony method

learning effect because the values of the evaluative criterion of successful samples are worse (and larger) than ones of failed samples. Figure 13.13 shows the square measure of close loops. From this figure, it also can be considered that the criterion of square measure of close loops is not appropriate for evaluating the learning effect because there is not significant distinction between success and failure in evaluation values.

However, the criterion of the monotony of close loops shown in Figure 13.15 distinguishes clearly the successful learning case except only a few samples, which is the highest curve of failure in the case of the criterion of distance between adjacent neurons shown in Figure 13.14. It can be explained from Figure 13.16 where the monotony criterion is effective in the result map layer, but the adjacent distance criterion is not satisfied because some discontinuous segments can be found in the figure. However, on the other hand, the distance between adjacent neurons of such discontinuous segments increases. Therefore, it is possible to make an evaluating index by combining two criteria: the distance between adjacent neurons and the monotony of close loops.

Figure 13.17 shows the correlation diagram of the criteria for all samples. As shown in Figure 13.17, the whole of successful samples are located together in the dashed line area. Therefore, it can be considered that the learning effect should be evaluated by combining two criteria, the monotony of close loops and distance between adjacent neurons, and the successful result can be obtained.

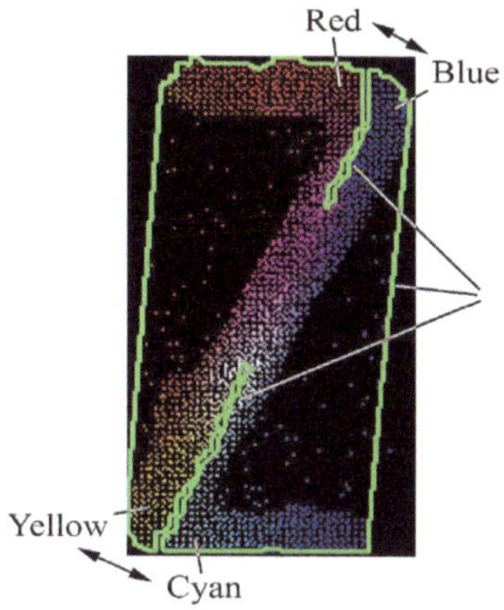

Figure 13.16 Example of the exception

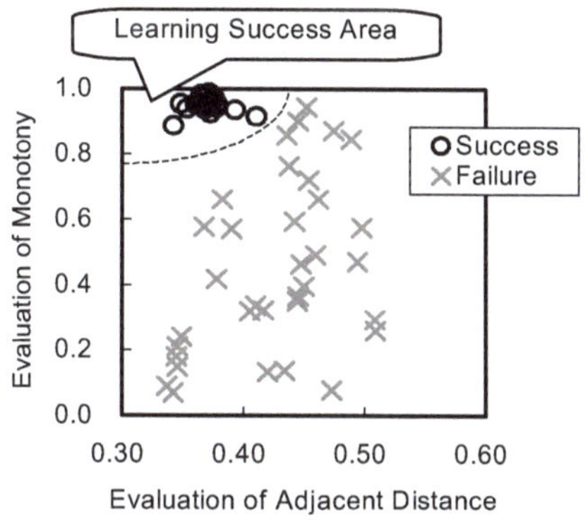

Figure 13.17 The correlation diagram

13.6 Conclusions

In this chapter, we applied the SOM approach to make master data, which is very important in the exterior inspection in the electronic device industry. Several key issues were discussed and practiced with experiments using real product images. Firstly, we successfully developed a SOM-based algorithm to learn an image's characteristics from one or several real images. Secondly, three sample selection methods of making the input of samples were described. These methods have been evaluated to be superior in the inspection accuracy of the exterior inspection. As a result, the random simplified index image methods can improve the inspection accuracy and take a smaller time to make master data comparing with the traditional method, so that the method can be considered to be superior in the master data making process. Thirdly, we discussed a theoretical problem on the evaluative criteria of the learning effect. For evaluating the learning effect an indicator called continuous weight is proposed. Four criteria were evaluated by using the indicator. The results of experiments show that combining the monotony of close loops and the distance between adjacent neurons can obtain a very good result. This fact can lead us to the optimal analysis of the learning parameter and development of a new judgement method in exterior inspection.

Additionally, it should be pointed out that the master data made by the SOM approach has a very different data feature compared with the common approach. The SOM master data has a 3D image, in which more neurons have been collected together around the region that represents the feature of the original image. Our further work is to develop a new judgment approach that can use such a characteristic.

Acknowledgements This research work was a cooperative effort including Mr. Wataru Fujiwara, Dr. Mituhiro Hoshino and Ikou Kaku. Their contribution is very much appreciated.

References

Fujiwara W, Hoshino M, Kaku I *et al.* (2001) Making the master data automatically in the exterior inspection by SOM approach. In: Proceeding of the 2001 Information and Systems Society Conference of IEICE, p 140

Fujiwara W, Hoshino M, Kaku I *et al.* (2002a) An effective method of exterior inspecting with self-organizing maps. In: Proceedings of forum on information technology, pp 315–316

Fujiwara W, Hoshino M, Kaku I *et al.* (2002b) A study on the effective method of exterior inspecting using a neural network approach. In: Proceedings of the 6th China-Japan international symposium on industrial management, pp 369–375

Iijima T (1973) Pattern Recognition. CORONA PUBLISH Co. 168

Kaku I, Fujiwara W, Hoshino M *et al.* (2003) An effective learning approach to automatic master data making in exterior inspection. In: Proceedings of the 16th international conference on production research

Kohonen T (1997) Self-organizing maps. Springer, Berlin Heidelberg New York

Sakaue K (1997) Investigation about the Evaluation Method of an Image Processing Performance. National Institute of Advanced Industrial Science and Technology

Sakusabe A (1991) A production of an image processing system [IM-21]. In: Proceeding of the 2nd Fruition Conference at the Development Section in Akita Shindengen, pp 91–105

Chapter 14
Application for Privacy-preserving Data Mining

Recently, data mining with capability of preserving privacy has been an area gaining a lot of researcher attention. In fact, researchers working on inference control in statistical databases have long raised a number of related concerns. In the literature, different approaches have been proposed, including the cryptographically secure multiparty computation, random perturbation, and generalization.

Nowadays, the main data mining tasks include association rule mining, classification, clustering and so on. According to different mining missions, there are different privacy-preserving data mining algorithms.

This chapter is organized as follows. Section 14.1 presents the application for privacy-preserving association rule mining. In Section 14.2 privacy-preserving clustering is discussed. In Section 14.3, we give a scheme to privacy-preserving collaborative data mining. In Section 14.4, we introduce the evaluation of privacy preserving and future work. Finally, the conclusion is presented in Section 14.5.

14.1 Privacy-preserving Association Rule Mining

Since its introduction in 1993 by Agrawal, association rule mining has received a great deal of attention. It is still one of most popular pattern-discovery methods in the field of knowledge discovery. The goal of association rule mining is to discover meaningful association rules among the attributes of a large quantity of data (Agrawal *et al.* 1993). One of the most important association rule mining algorithm is the Apriori algorithm, and many research efforts are based on it.

14.1.1 Privacy-preserving Association Rule Mining in Centralized Data

One crucial aspect of privacy-preserving frequent itemset mining is the fact that the mining process deals with a trade-off: privacy and accuracy, which are typically contradictory, and improving one usually incurs a cost in the other. One alternative to address this particular problem is to look for a balance between hiding restrictive patterns and disclosing non-restrictive ones. Oliveira and Zaiane (2002) proposed a framework for enforcing privacy in mining frequent itemsets. They combine, in a single framework, techniques for efficiently hiding restrictive patterns: a transaction retrieval engine relying on an inverted file and Boolean queries, and a set of algorithms to "sanitize" a database. In addition, they introduce performance measures for mining frequent itemsets that quantify the fraction of mining patterns, which are preserved after sanitizing a database.

The main idea of the algorithm is: let D be a transactional database, P be a set of all frequent patterns that can be mined from D, and $Rules_H$ be a set of decision support rules that need to be hidden according to some security policies. A set of patterns, denoted by R_P, is said to be restrictive if $R_P \subset P$ if and only if R_P would derive the set $Rules_H$. $\sim R_P$ is the set of non-restrictive patterns such that $\sim R_P \cup R_P = P$.

This process is composed of four steps as follows:

1. In the first step, the set P of all patterns from D is identified.
2. The second step, distinguishes restricted patterns R_P from the non-restrictive patterns $\sim R_P$ by applying some security policies. It should be noted that what constitutes as restrictive patterns depends on the application and the importance of these patterns in a decision process.
3. Sensitive transactions are identified within D. In this approach, a very efficient retrieval mechanism called the transaction retrieval engine is used to speed up the process of finding the sensitive transactions.
4. Finally, step 4 is dedicated to the alteration of these sensitive transactions to produce the sanitized database D'.

In general, the process of searching for sensitive transactions through the transactional database follows three steps:

1. Vocabulary search. Each restrictive pattern is split into single items. Isolated items are transformed into basic queries to the inverted index.
2. Retrieval of transactions. The lists of all transaction IDs of transactions containing each individual item are retrieved.
3. Intersections of transaction lists. The lists of transactions of all individual items in each restrictive pattern are intersected using a conjunctive Boolean operator on the query tree to find the sensitive transactions containing a given restrictive pattern.

The approaches to remove information (item restriction-based algorithms) have essentially four major steps:

1. Identify sensitive transactions for each restrictive pattern.
2. For each restrictive pattern, identify a candidate item that should be eliminated from the sensitive transactions. This candidate item is called the victim item.
3. Based on the disclosure threshold ψ, calculate for each restrictive pattern the number of sensitive transactions that should be sanitized.

4. Based on the number found in step 3, identify for each restrictive pattern the sensitive transactions that have to be sanitized and remove the victim item from them where the sensitive transactions to be sanitized are selected.

Then we can make D' public, where all the sensitive rules have been removed from access by other people for data mining to furthermore preserve privacy.

In Evfimievski *et al.* (2002), the authors present a framework for mining association rules from transactions consisting of categorical items where the data has been randomized to preserve privacy of individual transactions. While it is feasible to recover association rules and preserve privacy using a straightforward "uniform" randomization, the discovered rules can unfortunately be exploited to find privacy breaches. They analyze the nature of privacy breaches and propose a class of randomization operators that are much more effective than uniform randomization in limiting the breaches. Also they derive formulae for an unbiased support estimator and its variance, which allow us to recover itemset supports from randomized datasets, and show how to incorporate these formulae into mining algorithms.

The authors in Rizvi and Harista (2002) present a scheme, based on probabilistic distortion of user data, called mining association with secrecy konstraints (MASK). The scheme can extract association rules from large historical databases, simultaneously provide a high degree of privacy to the user and retain a high degree of accuracy in the mining results. The scheme is based on a simple probabilistic distribution of user data, employing random numbers generated from a predefined distribution function. It is this distorted information that is eventually supplied to the data miner, alone with a description of the distortion procedure.

14.1.2 Privacy-preserving Association Rule Mining in Horizontal Partitioned Data

Horizontal distribution refers to the cases where different database records reside in different places, while all the places have the same attributes. For this kind of data distribution, most research employs encryption technology. At present, we often use secure multiparty computation (SMC), based on cryptology, to achieve privacy preservation.

Following the idea of SMC, people have designed privacy-oriented protocols for the problem of privacy-preserving collaborative data mining. Clifton *et al.* (2002) proposed four efficient methods for privacy-preserving computations: secure sum, secure set union, secure size of set intersection, and scalar product. They also show how they can be used to solve several privacy-preserving data mining (PPDM) problems.

Secure sum is often given as a simple example of secure multiparty computation. Distributed data-mining algorithms frequently calculate the sum of values from individual sites. Assuming three or more panics and no collusion, the following method securely computes such a sum.

Assume that the value $u = \sum_{i=1}^{s} u_i$ to be computed is known to lie in the range $[0, n]$.

One site is designated the master site, numbered 1. The remaining sites are numbered $2, \ldots, s$. Site 1 generates a random number R, uniformly chosen from $[1, n]$. Site 1 adds this to its local value u_1, and sends the sum $(R + u_1) \bmod n$ to site 2. Since the value R is chosen uniformly from $[1, n]$, the number $(R + u_1) \bmod n$ is also distributed uniformly across this region, so site 2 learns nothing about the actual value of u_1.

For the remaining sites $l = 2, \ldots, s - 1$, the algorithm is as follows. Site l receives $V = (R + \sum_{l=1}^{i-1} u_j) \bmod n$.

Since this value is uniformly distributed across $[1, n]$, i learns nothing. Site i then computes $V = (R + \sum_{j=1}^{i} u_j) \bmod n = u_j + V \bmod n$ and passes it to site $l + 1$.

Site s performs the about step, and sends the result to site 1. Site 1, knowing R, can subtract R to get the actual result. Note that site 1 can also determine $\sum_{l=2}^{s} u_l$ by subtracting u_1. This is possible from the global result regardless of how it is computed, so site 1 has not learned anything from the computation. This method shows an obvious problem if sites collude. Sites $l - 1, l + 1$ can compare the values they send/receive to determine the exact value for u_l. The method can be extended to work for an honest majority. Each site divides u_l into shares. The sum for each share is computed individually. However, the path used is permuted for each share, such that no site has the same neighbor twice. To compute u_l, the neighbors of L from each iteration would have to collude. Varying the number of shares varies the number of dishonest (colluding) panics required to violate security.

SMC provides us a good research framework of conducting computations among multiple parties while maintaining the privacy of each party's input. However, all of the known methods for secure multiparty computation rely on the use of a circuit to simulate the particular function, which becomes the efficiency bottleneck. Even with some improvements (Gennaro *et al.* 1998), the computational costs for problems of interest remain high, and the impact on real-world applications has been negligible.

In Kantarcioglu and Clifton (2002), authors present a scheme aimed at addressing secure mining of association rules over horizontally partitioned data. The methods incorporate cryptographic techniques to minimize the information shared, while adding little overhead to the mining task.

14.1.3 Privacy-preserving Association Rule Mining in Vertically Partitioned Data

Vertical data distribution refers to the cases where all the values for different attributes reside in different places; each party has different sets of attributes but the key of each record is the same.

Vaidya and Clifton (2002) address the problem of association rule mining where transactions are distributed across sources. Each site holds some attributes of each transaction, and the sites wish to collaborate to identify globally valid association rules. However, the sites must not reveal individual transaction data. They present a two-party algorithm for efficiently discovering frequent itemsets with minimum support levels, without either site revealing individual transaction values. The absence or presence of an attribute is represented as 0 or 1. Transactions are strings of 0 or 1. The algorithm, based on the Apriori algorithm, can achieve association rules without sharing information other than through scalar product computation.

The complete algorithm to find frequent itemsets is:

1. L_1 = large one-itemsets
2, for ($k = 2; L_{k-1} \neq \phi; k{+}{+}$) do begin
3. $C_k = \textit{apriori-gen}(L_{k-1})$
4. for all candidates $c \in C_k$ do begin
5. if all the attributes in c are entirely at A or B
6. that party independently calculates $c.count$
7. else
8. let A have l of the attributes and B have the remaining m attributes
9. construct $\vec{X}$ on A′s side and $\vec{Y}$ on A′s side where $\vec{X} = \prod_{i=1}^{l} \vec{\mathrm{A}_i}$ and $\vec{Y} = \prod_{i=1}^{M} \vec{\mathrm{B}_i}$
10. compute $c.count = \vec{X} \times \vec{Y} = \sum_{i=1}^{n} x_i^* y_i$
11. end if
12. $L_k = \{c \in C_k | c.count \geq \min - \text{sup}\}$
13. end
14. end
15. Answer $L = \cup_k L_k$

In step 3, the function *apriori-gen* takes the set of large itemsets L_{k-1} found in the $(k-1)$th pass as an argument and generates the set of candidate itemsets C_k. This is done by generating a superset of possible candidate itemsets and pruning this set. Given the counts and frequent itemsets, we can compute all association rules with $support \geq min\_sp$.

Only steps 1, 3, 10 and 12 require sharing information. Since the final result $\cup_k L_k$ is known to both parties, steps 1, 3 and 12 reveal no extra information to either party. We now show how to compute step 10 without revealing information.

Secure computation of a scalar product is the key to our protocol. Scalar product protocols have been proposed in the SMC literature (Agrawal and Srikant 2000), however, these cryptographic solutions do not scale well to this data-mining problem. They give an algebraic solution that hides true values by placing them in equations masked with random values. The knowledge disclosed by these equations only allows for the computation of private values if one side learns a substantial number of the private values from an outside source.

We assume without loss of generality that n is even.

Step 1 A generates random $R_1, R_2, \ldots, R_n$. From these, $\vec{X}$, and a matrix C forming coefficients for a set of linear independent equations, A sends the following vector $\vec{X}'$ to B:

$$\begin{aligned}
&\left(x_1 + c_{1,1}{}^* R_1 + c_{1,2}{}^* R_2 + \ldots + c_{1,n}{}^* R_n\right) \\
&\left(x_2 + c_{2,1}{}^* R_1 + c_{2,2}{}^* R_2 + \ldots + c_{2,n}{}^* R_n\right) \\
&\ldots \\
&\left(x_n + c_{n,1}{}^* R_1 + c_{n,2}{}^* R_2 + \ldots + c_{n,n}{}^* R_n\right) .
\end{aligned}$$

In step 2, B computes $\vec{X}' \cdot \vec{Y}$. B also calculates the following n values:

$$\begin{aligned}
&\left(c_{1,1}{}^* y_1 + c_{2,1}{}^* y_2 + \ldots + c_{n,1}{}^* y_n\right) \\
&\left(c_{1,2}{}^* y_1 + c_{2,2}{}^* y_2 + \ldots + c_{n,2}{}^* y_n\right) \\
&\ldots \\
&\left(c_{1,n}{}^* y_1 + c_{2,n}{}^* y_2 + \ldots + c_{n,n}{}^* y_n\right) .
\end{aligned}$$

But B cannot send these values, since A would then have n independent equations in n unknowns $(y_1, y_2, \ldots, y_n)$, revealing the y values. Instead, B generates r random values, $R'_1, R'_2, \ldots, R'_r$. The number of values A would need to know to obtain full disclosure of B's values is governed by r.

B partitions the n values created earlier into r sets, and the R' values are used to hide the equations as follows:

$$\begin{aligned}
&\left(c_{1,1}{}^* y_1 + c_{2,1}{}^* y_2 + \ldots + c_{n,1}{}^* y_n + R'_1\right) \\
&\ldots \\
&\left(c_{1,n/r}{}^* y_1 + c_{2,n/r}{}^* y_2 + \ldots + c_{n,n/r}{}^* y_n + R'_1\right) \\
&\left(c_{1,(n/r+1)}{}^* y_1 + c_{2,(n/r+1)}{}^* y_2 + \ldots + c_{n,(n/r+1)}{}^* y_n + R'_2\right) \\
&\ldots \\
&\left(c_{1,2n/r}{}^* y_1 + c_{2,2n/r}{}^* y_2 + \ldots + c_{n,2n/r}{}^* y_n + R'_2\right) \\
&\ldots \\
&\left(c_{1,((r-1)n/r+1)}{}^* y_1 + c_{2,((r-1)n/r+1)}{}^* y_2 + \ldots + c_{n,((r-1)n/r+1)}{}^* y_n + R'_r\right) \\
&\ldots \\
&\left(c_{1,n}{}^* y_1 + c_{2,n}{}^* y_2 + \ldots + c_{n,n}{}^* y_n + R'_r\right) .
\end{aligned}$$

Then B sends S and the n above values to A, who writes:

$$\begin{aligned}
S = &\left(x_1 + c_{1,1}{}^* R_1 + c_{1,2}{}^* R_2 + \ldots + c_{1,n}{}^* R_n\right){}^* y_1 \\
&+ \left(x_2 + c_{2,1}{}^* R_1 + c_{2,2}{}^* R_2 + \ldots + c_{2,n}{}^* R_n\right){}^* y_2 \\
&\ldots \\
&\left(x_n + c_{n,1}{}^* R_1 + c_{n,2}{}^* R_2 + \ldots + c_{n,n}{}^* R_n\right){}^* y_n
\end{aligned}$$

Simplifying further and grouping the $x_i{}^* y_i$ terms gives:

$$\begin{aligned}S = &\left(x_1{}^* y_1 + x_2{}^* y_2 + \ldots + x_n{}^* y_n\right)\\&+ \left(y_1{}^* c_{1,1}{}^* R_1 + y_1{}^* c_{1,2}{}^* R_2 + \ldots + y_1{}^* c_{1,n}{}^* R_n\right)\\&+ \left(y_2{}^* c_{2,1}{}^* R_1 + y_2{}^* c_{2,2}{}^* R_2 + \ldots + y_2{}^* c_{2,n}{}^* R_n\right)\\&\ldots\\&+ \left(y_n{}^* c_{n,1}{}^* R_1 + y_n{}^* c_{n,2}{}^* R_2 + \ldots + y_n{}^* c_{n,n}{}^* R_n\right) .\end{aligned}$$

The first line of the R.H.S. can be succinctly written as $\sum_{i=1}^{n} x_i{}^* y_i$, the desired final result. In the remaining portion, we group all multiplicative components vertically, and rearrange the equation to factor out all the R_i values, giving:

$$\begin{aligned}S = &\sum\nolimits_{i=1}^{n} x_i{}^* y_i\\&+ R_1{}^* \left(c_{1,1}{}^* y_1 + c_{2,1}{}^* y_2 + \ldots + c_{n,1}{}^* y_n\right)\\&+ R_2{}^* \left(c_{1,2}{}^* y_1 + c_{2,2}{}^* y_2 + \ldots + c_{n,2}{}^* y_n\right)\\&\ldots\\&+ R_n{}^* \left(c_{1,n}{}^* y_1 + c_{2,n}{}^* y_2 + \ldots + c_{n,n}{}^* y_n\right)\end{aligned}$$

Adding and subtracting the same quantity from one side of the equation does not change the equation in any way.

Hence, the above equation can be rewritten as follows:

$$\begin{aligned}S = &\sum\nolimits_{i=1}^{n} x_i{}^* y_i\\&+ \left\{R_1{}^* \left(c_{1,1}{}^* y_1 + c_{2,1}{}^* y_2 + \ldots + c_{n,1}{}^* y_n\right) + R_1{}^* R_1' - R_1{}^* R_1'\right\}\\&\ldots\\&+ \left\{R_{n/r}{}^* \left(c_{1,n/r}{}^* y_{n/r} + c_{2,n/r}{}^* y_2 + \ldots + c_{n,n/r}{}^* y_n\right)\right.\\&\qquad \left. + R_{n/r}{}^* R_1' - R_{n/r}{}^* R_1'\right\}\\&+ \left\{R_{n/r+1}{}^* \left(c_{1,n/r+1}{}^* y_{n/r+1} + c_{2,n/r+1}{}^* y_2 + \ldots + c_{n,n/r+1}{}^* y_n\right)\right.\\&\qquad \left. + R_{n/r+1}{}^* R_2' - R_{n/r+1}{}^* R_2'\right\}\\&\ldots\\&+ \left\{R_{2n/r}{}^* \left(c_{1,2n/r}{}^* y_{2n/r} + c_{2,2n/r}{}^* y_2 + \ldots + c_{n,2n/r}{}^* y_n\right)\right.\\&\qquad \left. + R_{2n/r}{}^* R_2' - R_{2n/r}{}^* R_2'\right\}\\&\ldots\\&+ \left\{R_{(r-1)n/r+1}{}^* \left(c_{1,(r-1)n/r+1}{}^* y_{(r-1)n/r+1} + c_{2,(r-1)n/r+1}{}^* y_2 + \ldots\right.\right.\\&\qquad \left. + c_{n,(r-1)n/r+1}{}^* y_n\right) + R_{(r-1)n/r+1}{}^* R_r' - R_{(r-1)n/r+1}{}^* R_r'\\&\ldots\\&+ \left\{R_n{}^* \left(c_{1,n}{}^* y_1 + c_{2,n}{}^* y_2 + \ldots + c_{n,n}{}^* y_n\right) + R_n{}^* R_r' - R_n{}^* R_r'\right\} .\end{aligned}$$

Now A factors out the R_i from the first two components and groups the rest vertically, giving:

$$
\begin{aligned}
S = &\sum\nolimits_{i=1}^{n} x_i{}^* y_i \\
&+ R_1{}^* \left(c_{1,1}{}^* y_1 + c_{2,1}{}^* y_2 + \ldots + c_{n,1}{}^* y_n + R_1'\right) \\
&\ldots \\
&+ R_{n/r}{}^* \left(c_{1,n/r}{}^* y_{n/r} + c_{2,n/r}{}^* y_2 + \ldots + c_{n,n/r}{}^* y_n + R_1'\right) \\
&+ R_{n/r+1}{}^* \left(c_{1,n/r+1}{}^* y_{n/r+1} + c_{2,n/r+1}{}^* y_2 + \ldots + c_{n,n/r+1}{}^* y_n + R_2'\right) \\
&\ldots \\
&+ R_{2n/r}{}^* \left(c_{1,2n/r}{}^* y_{2n/r} + c_{2,2n/r}{}^* y_2 + \ldots + c_{n,2n/r}{}^* y_n + R_1'\right) \\
&\ldots \\
&+ R_{(r-1)n/r+1}{}^* \left(c_{1,(r-1)n/r+1}{}^* y_{(r-1)n/r+1} + c_{2,(r-1)n/r+1}{}^* y_2 + \ldots \right. \\
&\quad \left. + c_{n,(r-1)n/r+1}{}^* y_n + R_r'\right) \\
&\ldots \\
&+ R_n{}^* \left(c_{1,n}{}^* y_1 + c_{2,n}{}^* y_2 + \ldots + c_{n,n}{}^* y_n + R_r'\right) \\
&- R_1{}^* R_1' - \ldots - R_{n/r}{}^* R_1' \\
&- R_{n/r+1}{}^* R_2' - \ldots - R_{2n/r}{}^* R_2' \\
&\ldots \\
&- R_{(r-1)n/r+1}{}^* R_r' - \ldots - R_n{}^* R_r' \,.
\end{aligned}
$$

Party A already knows the n R_i values. Party B also sent n other values, these are the coefficients of the n R_i values above.

Party A multiplies the n values received from party B with the corresponding R_i and subtracts the sum from S to get:

$$
\begin{aligned}
\text{Temp} = &\sum\nolimits_{i=1}^{n} x_i{}^* y_i \\
&- R_1{}^* R_1' - \ldots - R_{n/r}{}^* R_1' \\
&- R_{n/r+1}{}^* R_2' - \ldots - R_{2n/r}{}^* R_2' \\
&\ldots \\
&- R_{(r-1)n/r+1}{}^* R_r' - \ldots - R_n{}^* R_r' \,.
\end{aligned}
$$

Factoring out the R_i' gives:

$$
\begin{aligned}
\text{Temp} = &\sum\nolimits_{i=1}^{n} x_i{}^* y_i \\
&- (R_1 + R_2 + \ldots + R_{n/r})^* R_1' \\
&- (R_{n/r+1} + R_{n/r+2} + \ldots + R_{2n/r})^* R_2' \\
&\ldots \\
&- (R_{((r-1)n/r)+1} + R_{((r-1)n/r)+2} + \ldots + R_n)^* R_r' \,.
\end{aligned}
$$

To obtain the desired final result ($\sum_{i=1}^{n} x_i * y_i$), party A needs to add the sum of the r multiplicative terms to Temp. In step 3, party A sends the r values to party B, and party B (knowing R') computes the final result. Finally party B replies with the result.

14.2 Privacy-preserving Clustering

Clustering has many important applications such as pattern recognition, image processing and marketing. In business, clustering can help marketers discover different groups among their customers and characterize customer groups based on purchasing patterns. In machine learning, clustering is an example of unsupervised learning. Clustering is a form of learning by observation, rather than learning by examples since it does not rely on predefined classes and class-labeled training examples.

14.2.1 Privacy-preserving Clustering in Centralized Data

Oliveira and Zaiane (2004) address the problem of protecting the underlying attribute values when sharing data for clustering. They propose a novel spatial data transformation method called rotation-based transformation (RBT). RBT can be seen as a technique on the border with obfuscation since the transformation process makes the original data difficult to perceive or understand, and preserves all the information for clustering analysis.

The major features of their data transformation are: (1) it is independent of any clustering algorithm; (2) it has a sound mathematical foundation; (3) it is efficient and accurate; and (4) it does not rely on intractability hypotheses from algebra and does not require CPU-intensive operations. They show analytically that although the data are transformed to achieve privacy, they can also get accurate clustering results by the safeguard of the global distances between data points.

14.2.2 Privacy-preserving Clustering in Horizontal Partitioned Data

Jha *et al.* (2005) present the design and analysis of a privacy-preserving k-means clustering algorithm, where only the cluster means at the various steps of the algorithm are revealed to the participating parties. The crucial step in their privacy-preserving k-means is privacy-preserving computation of cluster means. They present two protocols (one based on oblivious polynomial evaluation and the second based on homomorphic encryption) for privacy-preserving computation of cluster means.

Protocol based on oblivious polynomial evaluation.

Let F be a finite field. Party 1 has two polynomials P and Q with coefficients in F. Party 2 has two points α and β in F. Both parties want to compute $\frac{P(\alpha)}{Q(\beta)}$. In other words, we want to privately compute the following functionality:

$$((P, Q), (\alpha, \beta)) \to \frac{P(\alpha)}{Q(\beta)}.$$

We call this problem private rational polynomial evaluation (PRPE).

The protocol P_{PRPE} uses a protocol for oblivious polynomial evaluation, which is defined below.

Definition 14.1. Let F be a finite field. The oblivious polynomial evaluation or OPE problem can be defined as follows: Alice A has a polynomial P over the finite field F, and Bob B has an element $x \in F$. After executing the protocol implementing OPE B should only know $P(x)$ and A should know nothing.

They provide a protocol $P_{\text{PRPE}}((P, Q), (\alpha, \beta))$ for PRPE, which uses $P_{\text{OPE}}(P, \alpha)$ as an oracle. The protocol is shown below.

Step 1 Party 1 picks a random element $z \in F$ and computes two new polynomials zP and zQ. In other words, party 1 "blinds" the polynomials P and Q.

Step 2 Party 2 computes $zP(\alpha)$ and $zQ(\beta)$ by invoking the protocol for OPE twice, *i.e.*, invokes the protocol $P_{\text{OPE}}(zP, \alpha)$ and $P_{\text{OPE}}(zQ, \beta)$.

Step 3 Party 2 computes $\frac{P(\alpha)}{Q(\beta)}$ by computing $\frac{zP(\alpha)}{zQ(\beta)}$ and sends it to party 1.

Protocol based on homomorphic encryption.

Let (G, E, D, M) be a encryption scheme (where G is the function to generate public parameters, E and D are the encryption and decryption functions, and M is the message space, respectively) with the following properties:

- The encryption scheme (G, E, D) is semantically secure. Essentially, an encryption scheme is semantically secure if an adversary gains no extra information by inspecting the ciphertext.
- For all $m \in M$ and $\alpha \in M$, $m_1 \in E(m)$ implies that $m_1^{\alpha} \in E(m\alpha)$. Encrypting the same message twice in a probabilistic encryption function can yield a different ciphertext, so $E(m)$ denotes the set of ciphertexts that can be obtained by encrypting m.
- There is a computable function f such that for all messages m_1 and m_2 the following property holds:

$$f(E(m_1), E(m_2)) = E(m_1 + m_2)$$

The protocol is shown below:

Step 1 Party 1 encrypts x and n and sends the encrypted values $x_1 \in E(x)$ and $n_1 \in E(n)$ to party 2.

Step 2 Party 2 computes a random message $z \in M$, and encrypts $z \cdot y$ and $z \cdot m$ to obtain $z_1 \in E(z \cdot y)$ and $z_2 \in E(z \cdot m)$. Party 2 computes the following two messages and sends it to party 1:

$$m_1 = f(m_1^z, z_1)$$

$$m_2 = f(n_1^z, z_2) \ .$$

Step 3 Using the two properties of the probabilistic encryption scheme (G, E, D), we have the following:

$$m_1 = E(z \cdot x + z \cdot y)$$

$$m_2 = E(z \cdot n + z \cdot m) \ .$$

Therefore, party 1 can compute $z(x + y)$ and $z(n + m)$, and hence can compute $\frac{x+y}{m+n}$. Party 1 sends $\frac{x+y}{m+n}$ to party 2.

14.2.3 Privacy-preserving Clustering in Vertically Partitioned Data

Vaidya and Clifton (2003) present a method for k-means clustering when different sites contain different attributes for a common set of entities. Each site learns the cluster of each entity, but learns nothing about the attributes at other sites. They introduce the family of geometric data transformation methods (GDTM) to meet privacy preservation in clustering analysis.

Let V be a d-dimensional vector subspace, where each element v_i, $1 \leq i \leq d$, is the form $v_i = (a_1, a_2, \ldots, a_d)$, and each a_i in v_i is one observation of a confidential numerical attribute, and let $N = (\langle op_1, e_1 \rangle, \ldots, \langle op_d, e_d \rangle)$ be a uniform noise vector. They define a geometric transformation function f as d-dimensional space into itself, which transforms V into V' by distorting all attributes of v_i in V according to its corresponding ith element in N. Each vector v' of V' is the form $v' = (\langle a_1 [op_1] e_1 \rangle, \ldots, \langle a_d [op_d] e_d \rangle)$, and $\forall i$, $\langle a_i [op_i] e_i \rangle \in {}^s R$. They consider the following geometric transformation functions: translation, scaling, and rotation whose corresponding operations are *Add*, *Mult*, and *Rotate*.

All transformation data algorithms have essentially two major steps: (1) identify the noise term and the operation that must be applied to each confidential attribute. This step refers to the instantiation of the uniform noise vector N. (2) Based on the uniform noise vector N, defined in the previous step, transform V into V' using a geometric transformation function.

The Translation Data Perturbation Method

In this method, denoted by TDP, the observations of confidential attributes in each $v_i \in V$ are perturbed using an additive noise perturbation. The noise term applied to each confidential attribute is constant and can be either positive or negative. The set of operations $D_i(OP)$ takes only the value corresponding to an additive noise applied to each confidential attribute. The sketch of the TDP algorithm is given as follows:

TDP algorithm

Input: V, N
Output: V'

Step 1 For each confidential attribute A_j in V, where $1 \leq j \leq d$ do
1. Select the noise term e_j in N for the confidential attribute A_j
2. The jth operation $op_j \leftarrow \{Add\}$

Step 2 For each $v_i \in V$ do
For each a_j in $v_i = (a_1, a_2, \ldots, a_d)$, where a_j is the observation of the jth attribute do
$a'_j \leftarrow transform(a_j, op_j, e_j)$

End

The Scaling Data Perturbation Method

In the scaling data perturbation method (SDP), the observations of confidential attributes in each $v_i \in V$ are perturbed using a multiplicative noise perturbation. The noise term applied to each confidential attribute is constant and can be either positive or negative. The set of operations $D_i(OP)$ takes only the value $\{Mult\}$ corresponding to a multiplicative noise applied to each confidential attribute. The sketch of the SDP algorithm is given as follows:

SDP algorithm

Input: V, N
Output: V'

Step 1 For each confidential attribute A_j in V, where $1 \leq j \leq d$ do
1. Select the noise term e_j in N for the confidential attribute A_j
2. The jth operation $op_j \leftarrow \{Mult\}$

Step 2 For each $v_i \in V$ do
For each a_j in $v_i = (a_1, a_2, \ldots, a_d)$, where a_j is the observation of the jth attribute do
$a'_j \leftarrow transform(a_j, op_j, e_j)$

End

The Rotation Data Perturbation Method

The rotation data perturbation method (RDP) works differently from previous methods. In this case, the noise term is an angle θ. The rotation angle θ, measured clockwise, is the transformation applied to the observations of the confidential attributes. The set of operations $D_i(OP)$ takes only the value $\{Rotate\}$ that identifies a common rotation angle between the attributes A_i and A_j. Unlike the previous methods, RDP may be applied more than once to some confidential attributes. For instance, when a rotation transformation is applied this affects the values of two coordinates. In a 2D discrete space, the X and Y coordinates are affected. In a 3D discrete space or higher, two variables are affected and the others remain without any alteration. This requires that one or more rotation transformations are applied to guarantee that all the confidential attributes are distorted in order to preserve privacy. The sketch of the RDP algorithm is given as follows:

RDP algorithm
Input: V, N
Output: V'

Step 1 For each confidential attribute A_j, A_k in V, where $1 \leq j \leq d$ and $1 \leq k \leq d$ do
  1. Select an angle θ for the confidential attribute A_j, A_k
  2. The jth operation $op_j \leftarrow \{Rotate\}$
  3. The kth operation $op_k \leftarrow \{Rotate\}$

Step 2 For each $v_i \in V$ do
  For each a_l in $v_i = (a_1, a_2, \ldots, a_d)$, where a_l is the observation of the lth attribute do
    $a'_l \leftarrow transform(a_l, op_l, e_l)$

End

The Hybrid Data Perturbation Method

The hybrid data perturbation method (HDP) combines the strength of our previous methods: TDP, SDP and RDP. In this scheme, they select randomly one operation for each confidential attribute that can take the values $\{Add, Mult, Rotate\}$ in the set of operations D_i(OP). Thus, each confidential attribute is perturbed using either an additive, a multiplicative noise term, or a rotation. The sketch of the HDP algorithm is given as follows:

HDP algorithm
Input: V, N
Output: V'

Step 1 For each confidential attribute A_j in V, where $1 \leq j \leq d$ do
  1. Select the noise term e_j in N for the confidential attribute A_j
  2. The jth operation $op_j \leftarrow \{Add, Mult, Rotation\}$

Step 2 For each $v_i \in V$ do
For each a_j in $v_i = (a_1, a_2, \ldots, a_d)$,
where a_j is the observation of the jth attribute do
$a'_j \leftarrow transform(a_j, op_j, e_j)$
End

14.3 A Scheme to Privacy-preserving Collaborative Data Mining

In this section, we combine the data perturbation methods and the secure computation methods and propose a scheme to privacy-preserving collaborative k-nearest neighbor (k-NN) search in data mining (Zhu 2009).

14.3.1 Preliminaries

In this section, we first describe the cryptographic tools and definitions used here.

14.3.1.1 Homomorphic Encryption

A homomorphic encryption scheme is an encryption scheme that allows certain algebraic operations to be carried out on the encrypted plaintext, by applying an efficient operation to the corresponding ciphertext (without knowing the decryption key!). Let (e, d) denote a cryptographic key pair and $e(.)$ denotes the encryption function with public key e, $d(.)$ denotes the decryption function with private key d. A secure public key cryptosystem is called homomorphic if it satisfies the following requirements:

- Given that the m_1 and m_2 are the data to be encrypted, there exists an efficient algorithm to compute the public key encryption of $m_1 + m_2$, denoted as

$$e(m_1 + m_2) = e(m_1) \times e(m_2) .$$

- $e(m_1)^k = e(km_1)$
Because of the property of associativity, $e(m_1 + m_2 + \ldots + m_n)$ can be computed as $e(m_1) \times e(m_2) \times \ldots \times e(m_n)$, where $e(m_i) \neq 0$. That is,

$$e(m_1 + m_2 + \ldots + m_n) = e(m_1) \times e(m_2) \times \ldots \times e(m_n) .$$

14.3.1.2 ElGamal Encryption System

In cryptography, the ElGamal encryption system is an asymmetric key encryption algorithm for public key cryptography which is based on the Diffie–Hellman key agreement. It was described by Taher Elgamal in 1984 (Elgamal 1985). ElGamal encryption can be defined over any cyclic group G. Its security depends upon the difficulty of a certain problem in G related to computing discrete logarithms.

ElGamal encryption consists of three components: the key generator, the encryption algorithm, and the decryption algorithm.

The key generator works as follows:

- Alice generates an efficient description of a multiplicative cyclic group G of order q with generator g.
- Alice chooses a random x from $\{0, 1, \ldots, q - 1\}$.
- Alice computes $y = g^x \bmod q$ as her public key. Alice retains x as her private key, which must be kept secret.

The encryption algorithm works as follows: to encrypt a message m to Alice under her public key (G, q, g, y).

- Bob converts m into an element of G.
- Bob chooses a random r from $\{0, 1, \ldots, q - 1\}$, then calculates $c_1 = g^r$ and $c_2 = my^r$.
- Bob sends the ciphertext (c_1, c_2) to Alice.

The decryption algorithm works as follows: to decrypt a ciphertext (c_1, c_2) with her private key x.

- Alice computes $m = c_2/c_1^x$ as the plaintext message.

14.3.1.3 The k-nearest Neighbor Search

In the k-NN method, a number of patterns k within a region are fixed, whereas a region size (and thus a volume V) varies depending on the data. The k-NN probability density estimation method can be simply modified as the k-NN classification rule. The k-NN query is one of the most common queries in similarity search and its objective is to find the k nearest neighbors of points in horizontally partitioned data. The formal definition for k-NN search is given below (Shaneck *et al.* 2006):

Definition 14.2. In a distributed setting, given m horizontally distributed data sets $S_1, S_2, \ldots, S_m$, and a particular point $x \in S_j (1 \le j \le m)$ and a query parameter k, k-NN search returns the set $N_k(x) \subseteq S = \cup_{i=1}^m S_i$ of size k, such that, for every point $z \in N_k(x)$ and for every point $y \in S$, $y \notin N_k(x) \Rightarrow d(x, z) \le d(x, y)$, where $d(x, y)$ represents the distance between the point x and y.

The nearest neighbors of an instance are defined in terms of a distance function such as the standard Euclidean distance. More precisely, let point $x = (a_1(x), a_2(x), \ldots, a_r(x))$, where $a_i(x)$ denotes the value of the ith attribute of instance x. Then the distance between two instances x_i and x_j is defined as $d(x_i, x_j)$, where

$$d(x_i, x_j) = \sqrt{\sum_{q=1}^{r} (a_q(x_i) - a_q(x_j))^2} \,.$$

Here, we use the square of the standard Euclidean distance $d^2(x_i, x_j)$ to compare the different distances.

14.3.2 The Analysis of the Previous Protocol

In this section, we analyze the protocol given in Zhan and Matwin (2006) and point out its secure flaw in malicious adversaries.

For privacy-preserving k-NN search, a solution for privacy-preserving k-NN classification is developed in Zhan and Matwin (2006). There, the authors focus on how to prevent inside attackers from knowing private data in collaborative data mining in the semihonest model.

In vertical collaboration, each party holds a subset of attributes for every instance. Given a query instance x_q, we want to compute the distance between x_q and each of the N training instances. Since each party holds only a portion (*i.e.*, partial attributes) of a training instance, each party computes her portion of the distance (called the distance portion) according to her attributes set. To decide the k-NN of x_q, all the parties need to sum their distance portions together. For example, assume that the distance portions for the first instance are $s_{11}, s_{12}, \ldots, s_{1n}$; and the distance portions for the second instance are $s_{21}, s_{22}, \ldots, s_{2n}$. To compute whether the distance between the first instance and x_q is larger than the distance between the second instance x_q, we need to compute whether $\sum_{i=1}^{n} s_{1i} \geq \sum_{i=1}^{n} s_{2i}$. How can we obtain this result without compromising data privacy? In Zhan and Matwin (2006), the authors developed a privacy-oriented protocol to tackle this challenge.

Protocol 1 consists of three steps (Zhan and Matwin 2006). We briefly depict their idea in the following.

In step I, in order to compute $e(\sum_{l=1}^{n} s_{il})$ for $i \in [1, N]$, P_n generates a cryptographic key pair (e, d) of a semantically secure homomorphic encryption scheme and publishes its public key e. P_l generates N random numbers R_{il}, for all $i \in [1, N], l \in [1, n]$. Then forward transmission is as in Figure 14.1. In Figure 14.1 (a), when P_2 received the message $e(s_{i1} + R_{i1})$ from P_1, he computes $e(s_{i1} + R_{i1}) + e(s_{i2} + R_{i2}) = e(s_{i1} + s_{i2} + R_{i1} + R_{i2})$ and sends them to P_3, and so on. In Figure 14.1 (b), they send the random numbers R_{il} encrypted by the public key on the backward order.

In this protocol, if P_{n-2} and P_n collude to get the P_{n-1}'s private data $s_{i(n-1)}$, P_{n-2} only sends $e(-R_{i(n-1)})$ to P_n (shown as dashed line in Figure 14.1). P_n decrypts it and gets the random number $R_{i(n-1)}$, then gets the $s_{i(n-1)}$. Figure 14.2 is an example to explain the procedure of collusion attack when $n = 4$. In the Figure 14.2, P_4 and P_2 collude, and they can get R_{i1} and R_{i3}. From the forward transmission message, they can obtain the private data s_{i1} and s_{i3}. In Figure 14.2, we use dashed line to express the attacking step.

In step 2 of protocol 1, the procedure of computing $e(\sum_{l=1}^{n} -s_{jl})$ is similar to the step 1. This protocol cannot prevent a colluded attack and cannot provide the data privacy in data mining.

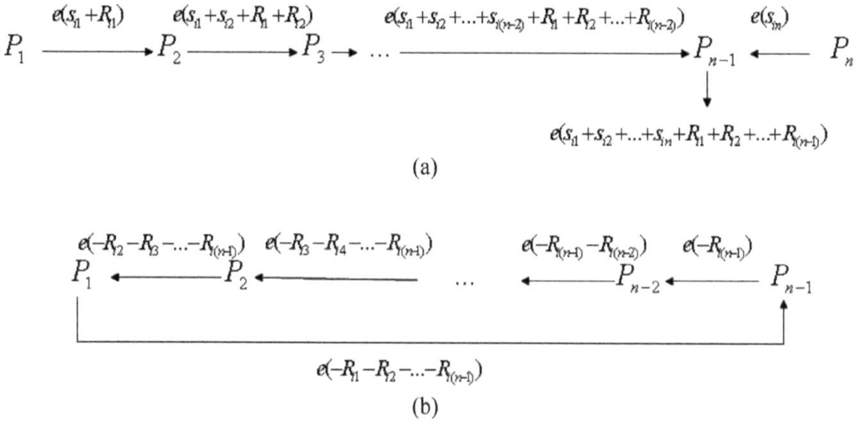

Figure 14.1 Step 1 of protocol 1 in Zhan and Matwin (2006): (a) forward transmission, and (b) backward transmission

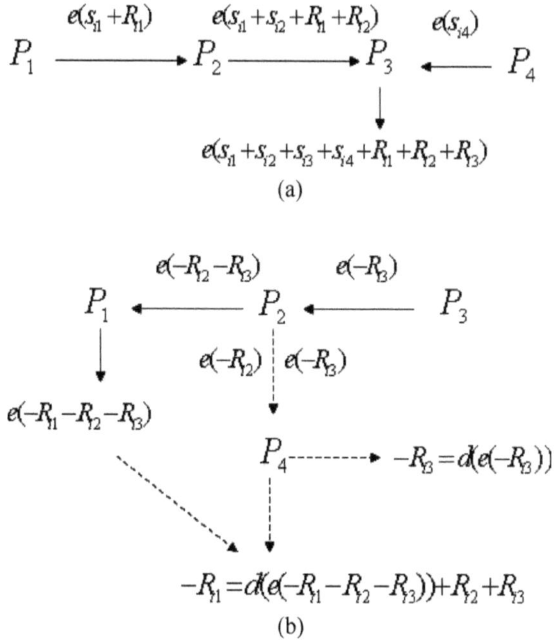

Figure 14.2 The protocol of four parties: (a) forward transmission, and (b) backward transmission

14.3.3 A Scheme to Privacy-preserving Collaborative Data Mining

The condition to conduct data mining is the same as in Zhan and Matwin (2006). In vertical collaboration, each party holds s subset of attributes for every instance. The notation is seen in Section 14.3.2. In this scheme, we use the ElGamal encryption system and symbols shown in Section 14.3.1.2. We define the operations as follows:

$$E_y(m_1) + E_y(m_2) = \left(g^{r_1+r_2}, m_1 m_2 y^{r_1+r_2}\right)$$

$$E_y(m_1) - E_y(m_2) = (m_1 - m_2)y^r \ ,$$

where r, r_1, r_2 are chosen to be random numbers from $[0, q-1]$.

14.3.3.1 Initialization

In the following, we restrict our discussion to one group of respondents and denote the l respondents in this group by $P_1, P_2, \ldots, P_l$. We assume that there is a private and authenticated communication channel between each respondent and miner. Each party P_i has a key pair (x_i, y_i) $(x_i \in [0, q-1], y_i \in G)$ such that $y_i = g^{x_i}$ in G, where G is a cyclic group in which the discrete logarithm is hard. Let g be a generator of G and $|G| = q$, where q is a large prime. Here, the public key y_i is known to all parties, while the private key x_i is kept secret by party P_i. Let $y = \prod_{i=1}^{l} y_i$ and $x = \sum_{i=1}^{l} x_i$. In this scheme, we use this public value y as a public key to encrypt respondent data. Clearly, $y = g^x$. So, decrypting these encryptions of respondent data needs this secret value x, which is not known to any party.

Parties may not trust each other, but all parties are aware of the benefit brought by such collaboration. In the privacy-preserving model, all parties of the partnership promise to provide their private data to the collaboration, but none of them wants the others or any third party to learn much about their private data.

14.3.3.2 Compute the k-NN

After the initial phase, each party has a public key y, where this can be done by the initiator (also miner). Encryption is under a homomorphic encryption scheme. The protocol of computing the k-NN is as follows. In the protocol, r_i is a random number from $[0, q-1]$ by party P_i privately, $i = 1, \ldots, l$.

Protocol: Compute the k-NN

Define array $e\,[1, \ldots, N]$

Note: collect the data

For $i = 1$ to N

    $e[i] = 0$

For $j = 1$ to $l - 1$ do
P_j Calculate $e_{ij} = E_y(s_{ij})$
P_j Send e_{ij} to P_l
P_l computes $e[i] = e[i] + e_{ij}$
End for
Note: P_l obtained $e[i] = e[i] + e_{il} = \left(c_i^1, c_i^2\right)$
End for

Note: Decryption and obtain the result

Define D, $D1$ array of $[1 \ldots N, 1 \ldots N]$
For $i = 1$ to N
For $j = 1$ to N

$$D[i, j] = c_i^2 - c_j^2 = \left(\prod_{k=1}^{l} s_{ik} - \prod_{k=1}^{l} s_{jk}\right) y^{\sum_{k=1}^{l} r_k}$$

End for
End for
$D1 = $ Permutation $\pi(D)$
P_l sends c_1^1 to $P_1, P_2, \ldots, P_{l-1}$

Note: $c_1^1 = c_2^1 = \ldots = c_N^1$

For $i = 1$ to $l - 1$
P_i computes $(c_1^1)^{x_i}$;
P_i sends $(c_1^1)^{x_i}$ to P_l;
End for

P_l computes and obtains $(c_1^1)^{\sum_{i=1}^{l} x_i} = g^{\sum_{i=1}^{l} r_i \sum_{i=1}^{l} x_i} = y^{\sum_{i=1}^{l} r_i}$
For $i = 1$ to N
For $j = 1$ to N

P_l computes and gets $D1[i, j] = \left(\sum_{k=1}^{l} s_{ik} - \sum_{k=1}^{l} s_{jk}\right)$;

If $D1[i, j] \geq 0$ then $D1[i, j] = +1$;
else $D1[i, j] = -1$;
End for
End for

Finally, P_l can compute k smallest elements as in Zhan and Matwin (2006) and then gets the k-NN for a given instance.

14.3.4 Protocol Analysis

By analyzing this scheme, we come to the conclusion that this scheme is correct and efficient.

14.3.4.1 Correctness

The correctness of the protocol can be verified as follows. We choose the ElGamal encryption system as encryption algorithm. In order to simplify, we assume that there are four parties to conduct collaborative data mining, *i.e.*, $l = 4$. We assume that there are N records in the database and each party holds a subset of attributes for every record. For record i, $P_j (j = 1, \ldots, 4)$ holds a subset of attributes s_{ij}.

In initialization phase, four parties have an agreement to conduct collaborative data mining and P_4 is a miner. $P_j (j = 1, \ldots, 4)$ has key pairs (x_j, y_j), where y_j is the public key and x_j is the private key. $P_j (j = 1, \ldots, 4)$ all can compute the $y = \prod_{j=1}^{4} y_j$ because every y_j is public.

In collecting the data phase:

Define array $e[1, \ldots, N]$
For $i = 1$ to N
    $e[i] = 0$;
    P_1 Calculate $e_{i1} = E_y(s_{i1}) = (g^{r_1}, s_{i1} y^{r_1})$ and send e_{i1} it to P_4;
    P_2 Calculate $e_{i2} = E_y(s_{i2}) = (g^{r_2}, s_{i2} y^{r_2})$ and send e_{i2} it to P_4;
    P_3 Calculate $e_{i3} = E_y(s_{i3}) = (g^{r_3}, s_{i3} y^{r_3})$ and send e_{i3} it to P_4;
    P_4 computes
        $e[i] = e_{i1} + e_{i2} + e_{i3} + (g^{r_4}, s_{i4} y^{r_4})$
            $= (g^{r_1+r_2+r_3+r_4}, s_{i1} s_{i2} s_{i3} s_{i4} y^{r_1+r_2+r_3+r_4}) = \left(c_i^1, c_i^2\right)$
End for

In computing the k-NN phase:

Define D, $D1$ array of $[1 \ldots N, 1 \ldots N]$
For $i = 1$ to N
    For $j = 1$ to N
        $D[i, j] = e[i] - e[j] = c_i^2 - c_j^2$
            $= (s_{i1} s_{i2} s_{i3} s_{i4} - s_{j1} s_{j2} s_{j3} s_{j4}) y^{r_1+r_2+r_3+r_4}$
    End for
End for
$D1 =$ Permutation $\pi(D)$;
P_4 sends $g^{r_1+r_2+r_3+r_4}$ to P_1, P_2, P_3;
P_1, P_2, P_3 decrypts $g^{r_1+r_2+r_3+r_4}$ using its private key x_i, respectively, and then send to P_4;
P_4 computes and gets $(g^{r_1+r_2+r_3+r_4})^{x_1+x_2+x_3+x_4} = y^{r_1+r_2+r_3+r_4}$;
For $i = 1$ to N
    For $j = 1$ to N
        P_4 computes and gets $D1[i, j] = (s_{i1} s_{i2} s_{i3} s_{i4} - s_{j1} s_{j2} s_{j3} s_{j4})$;
        If $D1[i, j] \geq 0$ then $D1[i, j] = +1$;
                else $D1[i, j] = -1$;
        End for
    End for

Finally, P_4 can compute k smallest elements as in Zhan and Matwin (2006) and then gets the k-NN for a given instance. When the protocol finishes, we can obtain the correct result.

14.3.4.2 Data Privacy and Security

Proposition 14.1. *This scheme can provide data privacy.*

Proof. This scheme consists of three phases.

Phase 1 is the initial phase. In this phase, initiator can obtain the public key of all participators and computes the public encryption key y. The corresponding private key $x = \sum_{i=1}^{l} x_i$ is not known to any individual party.

In phase 2, the initiator collects the data from other parties. Every party encrypts their private data using the public key y and no party can obtain the private data because they do not know the private key x.

In phase 3, the initiator computes the k-NN. The initiator stores the collected data from participators into the array e and then computes the difference between any two elements of the array e. Because no party knows the private key, the initiator cannot decrypt the encrypted data. Only when all the participators join to decrypt the data, can the initiator obtain the array $D1$ and compute the k-NN. Because $D1$ is a permutation of D, the initiator cannot find any private data from $D1$.

Therefore, this scheme can provide data privacy.

Proposition 14.2. *This scheme is secure in the semihonest model and can prevent colluded attack, the inside and outside attack.*

Proof. In this scheme, no one can obtain the private data if one of the participators does not join the decryption operation.

In the semihonest model, each party follows the rules of the protocol properly, but is free to use all his intermediate computation records to derive additional information about others' inputs. This scheme satisfies these conditions.

If some parties collude to obtain others' input data, this is impossible unless they obtain their private keys. For the same reason, this scheme can prevent the inside and outside attack.

14.3.4.3 Efficiency

The communication complexity analysis:

In this scheme, l denotes the total number of parties and N is the total number of records. Assume that α denotes the number of bits of each ciphertext and β stand for the number of bits of each plaintext. The total communication cost is $\alpha(l-1)N + 2\alpha(l-1)$.

The total communication cost of the protocol in Zhan and Matwin (2006) is $2\alpha lN + 2\alpha lN + \alpha N(N-1) + \beta(N-1) + \frac{3}{2}\alpha l^2 + \alpha(l-1)$.

Compared with the protocol in Zhan and Matwin (2006), the communication complexity of this scheme is lower.

The computation complexity analysis:

Comparing with the protocol in Zhan and Matwin (2006), the computation costs are included in Table 14.1. If we do not consider the effect of different public key systems, this scheme is more efficient.

Table 14.1 Comparison of the computation cost

Computation cost	Protocol in Zhan and Matwin (2006)	This scheme
Numbers of keys	One cryptographic key pair	0
Random numbers	$2N$	l
Number of encryption	$4lN$	$lN + l - 1$
Number of multiplication	$N^2 + 4lN + 3N$	$3l$
Number of decryption	$N(N-1)$	l
Number of addition	$2lN$	$l + N(N-1)$
Sorting N number	$g^{N\log(N)}$	$\frac{1}{2}g^{N\log(N)}$

In this section, we discussed the related research work on privacy-preserving data mining and pointed out the flaw of security, which cannot prevent from colluded attack. Then we presented a scheme to privacy-preserving collaborative data mining which can be used to compute the k-NN search based on the homomorphic encryption and ElGamal encryption system in distributed environment. This scheme is security in the semihonest model and is efficient.

14.4 Evaluation of Privacy Preservation

An important aspect in the development and assessment of algorithms and tools for privacy-preserving data mining is the identification of suitable evaluation criteria and the development of related benchmarks. It is often the case that no privacy-preserving algorithm exists that outperforms all the others on all possible criteria. Rather, an algorithm may perform better that another on specific criteria, such as performance and/or data utility. It is thus important to provide users with a set of metrics which will enable them to select the most appropriate privacy-preserving technique for the data at hand, with respect to some specific parameters they are interested in optimizing.

Verykios *et al.* (2004) proposed the following evaluation parameters to be used for assessing the quality of privacy-preserving data mining (PPDM) algorithms:

- The performance of the proposed algorithms in terms of time requirements, that is the time needed by each algorithm to hide a specified set of sensitive information, which mainly includes computational cost and communication cost.
- The data utility after the application of the privacy-preserving technique, which is equivalent to the minimization of the information loss or else the loss in the functionality of the data.
- The level of uncertainty with which the sensitive information have been hidden can still be predicted.
- The resistance accomplished by the privacy algorithms, to different data-mining techniques.

Wu *et al.* (2007) assessed the relative performance of PPDM algorithms:

In terms of computational efficiency, rule hiding is less efficient than data hiding, because one has to identify the items that contribute to the sensitive rule first and then hide the rule. For the privacy requirement, we think the hiding rule is more critical than hiding data, because after the sensitive rules are found, more information can be inferred. This is not to say that rule hiding is more accurate than data hiding. The selection of either hiding data or rule often depends on the goal of privacy preserving (hiding purpose) and data distribution. For instance, we can only hide data under a distributed database environment.

In general, clustering is more complex than classification (including association rules) because it often requires using an unsupervised learning algorithm. The algorithm used for the association rule and classification can learn from known results, thus, they are more efficient. However, the preserving power and accuracy are highly dependent on the hiding technique used or the algorithm used, not the data-mining task.

The inherent mechanism of blocking and sanitization is basically similar. The former uses a "?" notation to replace selected items to be protected, while the latter deletes or modifies these items from viewing; therefore, their complexity is almost the same. However, the privacy-preserving capability of blocking is lower than sanitization. Moreover, like sanitization, the blocking technique is NP-hard. Therefore, these two modification methods cannot be used to solve larger-sized of problems.

Most existing studies that use distortion methods focus on maintaining the level of privacy disclosure and knowledge discovery ability. It seems that efficiency and computational cost are not the most important issues for the distortion method. In general, data distortion algorithms have good effectiveness in hiding data. However, these methods are not without faults. First, the distorting approach only works if one does not need to reconstruct the original data values. Thus, if the data-mining task changes, new algorithms need to be developed to reconstruct the distributions. Second, this technique considers each attribute independently; as a result, when the number of attributes become large, the accuracy of data-mining results will degrade significantly. Finally, there is a trade-off between accuracy of data-mining results and data security using distortion methods. These methods may not be suitable for mining data in situations requiring both high accuracy and high security.

The generalization technique has been widely used in protecting individual privacy with the k-anonymity model in the past; however, it is relatively new to the data-mining community. Since generalization has the advantage of not modifying the true value of attributes, it may have a higher accuracy of data-mining results than data distortion techniques.

Cryptography-based secure multiparty computation (SMC) has the highest accuracy in data mining and good privacy-preservation capability as well; however, it has strict usage as it is only applicable to a distributed data environment. Two models of SMC are available: the semihonest model and malicious model. The semihonest models assume each party follows the protocol rules, but is free to later use what it sees during execution to compromise security. The malicious model, on the other hand, assumes parties can arbitrarily "cheat," and such cheating will not compromise either security or the results. How to prevent or detect a malicious party in a computation process is an unsolved issue. Not to mention that SMC has the burden of high communication cost, when the number of parties participating increases. Usually, the communication cost increases at an exponential speed when data size increases linearly. Also, different problems need different protocols and the complexities vary naturally.

14.5 Conclusion

Data mining is a well-known technique for automatically and intelligently extracting information or knowledge from a large amount of data. It can however disclose sensitive information about individuals, which compromises the individual's right to privacy. Moreover, data-mining techniques can reveal critical information about business transactions, compromising the free competition in a business setting.

Driven by one of the major policy issues of the information era, the right to privacy, privacy-preserving data mining (PPDM) has become one of the newest trends in privacy and security research. There has been great interest in the subject from both academic and industry: (a) the recent proliferation in PPDM techniques is evident; (b) the interests from academic and industry have grown quickly; (c) separate workshops and conferences devoted to this topic have emerged in the last few years. Therefore, PPDM is fast becoming an increasingly important field of study.

References

Ahmad W, Khokhar A (2007) An architecture for privacy preserving collaborative filtering on web portals. In: Proceedings of the 3rd International Symposium on Information Assurance and Security, pp 273–278, 29–31 August 2007

Agrawal R, Srikant R (2000) Privacy-Preserving Data Mining. In: Proceedings of the ACM SIGMOD Conference on Management of Data, pp 939–450, Dallas, TX, May 2000

Agrawal R, Imielinski T, Swami A (1993) Mining association rules between sets of items in large databases. In: P. Buneman and S. Jajodia (eds) Proceedings of ACM SIGMOD Conference on Management of Data, pp 207–216, Washington DC, May 1993

Clifton C (2005) What is privacy? Critical steps for privacy-preserving data mining. In: IEEE ICDM Workshop on Security and Privacy Accepts of Data Mining, Houston, TX, 27–30 November 2005

Clifton C, Kantarcioglu M, Vaidya J, Lin X, Zhu MY (2002) Tools for privacy preserving distributed data mining. ACM SIGKDD Explor Newslett 4(2):28–34

Elgamal T (1985) A public key cryptosystem and a signature scheme based on discrete logarithms. IEEE Trans Info Theory 31(4):469–472

Evfimievski A, Srikant R, Agrawal R, Gehrke J (2002) Privacy preserving mining of association rules. In: Proceedings of the 8th ACM SIGKDD International Conference on Knowledge Discovery in Databases and Data Mining, Edmonton, Alberta, Canada, pp 217–228, July 2002

Gennaro R, Rabin M, Rabin T (1998) Simplified VSS and fact-track multiparty computations with applications to threshold cryptography. In: Proceedings of the 17th Annual ACM Symposium on Principles of Distributed Computing, pp 101–111

Jha S, Kruger L, McDaniel P (2005) Privacy preserving clustering (LNCS 3679). Springer, Berlin Heidelberg New York

Kantarcioglu M, Clifton C (2002) Privacy-preserving distributed mining of association rules on horizontally partitioned data. In: Proceedings of ACM SIGKDDW Workshop on Research Issues in Data Mining and Knowledge Discovery (DMKD)

Li XB, Sarkar S (2006) A tree-based data perturbation approach for privacy-preserving data mining. IEEE Trans Know Data Eng 18(19):1278–1283

Oliveira SRM, Zaiane OR (2002) Privacy preserving frequent itemset mining. In: Workshop on Privacy, Security, and Data Mining at the 2002 IEEE International Conference on Data Mining (ICDM'02), Maebashi City, Japan

Oliveira SRM, Zaiane OR (2004) Achieving privacy preservation when sharing data for clustering. In: Proceedings of the International Workshop on Secure Data Management in a Connected World (SDM'04) in conjunction with VLDB 2004, Toronto, Canada. August 2004

Rizvi S, Haritsa JR (2002) Maintaining data privacy in association rule mining. In: Proceedings of the 28th International Conference on Very Large Data Bases, Hong Kong, China. August 2002

Shaneck M, Kim Y, Kumar V (2006) Privacy preserving nearest neighbor search. In: Proceedings of the 6th IEEE International Conference on Data Mining Workshops (ICDMW'06), 2006

Vaidya J, Clifton C (2002) Privacy preserving association rule mining in vertically partitioned data. In: Proceedings of the 8th ACM SIGKDD International Conference on Knowledge Discovery in Databases and Data Mining, Edmonton, Alberta, Canada

Vaidya J, Clifton C (2003) Privacy preserving k-means clustering over vertically partitioned data. In: SIGKDD '03, Washington DC, pp 206–214

Verykios V, Bertino E, Fovino IN, Provenza LP, Saygin Y, Theodoridis Y (2004) State-of-the-art in privacy preserving data mining. ACM SIGMOD Record 33(1):50–57

Wu CW (2005) Privacy preserving data mining with unidirectional interaction. In: Proceedings of IEEE International Symposium on Circuits and Systems (ISCAS 2005), 23–26 May 2005, pp 5521–5524

Wu X, Chu CH, Wang Y, Liu F, Yue D (2007) Privacy preserving data mining research: current status and key issues. Lecture Notes Comput Sci 4489:762–772

Zhan J, Matwin S (2006) A crypto-based approach to privacy-preserving collaborative data mining. In: Proceedings of the 6th IEEE International Conference on Data Mining Workshops (ICDMW'06), December 2006, pp 546–550

Zhu J (2009) A new scheme to privacy-preserving collaborative data mining. In: Proceedings of the 5th International Conference on Information Assurance and Security, Xi'an China, 2009

Index

MIX
Papier aus verantwortungsvollen Quellen
Paper from responsible sources
FSC® C105338

FSC
www.fsc.org

If you have any concerns about our products,
you can contact us on
ProductSafety@springernature.com

In case Publisher is established outside the EU,
the EU authorized representative is:
Springer Nature Customer Service Center GmbH
Europaplatz 3, 69115 Heidelberg, Germany

Printed by Libri Plureos GmbH
in Hamburg, Germany